Soil and Human Health

The intersection of soil health and human vitality is a critical yet overlooked aspect of global wellbeing. *Soil and Human Health: Food as Medicine* examines the soil–human health nexus, offering a comprehensive synthesis of research and insights that highlight the connections between soil processes, food systems, and human health.

The book explores key topics such as the One Health initiative, the role of soil processes in shaping human health outcomes, the microbiome's influence on health modulation, and the economic and cultural factors affecting food systems and consumption patterns. Case studies further illuminate the topic, providing real-world examples of how food systems impact human health across diverse contexts.

This book in the *Advances in Soil Sciences* series is an essential resource for researchers, practitioners, and policymakers seeking to advance the conversation on soil health, food systems, and their critical role in supporting human vitality.

Advances in Soil Science

Series Editors: Rattan Lal, B.A. Stewart

Soil Quality and Biofuel Production
R. Lal and B. A. Stewart

World Soil Resources and Food Security
R. Lal and B. A. Stewart

Soil Water and Agronomic Productivity
R. Lal and B. A. Stewart

Principles of Sustainable Soil Management in Agroecosystems
R. Lal and B. A. Stewart

Soil Management of Smallholder Agriculture
R. Lal and B. A. Stewart

Soil-Specific Farming: Precision Agriculture
R. Lal and B. A. Stewart

Soil Phosphorus
R. Lal and B. A. Stewart

Urban Soils
R. Lal and B. A. Stewart

Soil Nitrogen Uses and Environmental Impacts
R. Lal and B. A. Stewart

Soil and Climate
Rattan Lal and B.A. Stewart

Food Security and Soil Quality
R. Lal and B. A. Stewart

Soil Degradation and Restoration in Africa
Rattan Lal and B.A. Stewart

Soil and Fertilizers: Managing the Environmental Footprint
Rattan Lal

The Soil–Human Health-Nexus
Rattan Lal

Soil Organic Matter and Feeding the Future: Environmental and Agronomic Impacts
Rattan Lal

Soil Organic Carbon and Feeding the Future: Basic Soil Processes
Rattan Lal

Soil and Drought: Basic Processes
Rattan Lal

Managing Soil Drought
Rattan Lal

Soil and Human Health: Food as Medicine
Rattan Lal and Marium Husain

For more information about this series, please visit: https://www.crcpress.com/Advances-in-Soil-Science/book-series/CRCADVSOILSCI

Soil and Human Health

Food as Medicine

Edited by

Rattan Lal

Marium Husain

CRC Press
Taylor & Francis Group
Boca Raton London New York

CRC Press is an imprint of the
Taylor & Francis Group, an **informa** business

Designed cover image: Shutterstock

First edition published 2026
by CRC Press
2385 NW Executive Center Drive, Suite 320, Boca Raton FL 33431

and by CRC Press
4 Park Square, Milton Park, Abingdon, Oxon, OX14 4RN

CRC Press is an imprint of Taylor & Francis Group, LLC

ISBN: 978-1-032-72676-2 (hbk)
ISBN: 978-1-032-74541-1 (pbk)
ISBN: 978-1-003-46974-2 (ebk)

DOI: 10.1201/9781003469742

Typeset in Times
by SPi Technologies India Pvt Ltd (Straive)

Contents

Preface

The ancient concept of "Food is Medicine" is aimed at producing healthy and nutritious food and facilitating its access to the public to minimize the risks of food and nutritional insecurity by improving access and enhancing its retention via effective utilization. Around 400 BCE, Hippocrates emphasized the importance of a healthy diet by stating, "Let thy food be thy medicine and medicine be thy food." The Sanskrit scripture *The Charaka Samhita* states that "wholesome diet is essential for good health and to prevent diseases". An ancient Ayurvedic proverb states that "When diet is wrong, medicine is of no use; when diet is right, medicine is of no need". The Quran instructs Muslims to consume "lawful and good" food (halal and tayyib). Similarly, Judaism has many dietary laws (kosher) that suggest what is pure food.

The principal strategy is to enhance human health and well-being by improving economic and easy access to safe and nutritious food. However, realization of this goal also implies adoption of land use and soil/crop/animal management practices which protect and restore soil health of agroecoregions designated for food production from crops, trees, and livestock. The strategy is based on the "One Health" ideology which states that "health of soil, plants, animals, people, ecosystems, and planetary processes is one and indivisible". Therefore, the food–soil–health nexus must be promoted to enhance positive inter-dependence among them. Thus, there is an urgent need to focus on the protection and sustainable management of natural resources (soil, water, air, biodiversity) involved in the production of safe and nutritional food. In this context, food must meet the demand for essential macro- and micronutrients along with proteins, amino acids, fiber, and vitamins. Intake of a healthy and nutritionally rich diet can reduce disease risks and provide curative or medicinal benefits. However, not all chronic and non-communicable diseases can be addressed by choice of safe and nutritious food. While good food is not a substitute for conventional medicine to treat diseases, it is an important strategy for improving the overall well-being of humanity. Some soils, with elemental toxicity, and being habitats for some pathogens and viruses, can also adversely affect plant, animal, and human health. Some inputs into agroecosystems, such as amendments which may contain toxic elements or irrigation with sewage sludge which contain pathogens and heavy metals, can adversely affect quality and quantity of food produced and jeopardize human health and wellbeing. Therefore, soil-specific judicious management of natural resources is essential to produce safe, nutritious and healthy food.

The objective of this volume is to collate and synthesize some pertinent information on "Food is Medicine" and identify knowledge gaps and policy implications to promote the One Health concept among the general public, academicians/researchers, policy makers, and human health practitioners. The volume has been made possible by a strong cooperation among several stakeholders involved in completing the project. We thank the authors for sharing their knowledge and wisdom and their timely submission of revisions of the chapters. We also thank the staff of the CFAES Rattan Lal Center for Carbon Management and Sequestration (Lal Carbon Center) and of the OSU College of Medicine for their help and support. Thanks are also due to staff at the Taylor & Francis, especially Ms. Randy Brehm and Tom Connelly, for corresponding with authors, managing the flow of manuscripts, and supporting timely publication of the book.

Rattan Lal
Columbus, Ohio
January 31, 2025

Marium Husain
Columbus, Ohio
January 31, 2025

About the Editors

Rattan Lal, PhD, is a distinguished university Professor of Soil Science and Director of the CFAES Rattan Lal Center for Carbon Management and Sequestration at The Ohio State University, as well as an Adjunct Professor of the University of Iceland and the Indian Agricultural Research Institute (IARI), India. He received a BSc from Punjab Agricultural University, Ludhiana, India (1963); MSc from the Indian Agricultural Research Institute, New Delhi, India (1965); and PhD from the Ohio State University, Columbus, Ohio (1968). He served as Senior Research Fellow with the University of Sydney, Australia (1968–1969), Soil Physicist at IITA, Ibadan, Nigeria (1970–1987), and Professor of Soil Science at OSU (1987 to date). He has authored/co-authored over 1100 refereed journal articles and more than 575 book chapters, has written and edited/co-edited more than 100 books. He was included in the Thomson Reuters list of the World's Most Influential Scientific Minds (2014–2016), and he is among Clarivate's Highly Cited Researchers in Agriculture (2014–2022). He was ranked #1 globally and in the US among Agricultural Scientists (Plant Scientists and Agronomists) of the world by Research.com (2023, 2024). He has received an Honoris Causa degree from nine universities in Europe, USA and Asia; the Medal of Honor from UIMP, Santander, Spain (2018); the Distinguished Service Medal of IUSS (2018); and is fellow of five professional societies. Dr. Lal has mentored 125 graduate students, 15 undergraduate honors students, and 188 visiting scholars from around the world. He has also visited 102 countries. He was President of the World Association of Soil and Water Conservation (1987–1990), International Soil and Tillage Research Organization (1988–1991), Soil Science Society of America (2006–2008), and the International Union of Soil Sciences (2017–2018). He is Chair in Soil Science and Goodwill Ambassador for Sustainability Issues for the Inter-American Institute for Cooperation on Agriculture (IICA). He was a member of the 2021 United Nations Food Security Summit Science Committee and Action Tracks 1 & 3. He currently serves as a member of the Board for International Food and Agricultural Development (BIFAD) in the Department of Agriculture and as Chair of the Scientific Advisory Board for the Strategic Environmental Research and Development Program (SERDP) in the Department of Defense. Dr. Lal is laureate of the GCHERA World Agriculture Prize (2018), Glinka World Soil Prize (2018), Japan Prize (2019), US Awasthi IFFCO Prize (2019), Arrell Global Food Innovation Award (2020), World Food Prize (2020), India's Padma Shri Award (2021), Mahatma Ghandi Award for Environmental Protection (2022), and the Gulbenkian Prize for Humanity (2024).

Marium Husain, MD, is a medical oncologist who specializes in treating patients with soft tissue and bone sarcoma and serves as an assistant professor in the Division of Medical Oncology at the Ohio State University. She was motivated to focus on sarcomas after meeting a patient suffering from rhabdomyosarcoma of the liver during a medical mission trip to Haiti early in her career. He was unable to access basic treatment services and passed away. Ever since, it has been her mission to provide the best cancer care possible to her patients while offering a comprehensive approach that focuses on the whole person. Marium Husain has been honored to discuss her work at respected conferences around the country and her work research has been widely published in peer-reviewed journals such as BMC Cancer, Journal of Geriatric Oncology and Hematology and Medical Oncology. She is a member of the American Society of Clinical Oncology and the Connective Tissue Oncology Society. She believes the research opportunities as well as the support from the faculty and multidisciplinary teams at The James will enable us to achieve our goals. She says that her colleagues are some of the most committed and dedicated people she has met and that it's a privilege to be part of that team.

Contributors

Waqar Ahmad
School of Agriculture and Food Sustainability (SAFS)
The University of Queensland
QLD, Australia

and

Primary Industries and Natural Resources, Conservation and Ecosystem Management Section
Technical and Further Education (TAFE)
Strathfield, NSW, Australia

Mukkram Ali Tahir
Department of Soil and Environmental Sciences, College of Agriculture
University of Sargodha
Punjab, Pakistan

Bruce C. Ball
Formerly Scotland's Rural College
Edinburgh, Scotland, UK

Saul J. Bautista
Medical Education and Community Engagement
Ethos Farm to Health
Long Valley, New Jersey, USA

Sonia Blough
Freelance Food Systems and Climate Justice Researcher
Spain

Eric C. Brevik
School of Agricultural Sciences

and

School of Earth Systems and Sustainability
Southern Illinois University
Carbondale, Illinois, USA

Julia Chalif
Department of Gynecologic Oncology
The Ohio State University Comprehensive Cancer Center
James Cancer Hospital and Solove Research Institute
Columbus, Ohio, USA

Laura Chambers
Department of Gynecologic Oncology
The Ohio State University Comprehensive Cancer Center
Columbus, Ohio, USA

Muhammad Ansar Farooq
CFAES Rattan Lal Center for Carbon Management and Sequestration (Lal Carbon Center)
School of Environment and Natural Resources
College of Food, Agricultural, and Environmental Sciences (CFAES)
The Ohio State University
Columbus, Ohio, USA

and

Institute of Environmental Sciences and Engineering (IESE)
School of Civil and Environmental Engineering (SCEE)
National University of Sciences and Technology (NUST)
Islamabad, Pakistan

Jessica Fulton
College of Medicine
The Ohio State University
Columbus, Ohio, USA

Dennis J. Grencewicz
College of Medicine
The Ohio State University
Columbus, Ohio, USA

Shiva Jahanbakhshi
Department of Food Science and Technology
College of Food, Agricultural, and Environmental Sciences (CFAES)
The Ohio State University
Columbus, Ohio, USA

Cristina Laurenti
Department of Food System Sciences
The Research Institute of Organic Agriculture (FIBL)
Frick and Lausanne, Switzerland

Rafia Mubaraka
Department of Soil and Environmental Sciences, College of Agriculture
University of Sargodha
Pakistan

and

Department of Geography
University of Koblenz-Landau
Koblenz, Germany

Md Nayem Hasan Munna
CFAES Rattan Lal Center for Carbon Management and Sequestration (Lal Carbon Center)
School of Environment and Natural Resources, College of Food, Agricultural, and Environmental Sciences (CFAES)
The Ohio State University
Columbus, Ohio, USA

Chinmayee Priyadarshini
CFAES Rattan Lal Center for Carbon Management and Sequestration (Lal Carbon Center)
School of Environment and Natural Resources, College of Food, Agricultural, and Environmental Sciences (CFAES)
The Ohio State University
Columbus, Ohio, USA

Mansonia Pulido-Moncada
Aarhus University
Aarhus, Denmark

Bob Quinn
Quinn Institute
Big Sandy, Montana, USA

Raiza Rezende
Regenerative Healthcare European Association
Lisbon, Portugal

Noor Us Sabah
Department of Soil and Environmental Sciences, College of Agriculture
University of Sargodha
Sargodha, Punjab, Pakistan

and

CFAES Rattan Lal Center for Carbon Management and Sequestration (Lal Carbon Center)
School of Environment and Natural Resources, College of Food, Agricultural, and Environmental Sciences (CFAES)
The Ohio State University
Columbus, Ohio, USA

Amir Sadeghpour
School of Agricultural Sciences
Southern Illinois University
Carbondale, Illinois, USA

Nargis Akter Shapna
Department of Soil and Environmental Sciences
University of Barishal
Karnokathi, Bangladesh

Emanuela Simonetti
Kamut Enterprises of Europe
Bologna, Italy

Daniel Spakowicz
Pelotonia Institute for Immuno-Oncology
The Ohio State University Comprehensive Cancer Center
Columbus, Ohio, USA

Eirini Tsirimokou
Regenerative Healthcare European Association
Athens, Greece

Bilal Walker
Urban Seeds Grow – A New Jersey Non-Profit Corporation
Newark, New Jersey, USA

Caroline Wheeler
Department of Internal Medicine
The Ohio State University Comprehensive Cancer Center
Columbus, Ohio, USA

Ye Xia
Department of Plant Pathology
College of Food, Agricultural, and Environmental Sciences (CFAES)
The Ohio State University
Columbus, Ohio, USA

Munir Zia
Sustainability & Innovation Department
Fauji Fertilizer Company Limited
Rawalpindi, Pakistan

and

School of Biosciences
University of Nottingham
Loughborough, UK

Zuzanna Zielinska
Regenerative Healthcare European Association
Rotterdam, Netherlands

1 The One Health Concept

Rattan Lal and Marium Husain

"Let food be thy medicine and medicine be thy food." This quote is credited to Hippocrates, the ancient Greek physician, to whom is also attributed an oath that each medical student takes as they start their medical training. Whether these were Hippocrates' exact words is contested, yet the principle behind the statement has echoed throughout time and research: our diets and nutrition can influence our health as humans. Historically, people utilized different foodstuffs and herbs to treat illnesses. Diets included more diverse crops and there was less processing of foods. Humans had a different relationship with the natural world as well, where there was more interaction with livestock and the production of food was not as disconnected from daily living. Modern food systems are globalized and complex that has contributed to why this disconnection with our food and agricultural practices differs from historical traditions. The influence of environment on human health has always been a constant; our perception of this inextricable link is just a matter of our perspective.

The One Health model is an initiative by the World Health Organization (WHO) to adjust how we approach human health. It is "an integrated, unifying approach that aims to sustainably balance and optimize the health of people, animals and ecosystems." (Figure 1.1). The emphasis is that human health is dependent on the health of animals, plants and the environment. Soil, an important component of the environment, is the link among all ecospheres (Figure 1.2). It has strong interactive effects on all biotic and abiotic components associated with it. By protecting the health of soil and these other planetary members, we also ensure that human health is not only maintained but optimized beyond our current practice model. The One Health model also helps "to address the full spectrum of disease control – from prevention to detection, preparedness, response and management – and contribute to global health security." As we approach a growing human population of 8.2 billion in 2025 that is expected to reach 10.3 billion by mid-2080s [1], it is critically important to realize that for inequitable food access, approaching our food systems from a new/adjusted model, like One Health, can address the gaps in our current approach and improve health outcomes. Inadequate food access affected 733 million people in 2024 [2], and the vulnerability of the population to hunger and hidden hunger is being aggravated by climate change, political disturbances, and war which lead to severe disruptions in food production and supply chains. Conflict, climate change and other social/political issues are driving billions of people towards hunger and malnutrition by degradation of soil and the environment in war-prone and climatically vulnerable regions. Modern war can severely degrade soil health and undermine the quality (and quantity) of food grown on contaminated soils in polluted environments. Polluted soil and water damage human and animal health [3]. Thus, quantity and quality of food must be safeguarded by protecting and managing the soil and the environment in which crops are grown. Rapid urbanization is also aggravating risks of soil degradation [4] and affecting quality of food and human health.

This modified approach to food systems also includes adjusting our soil and agricultural practices. Indigenous peoples in the past implemented growing and cultivation practices that promoted diverse crops, crops that would have high yields and nutrition density based on the season, as well as an approach to livestock that have influenced the principles of the One Health model [5]. Current agricultural practices promote monocrops and short-term yields that maximize a profit within the globalized economic model [6]. Nutritional density has also been impacted by these practices, leading to decreased concentrations of micronutrients [7]. Initiatives like "soil to table"

DOI: 10.1201/9781003469742-1

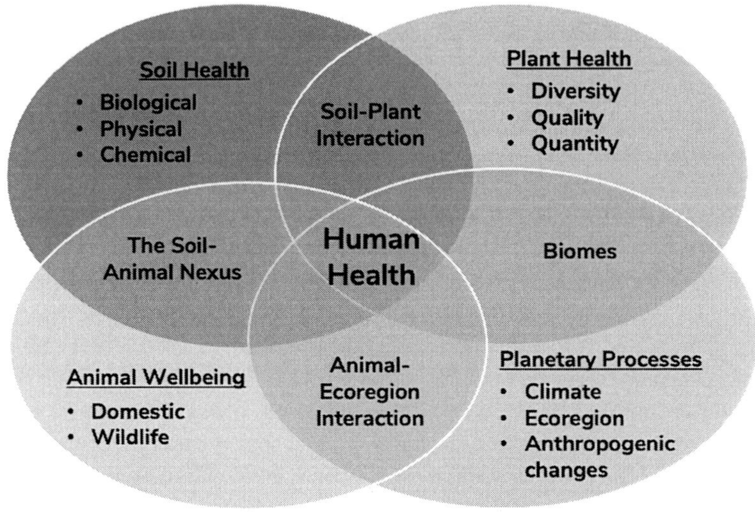

FIGURE 1.1 The One Health model based on an integrative unifying approach.

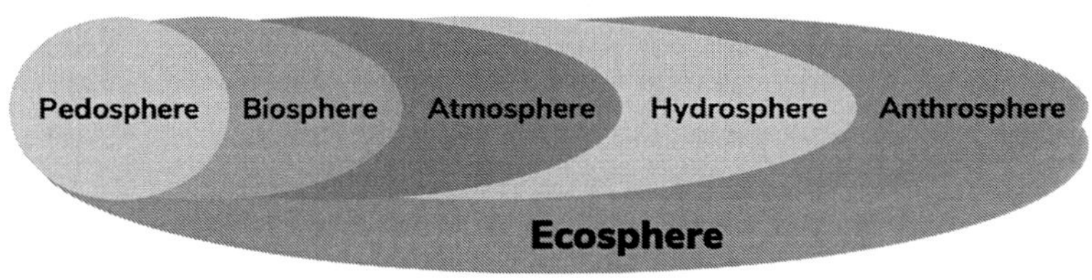

FIGURE 1.2 Pedosphere (soil) is the connecting link among all ecospheres.

attempt to redefine our approach to food access but do not address underlying soil management. Adjusting our approach to food systems can have significant impacts on human health and One Health.

This book delves into different aspects of how food can be used to transform human health and used medically. The authors provide evidence-based and personal insights into the food system. Chapter 1 introduces the concept of One Health and how our human health is not reflected only in what diseases we have, but also the foods we eat, the environments we live in, the animals we interact with, and the food and agricultural systems that define the modern era. Chapter 2 discusses growing diverse and nutritious food crops, specifically wheat, and how this can be used to improve sensitivity to modern wheat crops. Chapters 3 and 4 delve into soil practices and how our current agricultural practices impact our interaction with food and our health. Chapter 5 explores how the microbiome of the intestinal tract impacts human health, highlighting the importance of how food is not only cultural and social but biological. Chapter 6 discusses how micro and macro systems within and involving soil affect human health through micronutrient availability. Chapters 7 and 8 discuss the impact of soil and food systems on humans from a sociologic perspective, how our foods are transformed by culture and society. Finally, Chapters 9 and 10 focus on the deleterious impacts of current agricultural and soil practices that not only lead to human morbidity but also have harmful impacts on other living organisms. Chapter 11 concludes with action items and a path forward to transform our food and food systems to be a healing force, not only for our health as humans but also for all living beings on our planet. Chapters 12 and 13 are expert input from Dr. Lal on interdisciplinary strategies and urban agriculture for planetary health.

The objective of this book is to enhance awareness about the importance of the OneHealth model in developing and implementing a holistic approach to managing human health by protecting, restoring and sustainably managing the habitat and the environment in which people live.

REFERENCES

1. UN. 2024. World Population Prospects 2024: Summary of Results (UN DESA/POP/2024/TR/NO.9). https://www.un.org/development/desa/pd/sites/www.un.org.development.desa.pd/files/undesa_pd_2024_wpp_2024_advance_unedited_0.pdf
2. FAO, IFAD, UNICEF, WFP, and WHO. 2024. *The State of Food Security and Nutrition in the World 2024: Financing to end hunger, food insecurity and malnutrition in all its forms.* FAO; IFAD; UNICEF; WFP; WHO, Rome, Italy.
3. Certini, G., and R. Scalenghe. 2024. War is undermining soil health and availability more than urbanisation. *Science of The Total Environment* 908: 168124. doi:10.1016/j.scitotenv.2023.168124
4. Durdu, B., F. Gurbuz, H. Koçyiğit, and M. Gürbüz. 2023. Urbanization-driven soil degradation; ecological risks and human health implications. *Environmental Monitoring and Assessment* 195. doi:10.1007/s10661-023-11595-x
5. Johnston, L. 2022. Architects of abundance: Indigenous regenerative food and land management systems and the excavation of hidden history. University of Alaska FairbanksProQuest Dissertations & Theses, 2022.29322544.
6. McMichael, A.J. 1999. From hazard to habitat: Rethinking environment and health. *Epidemiology* 10(4): 460–464.
7. Bhardwaj, R.L., et al. 2024. An alarming decline in the nutritional quality of foods: The biggest challenge for future generations' health. *Foods* 13(6): 877.

2 Healing the Earth by Growing Food as Medicine

Bob Quinn and Emanuela Simonetti

INTRODUCTION

The idea of food as medicine has been gaining interest in recent years. There are many reasons for this and in this chapter, I hope to review just a few of them. I began my understanding of food as medicine from my firsthand experience in agriculture and food as a dry land farmer specializing in ancient wheat. After receiving degrees in botany and plant pathology at Montana State University and a PhD in plant biochemistry at UC Davis, I returned home in 1978 to our family farm near Big Sandy, Montana which my grandfather had started in 1920. I joined my father growing commodities of wheat and barley in a very industrial way. By 1983, I had started Montana Flour and Grains to help market our wheat directly to whole grain bakers in California and increase the profitability of our farm. Because of their requests, I started buying and selling organic grain and then started experimenting and converting my farm to organic production. By the end of 1988 we had stopped using chemicals on our farm and were no longer growing commodities, but were only growing food; good high-quality food as we termed it. In 1986, we introduced an ancient wheat to the health food market which we soon found had surprising healing effects on some people plagued with wheat sensitivities. This observation led to an extensive research program over many years which will be reviewed later in this chapter by Dr. Emanuela Simonetti as strong evidence for supporting the concept of food as medicine. Most of what you will read in this chapter, however, will be my own experiences and observations over the last seven decades from the production and processing side of agriculture, as well as my experience in markets and on farms throughout the world. Much of the story is based on my book *Grain by Grain: A Quest to Revive Ancient Wheat, Rural Jobs, and Healthy Food* which I wrote with my co-author, Liz Carlisle. There you can find more detailed information on my experiences and philosophy (Quinn, 2019).

FOOD AS MEDICINE – NOT A NEW IDEA

Whenever we hear a reference to food as medicine, we often think of Hippocrates from ancient Greece, considered the father of Western medicine (c. 460-370 BC). He is often loosely quoted as saying "Let food be your medicine and medicine be your food." It should not come as a surprise to anyone that similar ideas are found in many teachings throughout the ancient world including Confucius and Buddha.

Confucius (born in 551 BC in China) felt that food was a pillar of the state alongside the army and the people's trust. A hungry population is rebellious, and good food makes a healthy productive populace. He understood that acquiring, preparing, and eating food dominated the lives of most people. He said that food can never be too clean, and meat can never be sliced too thin. Confucius felt that profit motives and large-scale operations would result in depreciation of food quality. Eating fresh, local, and in-season food remains the best way to ensure quality. Knowing the source of our food increases human connection to the food and helps to ensure its quality (John, 2025).

Buddha (c. 563 BC (Nepal) – c. 483 BC) taught that troubles are caused by words flowing out of the mouth. Illness is caused by food going into the mouth. He also taught that using caution and moderation in what we consume is an important practice for good health. Before consuming any food,

DOI: 10.1201/9781003469742-2

we should determine if the food is fresh, if it is thoroughly cleaned, and how much is a reasonable amount to eat. The *Sutra of the Teachings Bequeathed by the Buddha* [*Fo Yijiao Jing*] states,

> When we eat, we should regard our food as medicine, for consuming too much or too little is not healthy. A regular and proper dose can support our bodies, cure our hunger, relieve our thirst, and prevent us from becoming ill. Like bees gathering honey, we take what we need, but we don't consume the whole flower.
> (Jing, 2017)

There are also many other similar teachings from the world's greatest religions which contain this important link between good food and good health. We would probably find many additional great teachers and philosophers of the past with similar ideas if we had access to all their writings.

REPLACING THE FOOD AS MEDICINE PHILOSOPHY IN THE MODERN WORLD

The modern world since WWII, however, has seen the rise of another philosophy. This was not really a teaching or a philosophy but more of a way of life and the dominant force of two enormous industries in America. The first is the agrichemical industry forming the heart of our new agro-industrial complex, and the second is Big Pharma forming the heart of the medical-industrial complex. Both of these, now in some cases owned by the same giant international corporations, such as Bayer, have come together to nearly erase the idea that food should be our medicine, and medicine should be our food.

On the one hand is modern medicine which has its origins in the acceptance of the germ theory of the mid-1800s. Through this, people began to understand the causal effect of infectious diseases and discover how to best treat them. The discovery of antibiotics and various vaccines added greatly to the treatment and prevention of major infections and epidemics. Slowly, the excitement over wonder drugs such as penicillin, which saved countless lives during WWII, gave way to dependence upon drugs as a normal way of life rather than something to be used periodically to help overcome a health crisis. Over the decades, infectious disease has decreased in importance while chronic disease is on the rise and is now the major cause of death in developed countries of the modern world.

In America today, according to CDC statistics, three-quarters of our population has at least one chronic disease. Over half of the population suffers from two or more chronic diseases (Watson, 2025). Most of the drugs taken today do not cure disease but focus on treating symptoms. The goal is to prolong life and enable people with these diseases to function at some acceptable level of performance. Big Pharma dominates the airways and media with a constant stream of the newest pill to treat the latest malady. Most medical schools spend less than a handful of hours in four years of instruction to cover the importance of nutrition in health care. In America the only traditional medicine which focuses on nutrition is veterinary medicine. In veterinary schools there is great emphasis on the study of nutrition.

During the 20th century, as major infectious diseases were addressed and became less prevalent, attention turned to chronic disease. In this case the focus was less on treating the root cause and more on treating the symptoms. The goal was reducing pain and suffering and increasing the ability of the sick to function. Very little effort was made to understand the source of the trouble and address the real cause of the disease in the first place. More money could be made by Big Pharma rolling out an ever-increasing number of new drugs. Most of the US population now take a handful of pills every morning. The ancient adage that food should be our medicine has been twisted into the idea of the modern age that pills should be our food.

THE FOCUS ON CHEAP ABUNDANT FOOD

Meanwhile, down on the farm, the way we have produced food over the last century has also drastically changed. We have gone from an agrarian society when 30.2% of the US population lived on a farm in 1920, the year my grandfather started our farm in Northcentral Montana, to less than

2% one hundred years later. Industrialization of agriculture has been the major source of this transition. My grandfather, Emmet Quinn, started farming with horses and mules in 1920. In 1927, he bought his first tractor, a small Cletrack crawler that ran on tracts instead of wheels. He added to that a small pull-type combine to harvest his wheat. He ran the tractor, his nephew ran the combine, and a hired man hauled the grain to his grain bin with a horse drawn wagon. This was a huge difference from my great grandfather's harvest crew just a couple miles down Little Sandy Creek from the Quinn Ranch. His harvest began with a worker driving a team of horses pulling a binder through the wheat fields. The binder cut and tied the grain into bundles and dumped them into piles of four or five. A couple more workers stacked 12 to 15 bundles in shocks by hand using a pitchfork. The bundles were placed in shocks with the butt end (the cut end) down and the top of the grain up for drying. They were then forked by hand into a wagon and hauled to a central place on the farm where a thrashing machine was set up. This was powered by a long belt driven by a large steam engine on wheels, which was also used for the heavy farm work such as the first breaking of sod with a moldboard plow. This machine required a couple of men to make sure the fire was maintained to run the steam engine and that other operations pertaining to the engine were going properly. During the thrashing operation, a couple of men would throw the bundles one by one into the trashing machine from the wagon when it arrived from the field. A couple of men would fill 100-pound burlap bags with clean grain from the thrashing machine, sew them shut, and load them on wagons to be hauled to the elevators in town or to storage buildings on the farm. This whole operation required about 15 men. So, when my grandfather reduced five implements to three, and a crew of 15 to three, and the time required for the whole harvest was reduced from weeks to days, To him it was a miracle.

The next miracle occurred about 20 years later when my father returned to the farm after his service in the Navy during WWII, after which he had spent a couple of years working for Westinghouse as an electrical engineer. He returned to the farm in 1948, when I was just a baby. He got a self-propelled combine to replace the tractor and pull-type combine his father had used. This meant that we now had just one combine operator and two truck drivers. They drove the trucks required to haul the grain away from the combines. This job fell to my sister and me as soon as we were big enough to see over the steering wheel and at the same time be able to push in the gas and brake pedals, as well as work the clutch and stick shift. Dad helped the newly organized REA (rural electric association) string wires to our farm and the farms of our neighbors so that we could enjoy the wonders of electric power that our city cousins had enjoyed for decades. This meant the grain could be elevated into storage bins on the farm by electric augers rather than shoveling it by hand. Steel bins replaced the wooden ones on the farm to better protect the grain from birds, rodents and rain.

The biggest change, however, came with the introduction of herbicides which could kill broadleaf weeds without harming narrow leaf grassy plants like wheat or barley. My father thought that was also a miracle, as weeds had been a plague for years and were only growing worse as crop diversity diminished. Everyone immediately adopted the new chemical weed control which seemed to have no downsides. In 1974 Roundup was introduced as another miracle drug for killing weeds. The active ingredient was glyphosate. Glyphosate works by killing living organisms which rely on a certain metabolic pathway to produce an essential amino acid. All plants except those genetically engineered to be resistant to this herbicide rely on this pathway to keep them alive. Many bacteria and fungi also have this pathway. No birds, fish or animals have this pathway in their systems, however, so they, including humans, are not affected by glyphosate directly. Because of this fact, the makers of Roundup pronounced it safe for humans. However, in the gut of every human live more bacterial cells than the number of human cells in their entire body. So, when humans ingest food or water contaminated with glyphosate, there is significant amount of damage done to the microbiome of the gut. The result is different for each individual, but indirectly, the damage to the gut microflora can be responsible for many unrelated diseases. It can even directly cause cancer. It took a lawsuit by a school grounds keeper in California, stricken with lymphoma, to require Monsanto to release documents that showed they knew the cancer-causing potential of glyphosate but refused to make that information public. In fact, their effort was made in the opposite direction, lobbying the EPA (Environmental Protection Agency) to increase the limits of glyphosate allowed on grain, and saying there was very small risk with little exposure to the poison.

In the beginning, glyphosate was used just for weed control in fields left fallow for the summer to conserve moisture for the following year, a common practice in the semi-arid Great Plains. In this region, and in other parts of the country which receive more moisture, it was also used for preplant, and postharvest weed control. This meant that the poisonous herbicide never touched a crop directly that would be eaten by animals or humans. The EPA limit was at 100-200 ppb (parts per billion). In 1993 efforts were made to start increasing this tolerance and now it is 30,000 ppb or 30 ppm (parts per million). This was done because Monsanto introduced GMO (genetically modified organisms) crops which were resistant to glyphosate. Now the herbicide could be used for complete weed control in the growing crop which was not affected by the herbicide, while all non-resistant weeds died. Previously, these crops were cultivated or sprayed with chemicals which only killed certain weeds. This success was soon followed by a push to use glyphosate as a desiccant to hasten the ripening of crops especially where fall rains or snow could damage them before they were able to be harvested. Before glyphosate these crops were cut with a swather and allowed to dry on the ground to speed the ripening process. This is risky because losses can occur if rain falls on the swathed grain and causes sprouting or molding. Many farmers were happy to adopt the new technology, and the sale and use of glyphosate continued to climb until today it is the most widely used herbicide in the world. A 2020 EPA report stated that 280 million pounds (140,000 tons) were applied to 298 million acres of crop land in America which is about 90% of the estimated 330 million acres of total crop land in America (Burd, 2020). No wonder nearly all our surface water, ground water, soil, and even rainwater, as well as most of the people living in America are contaminated with residual amounts of glyphosate.

Synthetic chemical fertilizers were introduced right on the heels of the new 2,4-D herbicide. By the mid 1950s my father was experimenting with its effectiveness and turned out to be an early adopter. The results of adding chemical NPK (nitrogen, phosphorous and potassium) to the fields were not as quickly seen, or readily adopted by farmers in our region, especially in dry years. By this time, in 1954, most of the land in our area had only been farmed for a little more than 35 years. Since we can only grow crops every other year because of lack of rainfall (around 12 to 14 inches) there had been only about 17 crops removed from the fields by this time. The natural nutrition in the soil was just beginning to be depleted so the effect of added chemical fertilizer was not always consistent and worth the extra cost and trouble to apply. However, within 10 years or so chemical fertilizer had become the norm for most farmers in the region.

With all these industrial changes brought to the farm belt, the driving goal for American agriculture pushed by the US government was being realized. That goal was the production of abundant cheap food for the country. The drive for more abundant food continued to produce results. The surplus of foods, especially grains, could now be exported bringing new income to industrial agriculture. The industrialization of almost all forms of agriculture, spurred by the enormous growth of the agrichemical industrial complex and ever larger farm machinery, continued to increase the amount of production and increase efficiency of production and processing. This combined to drive down prices and decrease the percentage of the US spending on food since WWII from 23% of the average American budget to nearly one-third that at about 7.2% in 2023, according to the Center for Economic and Policy Research (Baker, 2024). Claims were repeatedly made that we had the cheapest, safest and most abundant food supply in the world, and that our new mandate was to feed the world.

So, what could possibly be wrong with such a rosy picture? In recent years, however, there has been a startling awakening to and understanding of the high cost of cheap food.

THE HIGH COST OF CHEAP FOOD

The High Cost to Our Farms and Farmers

Since WWII, in the United States we have seen a rise in abundance of cheap food. After WWII, and especially after seeing hunger in Europe and remembering the challenges of the Great Depression, it became a matter of national security as well as national pride to ensure we had an abundant supply of cheap food in America. All kinds of programs and incentives went into achieving this goal. By all measures, the goal was achieved and was a resounding success. The abundance seems to have no drawbacks, but it is tied to cheap food and now we are starting to see that there is a very high cost to cheap food. The burden of this cost extends well past the checkout counter where the token cost of the food is paid. The rest of the cost is paid out of four main areas: our farms, our communities, our planet, and finally, our health.

Even though we as farmers are heralded as a wonder because of all our production, there is a dark side to all the accolades. On the one hand, according to the 2022 USDA (United States Department of Agriculture) Census of Agriculture summarized by the American Farm Bureau, the average US farmer feeds 1665 people (AFBF, 2025). In 1940, that number was closer to 19. In 1920 the percentage of Americans farming was nearly 30%. Today that number is closer to 1.5%. The downside of all this improvement is the loss of farms and farmers, and the economic hardship most farmers find themselves in today. In my neighborhood over three-quarters of the neighbors I grew up with are gone. They sold out or went broke. Most of the land has been purchased by the neighbors who remain, and some has been purchased by absentee landlords. For the last 50 years most of my neighbors have discouraged their children from returning to the farm, saying, "There is no future in farming here. You are better off to go to college, get an education and a job in town." Meanwhile, most of those who are left on the land are not in good financial position. This is a very high cost paid by the farmers and their families and workers who grow our food and yet are not able to be fairly compensated for their labor and efforts. Most farm families have at least one member or more with jobs in town to supplement the scanty farm income. That was almost unheard of when I was growing up.

Nationally 14.3 cents of each dollar spend on food sold in America went to the farmer according to a 2018 report of the farmer's share of the retail food dollar put out by the National Farmers Union. For wheat farmers, the number was far worse: 3.1 cents out of every dollar spent on bread goes to the grower (NFU, 2024). The real bad news is that this 3.1 cents is not enough to pay for all the chemical inputs the industrial farmer is now buying, as well as the cost of the machinery, plus the cost of the land. Because the cost of production exceeds the value of that production, many farmers are going broke. Over the last few decades, the majority of net income received by farmers in Montana has been provided by government subsidy programs. The depression caused by this situation has contributed to Montana becoming the leading state for suicide in America. I would say this is a very high cost for the cheap food that is being purchased in the grocery stores. Of course, this problem is of little concern to the agricultural industrial complex. As one farm goes out of business and disappears, there is always capital to fund the next farmer or an investor to take a turn until their money is gone and the cycle repeats itself. The drumbeat of the chemical companies is that their inputs are necessary to feed the world and keep the US leading the world's crop production. This philosophy dominates the teaching in high school agriculture classes, universities, the discussions by the country's largest farm organizations, and the political scene in Washington DC.

The High Cost to Our Communities

When a large number of farmers leave the community, the economy of the community suffers. The population of our little town of Big Sandy has dropped from nearly 1,000 in 1960 to about 600 today. As farmers and their families left, businesses on Main Street closed, so that now nearly half the buildings are closed and empty. When those businesses close, more people lose their jobs and are forced to move. So, the cycle continues. Our school population is down about two-thirds from when I graduated. The five grain elevators that used to buy the grain from local farmers have all closed. Now farmers are forced to buy bigger trucks and haul grain an additional 35 to 50 miles to the nearest elevator. Some drive a 100 miles for a little better price. This is all in the name of consolidation to increase efficacy and keep costs down. But what is happening is that corporate profits for out-of-state giants are ever increasing. This is mostly at the cost of jobs lost in the rural communities when small business owners have to close their shops, as farmers lose their farms and move away. So, another segment that is paying for the high cost of cheap food, is our small rural communities. These communities, which used to contribute to the life blood of America, are having their resources sucked out of them like a kid guzzling a milkshake through a straw.

The High Cost to Our Planet

Another entity that is paying a high cost for cheap food is our planet, due to the pollutions from the contamination of agricultural chemicals. Probably the most well known and problematic example is the dead zone in the Gulf of Mexico which is approximately 6,705 square miles (about the size of New Jersey) according to a 2024 report by the NOAA (National Oceanic and Atmospheric Administration) (Jessmer, 2024). This dead zone is produced by the runoff of chemicals used in agricultural fertilization programs and pollution from industry and our cities (chiefly nitrogen and phosphorus). These chemicals enter the waterways of the Mississippi watershed, which makes up 40% of the land of the continuous 48 states, and are deposited at the river's mouth in the Gulf of Mexico. The excess of nutrients stimulates an overgrowth of algae. When the algae die and decompose, the oxygen levels in the water are depleted. The resulting low oxygen levels cause fish and shrimp to leave the area, producing what we refer to as a dead zone.

Another alarming report of the Water and Land News, posted in April of 2019, reviewed the results of 55,000 water wells in Iowa tested between 2002 and 2017 (WLN, 2019). They found that "22% had nitrate levels at or above 5 ppm, the threshold at which many more recent studies have found an increased risk of colon, bladder, ovarian, kidney and rectoral cancer, as well as birth defects." Twelve per cent were over 10 ppm which is the legal EPA limit for drinking water. So it is not only our oceans being contaminated, but also the drinking water from wells across rural America.

What is even more astonishing are reports of glyphosate being found in the rain as early as 2004 (Chang et al., 2010). Researchers found more than 60% of 80 rain samples collected in Iowa, Mississippi, and Indiana in 2004, 2007, and 2008 were contaminated with glyphosate. We have tested the rain falling on our farm over the past several years and can confirm this report. We are now conducting research on this subject on our newly founded, non-profit, 700-acre, regenerative organic research, education and health institute, the Quinn Institute. We want to better understand the nature and origin of the contamination in the rain. In 2024, we had measurable contamination in 4 of 20 rain events which ranged from 1.7 ppb to 12 ppb. Glyphosate was detected in an additional 6 of the 20 samples. This means that half (10 out of 20) of the rainstorms we had in 2024 had detectable amounts of glyphosate in the rain falling from those storms. In past years, we have found that an organic farm 20 miles west of us has had up to four times the contamination of glyphosate that we have had in the same storm. That farm is surrounded by chemical farming for more than 10 miles in all directions, while we have chemical farming neighbors on only two sides of us, as we have the Bear's Paw Mountains to the northeast and the Missouri River Breaks to the southwest which precludes farming of any kind. Over the past seven years or so that we have done this testing, we have observed less contamination in dry years when there is less glyphosate used in the fields because no weeds were growing. We hope to expand this research in the future to include dust sampling and to compare our farm with a site in the middle of a large pasture near us where no glyphosate is being sprayed. We hope to get a better understanding of what is happening.

It is interesting to note that the limit of glyphosate contamination for organic wheat in America which is set by law at 5% of the EPA limit, is 1500 ppb (1.5 ppm) compared to 10 ppb in much of Europe. As mentioned earlier, Monsanto made sure the limit was raised a few years ago to a limit of 30 ppm before they started recommending the use of glyphosate in their Roundup herbicide on GMO food crops resistant to Roundup and also as a preharvest desiccant. Its use is the central feature of most regenerative farming systems which are not organic. So, not only is its use widespread, but it now has also been accepted by most of the regenerative movement which focuses on no-till farming. Many of these regenerative advocates say no-till is the only way to farm because they do not disturb the soil with any tillage so there is no mechanical disturbance. However, there is significant chemical disturbance in the soil where poisons are applied. Unfortunately, this is mostly ignored. Hence, a certain legitimacy is being lent to glyphosate in the excitement over no-till and regenerative agriculture. As a result, the hundreds of millions of pounds applied annually to crop fields in the US continues with few questions being asked. No wonder glyphosate contamination is now found in ground water, surface water, soil and even in rain across the country.

It is hard to estimate the high cost of this contamination to the economy or the health of our people. It is even harder to estimate the extremely high cost of the cleanup of such pollution on the earth. This is a cost that should not be borne by the taxpayers but by the makers of these defective products that do not stay where they are placed, even when used according to the manufacturer's instructions. In addition, these products do not quickly break down into harmless byproducts as the manufactures claim but are now found throughout our environment.

Another pressure on the planet is the enormous amount of greenhouse gases released to the atmosphere by the chemical production of nitrogen fertilizer. An article in *Nature* from 2022 reported that the synthetic nitrogen fertilizer supply chain was responsible for estimated emissions of 1.13 $GtCO_2$ in 2018, representing 10.6% of agricultural emissions and 2.1% of global greenhouse gas emissions. This included production, field emissions and transportation (Menegat et al., 2022).

In like manner, as the organic matter of the soil has declined by estimates up to 50% due to the current industrialized extractive method of agriculture, the resulting release of CO_2 to the atmosphere is astounding. One report declares that emissions from soil alone account for 15% of the entire global increase in climate warming, caused mostly by the increase of carbon dioxide from the loss of organic matter in the soil (Kopittke, 2024).

The High Cost to the Health of Our People

The last group or problem situation that pays a high price for the cost of cheap food is certainly paying the greatest price and is suffering the most. It is the health of our people. As mentioned before, today in America according to the CDC (Centers for Disease Control) statistics, approximately 76% of our population suffers from at least one chronic disease. This means that over three-quarters our population is sick with some type of debilitating chronic disease, most of which lead to an early death. Right after WWII, we were a much healthier nation, and it is estimated that the average household spent 4–5% of their income on health care. Today, that number exceeds 18%. The CDC goes on to say that 60% of the cause of these problems is connected to our diet. This problem might be divided into five main categories. First, there can be a problem with eating too much of the wrong thing such as sugar and fats. Second, there can be a problem with eating food low in nutrient value due to poor production systems, worn-out soils, as well as harvesting unripe crops before they reach peak nutrient levels. Or third, nutrition can be lost from fresh food during shipping over great distances for many days, or sitting on store shelves for long periods of time. Nutrients can be stripped from the food by over processing and/or trouble-causing additives can be introduced. This fourth and perhaps greatest source of trouble with our food comes from excessive processing when not only is nutrition lost but also countless additives are added to the food to alter taste, texture, or shelf life. Some additives are supposed to increase nutrition by adding minerals, vitamins, carbohydrates or protein. Many of these additives, however, cause more problems than they are trying to solve and, in many cases, they aggravate existing problems or create new ones. Finally, the fifth problematic category is the rise of traces of agrichemicals in our food, especially glyphosate, the most widely used chemical in America and throughout the world. As a result, today most Americans have significant traces of glyphosate in their body that can be measured in urine and hair samples.

Once again, it is a great irony that so many methods used to lower the cost of food at the checkout counter in so many ways, are actually costing the buyers, or those theylove, a price that most cannot afford to pay, and many would be happy and thankful to avoid if they could.

THE HEALING EFFECT OF GROWING FOOD AS MEDICINE

Healing Our Farms and Farmers

It may come as a surprise to some that the solutions to all the serious problems of the high cost of cheap food described above could be mediated or solved completely by a change in our food production and processing systems. I believe that the solution lies in starting with good seed, placing that seed in healthy soil, harvesting the resulting crops at the peak of nutrition, and finally processing those crops as little as possible, as near to the farms as possible, and using that food to feed the local population first before shipping it out of the region to feed others. I have come to this conclusion from my personal experience over the past 70 years. It is interesting to see how these conclusions are becoming more and more validated in recent years by research publications.

In taking the high-cost challenges one by one, I will start with my experience operating our wheat and cattle ranch. After completing my college education, I decided to move back to Montana to raise my family the way I was raised, by working together on the farm rather than roaming the streets of suburban Northern California where we were living after I finished college. Our farm was 2,400 acres at the time, which was about the average size of a dryland wheat and cattle ranch in Montana. It was big enough to support one family comfortably but not big enough for two. My parents were also still helping with the farm and being supported by it. I tried to figure out different ways to add value to the farm so my wife and/or I would not have to go to town to get a supplementary job, which was starting to be the rule rather than the exception for farm survival. I did not have much luck with new ideas, although I tried experimenting with a couple of non-starters like creating a soil-testing lab in my basement and selling prairie dog puppies to pet stores in Japan. In the summer of 1983, a fourth cousin of mine, Steve Lancaster, from Orange County in California contacted me and drove up for a visit. We hit it off and became friends. Since he was currently looking for a new job, we came up with the idea for starting a business together to provide high-protein wheat to whole grain bakers in California. He would be responsible for customer service and marketing in California, and I would be responsible for quality control and wheat procurement in Montana. Within a week, he had found our first customer, Food for Life in the LA area. Jim, the owner, was willing to pay us a dollar a bushel over the regular price of wheat which at time was about $ 3.00/bushel. So, I had just increased the value of the wheat produced on our farm by a third. That was a huge advantage for our farm. We were able to sell directly to the end user and put into our pockets the margins that usually went into the pockets of out-of-state grain companies.

This business worked very well the first year. Then, in the summer of 1984, Jim asked me to find organic wheat for his bakery. I was taken by surprise. I had heard of organic farming while I was in California and even knew that there were some organic farmers in Montana, but I did not believe in organic farming. After all I had been brought up using chemicals on our farm. I had been taught in all my agriculture classes in high school and all the way through college that not only were chemicals essential to modern agriculture, but also that they were seen as the hope for feeding the world. However, I did not want to let my prejudice stand in the way of keeping my best customer satisfied. There were not many organic farmers in Montana in 1984 nor was there much organic wheat. After a few phone calls, however, I found a farmer in the northeast corner of Montana who had some organic wheat in his bin that he was willing to sell me. I drove the 300 miles from

Big Sandy to Plentywood with my truck and had the farmer sign an affidavit saying he was following the California Organic law. There was no certification required in those days. A self-declaration was enough. I paid him and loaded the grain into my truck. I headed home, cleaned and bagged the grain and put it on a truck to California. Jim loved it and wanted more. Pretty soon I had met and become acquainted with almost all the organic farmers in Montana. They became friends and started inviting me to their winter gatherings. I was impressed with their stories of being able to grow their own nitrogen using legumes, eliminating the need to buy chemical fertilizer. They used rotations to break up cycles of weeds, pests and disease so they did not need to buy chemical pesticides and herbicides. They told me how the tilth of their fields had increased, and the fields felt softer to walk on after using these sustainable (now referred to as regenerative) organic principles for a few years. I was amazed by what I heard. I decided we needed to experiment on our farm.

We had 20 acres that had been in alfalfa for three years and had not had any chemicals applied to it during that time. We worked the alfalfa into the ground during the summer of 1986 and then tested the soil for nitrogen. We added enough urea to the neighboring 20 acres so that the measurable nitrogen would be the same in both the organic and non-organic fields. Then we planted winter wheat in both fields that fall. The organic field was planted with seven-inch spaced drills normally used for our spring grain, instead of the 14-inch spaced drills used for winter wheat to catch more snow with the deeper furrows it made. My new organic friends told me that one of the best controls for weeds was getting a quick ground cover by the cash crop planted in the field. Having seven inches between the rows would allow the wheat to cover the ground twice as fast as having 14 inches between the rows. That is exactly what happened. The next spring, the wheat in the organic field covered the ground quickly, not allowing weeds the light they needed to push through the wheat canopy. Meanwhile the neighboring chemical field was sprayed with 2,4-D as usual in May which killed all broadleaf weeds in the field at that time. However, we then received some unusual July rains. In our region most of the rain normally ends by the late June. Because there was so much space for sun to shine between the rows on the chemical field, weeds began to grow and by harvest time near the first of August, there were several serious patches of weeds in the chemical field while the organic field was clean, having almost no weeds at all. But what about the yields? We found at harvest time, after a normal year of rainfall, that the organic field yielded 33 bushels/acre (one bushel = 60 lbs.) and tested at 15.3% protein. while the chemical field yielded 34 bushels/acre and tested at 16.0% protein. Both of those results are so close as to not be significant. I was thrilled. My experiment had worked. My father was astonished. Here he had been spending tens of thousands of dollars on chemicals annually for years and the organic field had matched the quantity and quality by using no chemical inputs at all. The organic crop had benefited by the previous three years of alfalfa.

After some back-and-forth discussion, we agreed to increase our organic experiment to 15% of our land for the next year. That year, 1988, was a drought year, complete with a plague of grasshoppers which often accompanies a drought in our area. Instead of summer fallowing our two test plots to conserve moisture that year as we normally would do, we planted them both to durum. This was a mistake we plainly saw after it did not rain. I had pushed for the planting of those test plots because I was curious to see how fast the nitrogen produced by three years of alfalfa would be used up compared to the chemical nitrogen applied in the form of urea. The yield on the chemical field was zero, while the yield on the organic field was four bushels/acre. That was not a lot but it paid for most of the expense of seeding and harvesting, considering we were receiving twice the price for organic than the price of non-organic wheat. Over the years, we have seen that the results of these first two years of experiments were typical of the comparison of our organic grain yields with country averages for non-organic grain. In average rainfall years, the yields of organic and non-organic were about the same while in very wet years, non-organic pulled ahead of organic by 30 to 40%. In dry years, however, when yields went below 10 bushels per acre, the non-organic went to zero very quickly while the organic fields held up much better. Organic crops have more tolerance to drought.

Meanwhile the fields which were close to the pastures were under heavy attack from grasshoppers that year. The non-organic field was sprayed with malathion which kills everything in short order. But within 10 days or less, the malathion had dissipated and a new horde of grasshoppers arrived. The crop was poor, and I could not afford to spray again. At harvest time there were about as many grasshoppers in the combine tank as there were kernels of wheat. My organic friends told me to spread wheat bran laced with Nosema locusta, a biological control, around the edge of the fields where the grasshoppers were coming into my organic crops. We did that and as the grasshoppers ate the bran, they became infected with the protozoa in the product which are slightly smaller than a bacterium. That infection soon killed the grasshoppers and as they lay dying, their grasshopper friends came and ate them, reminding me of the similarity to politics in America today. In this way the infection spread throughout the incoming grasshoppers and although they ate about 30 to 40 feet of the crop to the ground along the edge of the field, what was left had almost no grasshoppers at harvest time.

This time I was the one who was astonished. After thinking over my experiences of the previous two years, I went to my father and said,

> Dad we have done everything the chemical companies and their advocates told us to do and for the last two years, our organic test plots or fields have equaled or surpassed the chemical fields in yield and protein. I am done with chemicals. I am going to figure out how to farm without them from now on.

My father listened carefully to what I had to say and simply said to me, "OK". I will always be thankful for his willingness to take a chance with me. From that point on I was a total convert to organic farming. We had a steep learning curve and, in fact, we are still learning and trying new things after almost 40 years. The most immediate result was that we started making money. We paid off our operating note in three years and have never had to have one since. An operating note is what the farmer borrows in the spring to buy all the inputs required to plant the crop in hopes that there will be enough in the fall to pay it back when the crops are sold. No chemical farmer I know can operate without one. But, by growing most of our inputs, thereby reducing cash outlays to the chemical companies which reduced our expense by nearly two-thirds, and increasing the value of what we sold by nearly 50 to 100%, we were now able to realize a net income from our farm far above what we received from the direct government subsidy payments that were coming to us in those days.

Besides all this good news, however, the best part for me was that farming had become fun again. My whole farm was my laboratory as I tried new experiments to increase soil health and thereby plant health and vigor. So, by breaking out of the industrial "rush to the bottom game" by seeing how many bushels of grain we could raise per acre and how cheap we could sell that grain before our farm joined so many others in bankruptcy or a forced sale, we chose a different road. This was the road to nutrition and production of a quality of food that heals and nourishes our people at the same time. It may cost a few dollars more at the grocery store, but it makes a lot more sense at the doctor's office. The little extra paid to the farmers for a fair return for their effort will be made up many times by reducing the cost of health care by improving health of our people, and in the savings of billions of dollars in cleanup costs by reducing chemical pollution. But before getting to those topics, let's consider the effect on our communities.

Healing Our Communities

As I mentioned earlier, when farmers and their farms suffer, their communities suffer as well. Each are dependent upon the other. If this is true in a negative sense, it is also true in a positive sense. As the financial security of our farms and farmers increase, they have more money to spend in town and can better support the businesses in the community. When those businesses are supported, then the local business owners can also better support each other, their employees and their customers in a continuous cycle. It is often said that each dollar spent in the community will cycle seven times, so the effect is significant. But that cycle can be short-circuited when dollars are sent out of the community to international agrichemical companies for example.

Another way to help the local economy and the nutrition of food eaten is to minimally process it closer to home by locally owned facilities. This not only creates jobs locally but also makes more nutritious food. Again, local ownership guarantees more money stays in the local community to help the local economy rather than going out to large out-of-state companies that seldom help the local communities with their extracted profits.

In addition, the cost of machinery to minimally process food is often much cheaper than machinery used for highly processed foods. For example, a flour mill which merely grinds the grain is much cheaper to build than one which sifts out the bran and germ to make while flour. These bran and germ byproducts, which are actually very valuable sources of human nutrition, are often incorporated into animal rations. So, the pigs are eating better than we are. Of course, most white flour is fortified to replace some of the nutrition which has been removed but it is not the same as what existed in the whole kernel of wheat in the first place. Very rarely is any substantial bran added back to white flour to make industrial wholewheat flour. Bran is an import source of fiber which can sustain gut health.

An oil-pressing facility which merely crushes oil seed to extract the oil in a cold-press operation is much cheaper to build and operate than one which requires solvents to extract every last drop of oil from the seed. These solvents must be removed and disposed of themselves without contaminating employees or the environment. Unfortunately, the refining methods of these types of oil processors remove vitamins and minerals that have significant nutritious value from the oil and should not be removed from our food.

I have been involved in this idea of minimal processing firsthand by creating three different businesses which added value to the crops I produced on my farm, and in turn added part of that value back to the local community by increasing jobs and spending money earned locally. My three businesses were Montana Four and Grains that started in 1983 and is now located in Ft. Benton. As I mentioned before, we started selling the high protein wheat from our farm to whole grain bakeries in California which led to finding and selling organic grain. Because of customer demand, we later added a stone flour mill and a cleaning plant. We were trying to promote eating healthy whole foods as opposed to refined foods. We sold whole grain and whole wheat flour that we were grinding on a stone mill, and later on a micronizing mill that cut the grain into fine flour with knives. Of course, these mills are not as efficient as the monster flour mills which now predominate in America. But not long ago every community had a small flour mill which served them. Fewer miles were put on the grain getting to the mill, and on the flour getting to the local stores. In the case of whole grain flour, the longer it sets, the more nutrition is lost. So faster turnover is better which gives another reason to shorten the supply line.

The next business I started was an organic vegetable oil business. It was called The Oil Barn and was located in a repurposed cow barn on our farm. I grew high oleic safflower and bought presses which could produce cold-pressed unrefined oil. Oleic acid is a monounsaturated fatty acid which is good for your heart and very stable, hence a good choice for high temperature cooking. The oil was not refined. It was only pressed and filtered through paper and cloth filters to remove plant particles. The end product was not clear, as are most of the refined vegetable oils in the stores. It was yellow. The color was due to the presence of soluble vitamins (especially vitamin E) in the oil. Because it was cold pressed and not refined, the oil-soluble nutrition that was in the seed ended up in the oil. High-temperature cooking will destroy much of the vitamin E in the oil but if you use it for salad dressing, the vitamin E is preserved for the benefit of those who eat it. We also sold our oil to cosmetic companies who used it in their skin products. Safflower oil is known to be good for your skin and customers certainly do not want oil that is contaminated with agrichemicals or solvents used in refining oil which might be absorbed through the skin. Our organic non-refined safflower oil was very popular. You may not normally think of good nutrition for your skin but that can also be another link to good health, adding to the importance of the nutrition that enters your stomach.

We bottled the oil for retail sale and also sold it to cafes and restaurants who loved the mild but unique flavor that it gave to their fried foods. We supplied the kitchens of Montana State University and the University of Montana and were able to have some of the used oil returned to us. After dewatering, cleaning and filtering the oil, we were able to put it directly into our tractors as straight vegetable oil. This served as a replacement for diesel and completed the energy cycle back to the farm, reducing our carbon footprint to less than zero. In this way we could also say we were selling the oil twice. To me, this is the best way to end the food verses fuel debate raging in some parts of our county. We solved the dilemma for the best use of farm ground, by using it for both at the same time. We first used the vegetable oil for food and then the used oil went for fuel. Having this system set up regionally would save miles of transportation and keep a lot of fuel money in the local community rather than it being sent to international oil companies. In addition, we are reducing the net greenhouse gasses required to produce the oil crop. More carbon is taken up by the growing safflower plant than is released by burning a portion of the oil produced by that plant in diesel engines in machinery required to cultivate and plant the crop as well as harvest the seeds. This contributes directly to the health of our planet and indirectly to the health of the people living on it which we will discuss more in the next sections.

My last business venture was producing an organic whole grain snack. The company was Big Sandy Organics located in my hometown of Big Sandy. The purpose of this business was to develop a health snack with grain we produced on our farm. The snack only had three ingredients, whole grain KAMUT® ancient wheat fried in our high oleic safflower oil and sprinkled with just a pinch of salt from the ancient seas of Utah. This salt was buried deep beneath the ground and protected from the pollutions that plague our modern oceans, the source of much of the salt we eat today. This product was simple and easy to make and tasted great. It has been a hit as a wholesome snack and is especially appreciated by schools looking to replace sugary sodas and candy from the vending machines in the high schools or the sugary snacks often given to grade school students. Most understand now that high use of these sugary treats aggravates diabetes and all kinds of other health problems. The whole grain treats, on the other hand, fit perfectly with government guidelines for high nutrition, high fiber, whole grain snack recommendations for the schools. Because the high fiber in these snacks automatically reduces constipation, teachers have noticed discipline problems decrease and attention spans increase. Midmorning hunger pangs are also lessened. This is still another example of the benefits of linking good food with good health.

All three of these businesses have been sold now, mostly to employees. My goal was never to create national or international empires but to be a model for local and regional economic benefits, with an emphasis on high nutrition for good health. Imagine the benefits to communities everywhere if this model were expanded to other crops and other products all over the country.

Healing Our Earth

The benefits of growing food as medicine for our planet might not be the first thing that comes to your mind. However, they are important to the whole picture when you consider all the negative effects that industrial chemical agriculture has had on our environment. These effects could be significantly reduced or eliminated by changing our dependance on chemical agriculture to regenerative organic agriculture. Regenerative organic systems focus on feeding and nourishing the soil in a natural way of soil building which resembles what we see in nature. In nature we have huge diversity of organisms. Each give and take from their environment in a way which nourishes and sustains the whole system. There are no chemicals applied to feed or protect the plant in nature. Unfortunately, many of the synthetic chemicals used in chemical agriculture find their way into surface or ground water and contaminate our air and soil. If the required nutrition was supplied to the plant by a healthy soil, there would be no chemical pollution of either surface or ground water. There would be no high nitrate water in the wells of Iowa. There would be no dead zones in the Gulf of Mexico. Plants grown in healthy soil are healthier themselves and produce more tasty and nutritious food. Healthy plants are better able to withstand attacks of disease and insects which reduce the need for poisons to be applied to destroy those attackers. Organic rules do not allow chemical pesticides and insecticides. The naturally occurring ones that are allowed generally break down in the soil or on the plants after application much faster than chemical ones used in chemical agriculture systems. The organic inputs are also less problematic for human health. The diversity of regenerative organic farms also plays an important role in breaking up cycles of weeds, insects, and disease which are usually associated with monocultures. Again, this reduces the need for harsh chemical poisons in the form of herbicides, pesticides and insecticides.

Another pressure on the planet is the enormous amount of greenhouse gases released into the atmosphere by the chemical production of nitrogen fertilizer. This product is not used in organic agriculture. Hence, every acre converted from chemical to organic systems would significantly reduce the production of chemical nitrogen fertilizers and the greenhouse gases released into the atmosphere by that process.

Besides all the decreases in negative effects due to regenerative organic systems, there is also a significant increase in positive effects. This is mainly due to carbon sequestration which is significantly increased in organic systems. Over the last hundred years, chemical and extractive agriculture have reduced the average carbon in the soils of America by nearly 4%. Organic systems have been shown to reduce this loss over time. Some say if the whole world went to regenerative organic systems almost all the excess carbon now in the atmosphere could be returned to the soil. Think how this would affect the health and prosperity of the world if the extreme weather we have been recently experiencing returned to more normal conditions, reducing violent storms, hurricanes, tornadoes, and typhoons.

Healing Our People

Imagine the savings in our health care programs as well as the well-being of our people if we were to reduce chronic disease back to pre-WWI levels. Think how much money you would have to pay a little more for good food, if you did not have to pay for treating chronic disease. By most estimates, your savings due to better health would be much more than the small increase in your food bill for better food. This could be done with three guiding principles: avoiding foods which are bad for us, seeking out foods which are good for us and eating them in proper proportions, and adding enough daily exercise and activity to our routine to strengthen and help maintain the health of our physical bodies.

The foods which are bad for us are well known such as those high in refined sugar, industrially produced fats, artificial chemicals from processing aids or food additives, as well as poisonous pesticide residues from agrichemicals being applied to fields or crops. It is important to consider the whole list when choosing what to avoid and not rationalize away the acceptance of some problems by avoiding others. For example, organic soda pop, candy bars or any other organic food high in sugar might free you from the worry of the agrichemical residues in your food and most artificial processing aids or food additives, but it will not remove the negative effects of too much refined sugar. However, by avoiding excess sugar and fats altogether, while at the same time being oblivious to the presence and potential negative effects of artificial additives, processing aids and poisonous residues from agrichemicals or foods produced from GMO seeds is not the answer either.

As for seeking good food, this means, in its simplest form, seeking out food with high nutritional value. Unfortunately, we do not have a universally accepted, quick, reliable and reasonably priced test for nutrition. We do have an increasing amount of nutritional information and lists of compounds known to play a role in nutrition but there has been little work studying how those compounds might work together synergistically to produce a superior nutrition in the body. Then there is the fact that everyone's body responds to different nutrient levels differently. A diet which might make one person sick does not affect another in the same way. However, it is understood that the more nutrition a person receives the better health they have.

It is now recognized that good taste and aroma can be indicators of good nutrition. So, if foods are not laden with artificial additives which mask natural taste and small, these two important attributes might be used to indicate the nutrient level in food. Many compounds contributing to good taste and pleasing aromas are polyphenols or other secondary plant compounds which are also strong antioxidants or responsible for reducing inflammation. Both contribute to better health. Many declare that the flavors and aroma of organic fruits and vegetables are superior to non-organic. Some claim grass-fed beef tastes better that that from a feed lot. Some claim heirloom and ancient grains have better flavor and aroma than modern grains. All these observations may be contributing factors that could be used to make good food choices until a comprehensive test for nutrition is developed.

Although many of the fine points are in dispute, what is agreed upon by all is that you are what you eat, and if you want to avoid and reduce chronic disease, you need to choose good food with as much nutrition and as little contamination as possible.

THE RESULTS OF RESEARCH

Much of what I have shared with you thus far has been based mostly on my personal observations and experiences but there is a growing amount of published research which documents the connection between food and health. I would like to divide this topic into the effects of three main categories of production: first, starting with the right seed demonstrated by clinical studies comparing modern and ancient wheat; second, using the right cropping system; and third, selecting the best animal production system.

For the seed topic, I would like to introduce Dr. Emanuela Simonetti, Scientific Director of Kamut Enterprises of Europe, who in the next section summarizes a portion of the research we have supported at Kamut International and Kamut Enterprises of Europe for nearly 20 years.

CLINICAL STUDIES ON CHRONIC DISEASE COMPARING MODERN AND ANCIENT WHEAT

Emanuela Simonetti

The first research studies on ancient wheat arose from the need to explain their higher digestibility and tolerability reported by consumers since their introduction to the market. Before the beginning of the current century, very few studies looked at the difference between ancient and modern wheat varieties. Recently however, there has been an exponential growth of research on these topics, attesting that the rediscovery of ancient grains has not involved farmers, processors, and consumers only, but also the scientific and health professional world.

Among the ancient wheat varieties, KAMUT® brand khorasan wheat is one of the most studied and can be considered a case study. This is because the trademark ensures a level of standardization and uniformity difficult to achieve with other ancient grains.

The KAMUT® trademark is used in marketing foodstuffs of a particular variety of ancient khorasan wheat called QK-77, which was registered by the USDA in 1990. Khorasan is the common name for *Triticum turgidum* ssp. *turanicum*, a tetraploid wheat which is a close relative of durum wheat. Any wheat sold under the KAMUT® trademark must possess several quality specifications related both to growing conditions and nutritional characteristics. For example, KAMUT® brand wheat must be grown only as a certified organic grain, have a protein range of 12%–18% and contain minimum 400 ppb of selenium. In addition, KAMUT® brand wheat is a good source of fiber, zinc, phosphorus, magnesium, vitamin B1 (thiamin), and vitamin B3 (niacin). It is also high in many minerals such as selenium, copper, manganese, and molybdenum. KAMUT® brand wheat is characterized by a particular phytochemical content, including an extremely diverse profile of polyphenols (Dinelli, 2009). Moreover, there are several studies that have highlighted a higher content of carotenoids, in particular lutein which is the main carotenoid of wheat, in samples of KAMUT® brand wheat flour and bread compared to the same products prepared with spelt wheat, durum wheat, and bread wheat (Abdel-Aal, 2007; Pasqualone, 2011).

The health properties of the ancient KAMUT® brand wheat have been studied in cell cultures, animal models, and human clinical trials. We will discuss findings from each of these in turn.

Cell cultures have been used to evaluate the biological effect of both single components of grain and whole wheat-based food after *in vitro* digestion. Truzzi et al., 2020 studied the effects of phenolic compounds extracted from different ancient and modern wheat varieties. They showed that total polyphenol extract from KAMUT® brand wheat increased proliferation of intestinal human cells (Caco-2) and human monocyte cells (U937), enhanced mouse fibroblast (L929) migration, and was less toxic to U937 cells at all tested concentrations, compared to other durum wheat varieties. These results suggested that the properties of wheat polyphenols are influenced by the specific wheat genotype.

In another study, undigested protein extracts from modern and ancient wheat were administered to peripheral blood mononucleated cells (PBMC) of healthy subjects and non-celiac gluten sensitivity (NCGS) patients (Valerii, 2015). Protein extracts of modern wheat caused higher production of CXCL-10, a pro-inflammatory chemokine involved in the loss of the intestinal epithelial barrier.

Rather than using single components of grain, other studies supplemented cells with real foods that are part of the human diet in order to look at the effects of the food matrix.

In her first study, (Valli et al., 2016), Valli, supplemented cultured hepatic cells (HepG2) with *in vitro* digested cookies prepared with three different whole grain flours: KAMUT® khorasan wheat grown in North America, khorasan wheat grown in Italy and modern durum wheat and then exposed the cells to an oxidative and inflammatory stress. The highest protective effect against oxidative and inflammatory stress was achieved by North American KAMUT® khorasan wheat-based cookies, followed by the Italian khorasan wheat-based cookies and then by the modern durum wheat-based cookies. Although the wheat variety appeared to be the main factor in the observed results, the growing location also played a key role.

In another study (Valli, 2018), the same kind of hepatic cells (HepG2) were used to test the biological effects of bread made with ancient, heritage, and modern wheat. In this case, the wheat samples were organically grown in the same location and growing season to minimize differences due to environmental factors. Each type of bread was digested *in vitro* and supplemented to liver cells and then assessed for antioxidant and anti-inflammatory effects. Supplementation with the *in vitro* digested fraction of ancient wheat-based bread exerted positive effects, particularly as pretreatment before exposure to an inflammatory stress, confirming the potential health benefits of ancient grains.

The antioxidant and anti-inflammatory properties of KAMUT® brand wheat observed in cell cultures have been also confirmed in animal models. Rats fed with modern whole-grain durum wheat bread or whole-grain KAMUT® brand wheat bread were examined in the basal condition and after the administration of doxorubicin, a pro-oxidative agent (Gianotti, 2011). Rats fed whole-grain KAMUT® brand wheat bread showed a lower oxidative status in plasma and in liver and a different activity of glutathione peroxidase and thioredoxin reductase in liver compared to rats fed the modern whole-grain durum wheat bread. In addition, the histological evaluation showed foci of lobular inflammation in the hepatic tissue in response to doxorubicin only in rats fed modern durum wheat bread (Benedetti, 2012). The observed antioxidant protection was attributed to a better antioxidant profile in terms of selenium content and polyphenol composition of KAMUT® khorasan than durum wheat.

Using the same experimental design, another study showed that KAMUT® brand wheat pasta counteracted the oxidative stress induced by doxorubicin administration (Carnevali et al., 2014). Moreover, KAMUT® brand wheat pasta-fed rats showed normal duodenum morphology, while modern durum wheat pasta-fed rats showed flat mucosa, an unusual shape and shortening of the villi and a high lymphocyte infiltration.

These studies evidenced for the first time a clear difference in inflammation caused by modern wheat compared to the lack of inflammation or even protection from inflammatory pressure seen in rats fed ancient wheat and paved the road to the human clinical trials on patients suffering from non-infectious chronic inflammatory disease.

The human clinical trials were nutritional intervention studies where participants replaced all other wheat cereals with either KAMUT® brand wheat or modern wheat products, both prepared from organic semi-whole or whole wheat, using identical processing procedures. The analyzed populations covered a wide spectrum including healthy individuals and patients with various diseases, such as acute coronary syndrome (ACS), irritable bowel syndrome (IBS), type 2 diabetes mellitus (T2DM), non-alcoholic fatty liver disease (NAFLD), pre-hypertension and fibromyalgia. The impact of the replacement diet with ancient wheat products was assessed measuring blood markers related to risk factors of oxidative stress and inflammation and traditional risk measurements. Volunteers were instructed not to eat any wheat products except for the ones provided for the experiment. They were not told anything about the nature of the wheat products given them. They were also told not to change their dietary or exercise routines during the experimental period.

The first study (Sofi, 2013) focused on cardiovascular health. It investigated the impact of consumption of KAMUT® khorasan wheat in 22 healthy adults with no prior clinical manifestation of cardiovascular disease, but with high risk factors for the disease. The study was a randomized, single-blinded, cross-over study with eight weeks on each diet. Results indicated that consumption of KAMUT® khorasan wheat products significantly reduced metabolic risk factors in the blood such as total cholesterol (−4%), cholesterol LDL (−8%), glucose (−4%), and increased mineral levels, particularly potassium (+5%) and magnesium (+2%). Similarly, redox status was significantly improved only after the KAMUT® khorasan intervention phase, as measured by a reduction in both thiobarbituric acid reactive substances (−21.5%) and carbonyl levels (−17.6%). Additionally, a decrease in the levels of markers of inflammation in the blood, such as TNFα (−35%) and l'IL-6 (−24%), IL-12 (−28%), MCP-1 (−17%), MIP-1β (−15%), and VEGF (−11%) were observed after a KAMUT® khorasan wheat-based diet. These positive results were not observed in the same participants after the consumption of modern wheat products.

The following study had the same experimental design and involved 22 patients with clinical manifestation of a cardiovascular disease, the Acute Coronary Syndrome (ACS), and under strict pharmacological therapy (Whittaker, 2015). Consumption of KAMUT® khorasan wheat products, compared to consumption of modern wheat products, significantly reduced key factors of cardiovascular risk including total cholesterol (−7%), cholesterol LDL (−8%), glucose (−8%), and insulin (−25%). Other results indicated a better antioxidant status with significant reductions in reactive oxygen species (ROS) and lipoperoxidation. There was also an increase of magnesium in the blood and a decrease in inflammation indicating cytokines such as TNFα (−34.5%). No significant differences from the baseline in the same patients were observed after the modern wheat intervention phase. Therefore, the results of this study indicated that a KAMUT® khorasan wheat-based diet can give an additional protection in ACS patients which are under a stringent drug therapy (secondary prevention).

Other findings confirming the positive effects of KAMUT® khorasan wheat on cardiovascular health came from a study on 63 pre-hypertension patients (Cicero, 2018). Results indicated that consumption of KAMUT® khorasan wheat improved blood triglycerides (−14.5%) and glucose (−31.6%), diurnal (−30.2%) and nocturnal (−36.8%) systolic blood pressure, and endothelial reactivity (+2.3%) compared to consumption of modern wheat.

Another study (Sofi, 2014) was conducted on 20 people suffering from irritable bowel syndrome (IBS). The results indicated that after consuming KAMUT® khorasan wheat products the patients experienced a significant improvement in symptoms related to IBS such as a decrease in the intensity and frequency of abdominal pain, less stomach bloating, less troublesome bowel movements, and less tiredness. The inflammatory profile of the patients was also improved with a reduction of cytokines including Interleukin 6 (−36.2%), Interferon γ (−33.6%), Monocyte Chemotactic Protein 1 (MCP-1) (−38,9%) and Vascular Endothelial Growth Factor (VEGF) (−23.7%). No significant differences were observed after consuming modern wheat products.

The beneficial effect of a KAMUT® khorasan wheat-based diet on symptoms of certain disease characterized by chronic widespread pain was also confirmed in another study involving 20 participants with fibromyalgia (Pagliai, 2020). In another double-blinded trial involving 21 patients with type 2 Diabetes Mellitus (T2DM) (Whittaker, 2017), the consumption of KAMUT® khorasan wheat products had a positive impact on blood insulin (−16.3%), glucose (−9.1%), and insulin resistance (−24.3%). In addition, there was a significant decrease in risk factors of T2DM patient vascular complications such as total cholesterol (−3.7%), LDL-cholesterol (−3.4%), inflammation indicating cytokines, such as VEGF (−33.6%) and IL-1ra (−27.2%), and a significant increase of total antioxidant capacity (+6.3%).

Another study was a randomized, double-blinded, parallel-arm study over a 3-month period with 40 patients suffering from Non-Alcoholic Fatty Liver Disease (NAFLD) (Dinu, 2018b). Consumption of KAMUT® khorasan wheat products produced a significant improvement in several key markers in the blood, such as total cholesterol (−6%) and liver function enzymes, such as alanine aminotransferase (ALT) (−12%), aspartate aminotransferase (AST) (−14%) and alkaline phosphatase (ALP) (−8%). Additionally, only in the NAFLD patients that consumed KAMUT® khorasan products, a significant improvement in markers of inflammation in the blood were observed, such as TNFα (−50%), IL-1 ra (−37%), IL-8 (−24%), INF-γ (−24%). Finally, although both kinds of products demonstrated positive effects on liver steatosis grading, Doppler Perfusion Index values, fat mass, and Reactive Oxygen Species production, the improvements were more evident after consumption of KAMUT® khorasan wheat products.

Lastly, two other trials involved healthy subjects. In a randomized, non-blind, parallel arm study with 30 healthy volunteers (Trozzi, 2019), after 16 weeks of a KAMUT® khorasan products-based diet, there was a greater reduction of fat mass and insulinemia than in modern wheat group. In a randomized single blinded crossover trial involving 20 male basketball players (Spisni, 2020), four-week consumption of KAMUT® khorasan products was associated with a reduction of monocyte chemoattractant protein-I and an improvement in a self-rated health status compared to baseline.

Besides KAMUT® brand wheat, other kinds of ancient wheat have been included in clinical trials. A study involving 22 type 2 diabetic people showed that the incorporation of whole emmer wheat flour in their diet for six weeks compared to modern whole wheat bread flour decreased total cholesterol, triglycerides and LDL-cholesterol (Yenagi, 2001).

Forty-five clinically healthy subjects were included in a randomized, double-blinded crossover trial aimed at evaluating the effect of a replacement diet with bread prepared with ancient bread wheat varieties (Verna, Gentil Rosso and Autonomia B) versus a modern bread wheat variety (Blasco) on cardiovascular risk profile (Sereni, 2016). After eight weeks of intervention, consumption of bread made with the ancient varieties lead to significant reduction of total cholesterol, LDL cholesterol and blood glucose and increase in circulating endothelial progenitor cells, whereas no significant differences were obtained with the modern variety.

The main effects observed in human clinical studies after consumption of products based on ancient wheat varieties appear to be antioxidant and anti-inflammatory, confirming the results obtained with cell cultures and animal models. In addition, there were also benefits on metabolic and clinical parameters and improvement of symptoms and sleep quality. The lifestyle habits and the medicinal therapy were maintained constant throughout each trial so that the observed impact on health was ascribable to the replacement diet and not to other interfering factors.

Overall, these human clinical trials showed that our food can be our medicine, in some cases with health effects on blood parameters and symptoms comparable to drugs.

The mode of action underlying these effects is still unclear, and it is believed that it is attributable not only to one wheat component, but rather that there is a synergy of multiple factors playing a significant role.

KAMUT® brand flour showed a higher percentage of amylose molecules, that are digested more slowly than amylopectin molecules, compared to modern wheat (Sofi, 2013). This could explain the lowering effect of blood glucose and insulin levels and the maintenance of satiety for longer periods reported by the volunteers that followed a KAMUT® brand wheat-based diet compared to modern wheat. A better composition in terms of mono/polyunsaturated fatty acids versus saturated fatty acids in ancient wheat compared to modern wheat could explain their effect on decreasing total and LDL cholesterol (Hidalgo, 2009). Moreover, the increased levels of minerals in the serum of studied subjects after eating ancient wheat is a direct consequence of a higher concentration of many minerals with anti-inflammatory action (e.g. selenium and magnesium) in these varieties (Fan, 2008; Murphy, 2008). This is related to the greater height of ancient wheat varieties (up to 1.8 m), compared to dwarf modern wheat varieties, which corresponds to a more developed root system capable of optimizing the absorption of nutrients from the soil. Moreover, the decreased concentrations of minerals in modern wheat varieties may be partially due to the "yield dilution phenomenon" (i.e., increased starch accumulation which dilutes the other wheat components) (Shewry, 2020) and to the intentional selection of modern wheat varieties with lower ash content to produce whiter flour.

Moreover, higher content or biodiversity in phytoactive compounds like polyphenols and carotenoids in ancient wheat compared to modern wheat, as said before, can explain at least in part their antioxidant and anti-inflammatory effects. For example, phenolics may activate one or more adaptive cellular stress response pathways, like the Nrf-2/ARE pathway which plays a central role in the upregulation of genes implicated in the protection of the cell from oxidative stress (Lee & Johnson, 2004), and the NF-κB pathway implicated in the regulation of inflammation (Vauzour, 2012). Effects and possible mechanisms of ancient wheat components on glycemic, lipid and mineral profiles, as well as on inflammatory and oxidative states are shown in Figure 2.1.

Another difference between ancient and modern wheat varieties is related to gluten, the most important and abundant wheat protein. Even if all the wheat varieties show toxic peptides for celiac people, the modern wheat varieties have been selected to have high gluten strength (GS). Investigations of the *in vitro* effects on cell lines of total protein extracts from landrace and modern bread wheat varieties showed that the latter, characterized by increased GS, were more cytotoxic and immunogenic than the landrace wheat proteins (Truzzi, 2021). Given the increasing prevalence of both celiac disease (CD) and non-celiac wheat sensitivity (NCWS), overlapping with the release of modern varieties starting from the so-called Green Revolution, for some experts these disorders are definitely the consequence of modern wheat cultivars (de Longeril, 2014). Beside gluten, other kinds of proteins such as ATIs (amylase-trypsin inhibitors) which are also involved in CD and NCWS have been shown to have different activities between modern and ancient wheat varieties (Simonetti, 2022).

Lastly, the positive effects of an ancient wheat-based diet could be at least in part related to modification of the gut microflora. A study carried out in 30 healthy volunteers (Saa, 2014) and another in 20 patients suffering from fibromyalgia (Baldi, 2022) showed that consumption of KAMUT® khorasan wheat products, compared to consumption of modern wheat products, was characterized by the release in the gut of more Short Chain Fatty Acids (SCFA) and phenolic compounds, capable of mitigating metabolic inflammation and providing other important benefits for human health, and by a better composition of gut microflora.

Overall, the ancient wheat varieties have unique nutraceutical, phytochemical and sensorial characteristics compared to the modern wheat varieties. The human clinical studies, although requiring further confirmation in a greater number of patients and with a higher number of tested wheat varieties, support the antioxidant and anti-inflammatory properties of the ancient wheat varieties shown in *in vitro* studies and show that all the kinds of wheat do not have the same impact on human health.

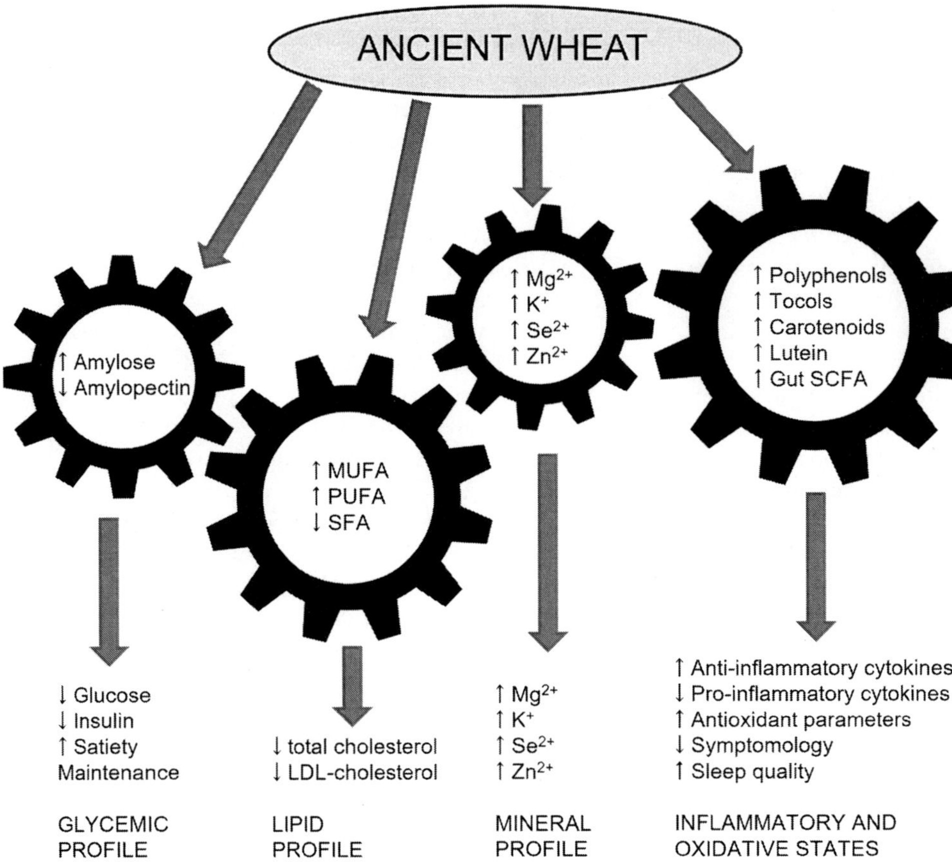

FIGURE 2.1 Schematic overview of possible mode of action of ancient wheats components on glycemic, lipid and mineral profiles, and on inflammatory and oxidative states.
Source: Adapted from (Dinu 2018a).

These studies have shown the medicinal value of ancient wheat varieties and the role of these nutrient-dense foods in preventing or mediating chronic disease. Although the observed effects are concordant in different cohorts of subjects, the mechanisms of action are not yet completely understood and deserve further studies.

CHOOSING THE RIGHT CROPPING SYSTEM

After choosing and planting the right seed, most of the time you will have a nutrient advantage with an heirloom variety; it is equally or maybe even more important to know how that seed grew into a new crop. Many research studies are now available which demonstrate the superior nutrient levels of organically grown food. Many of these studies were reviewed and became part of the book, *What Your Food Ate* by David Montgomery and Anne Biklé. They summarized their findings by saying: "In connecting the dots between agriculture and medicine, soil health emerged as an under-appreciated foundation of human health." They reported substantial declines in mineral micronutrient concentration in fruits vegetables, grains, meat, and dairy products in the decades since the rise of today's industrial agriculture. This is due to excessive tillage as well as agrichemicals, both of which disrupt community of soil life involved in micronutrient acquisition and phytochemical production. The most damaging industrial farming practices on soil life is deep tillage and adding high rates of nitrogen. When you add in soil disturbance due to chemical pesticide application, the problem is compounded. So, while nutrient factors in the food have declined, negative nutrient factors or anti-nutrient factors such as pesticide residues in the food have increased (Montgomery, 2022).

On the other hand, organically grown food has elevated levels of certain vitamins, minerals phytonutrients such as polyphenols which are very important antioxidants and reducers of inflammation. These properties help to mediate and prevent a host of chronic diseases. Organically grown food also has the advantage of not using chemical herbicides, insecticides, or fungicides. This means there will be no residues from the direct application of these poisonous compounds to crops. Of course, we are all exposed to micro amounts of these compounds by the contaminated air, soil, ground water, and even rain that falls on us but at least we can avoid the higher concentrations often found on food coming from intensive chemical, industrial agriculture where pesticides are often applied directly to the plants which produce the food we eat. We have often heard the quote that healthy soil produces healthy food, and it is true.

For those concerned about the degradation of soil using excess tillage but who still want their food grown without chemicals, regenerative organic agriculture provides an answer for that concern. Regenerative organic systems combine the legal definition of organic with a focus on soil and soil health including a reduction of tillage and an increase in soil life.

Nothing takes the place of putting good seed into healthy soil to produce a nutrient dense crop. It is hard to imagine a more important concept. Hopefully in the future there will be more understanding and appreciation of this idea and more research to help perfect regenerative organic systems in every part of our planet.

CHOOSING THE RIGHT ANIMAL PRODUCTION SYSTEM

Most of the industrial animal and bird production systems now common in America use some kind of CAFO (confined animal feeding operation). This means the animals or birds are confined indoors. Because they are not outside on pasture, they never get to choose the plants or food that they eat. There has been a lot of research on animal nutrition and now there are whole books dedicated to that subject and reviewing the research. One of the best is *Nourishment* by Fred Provenza.

> Renowned animal behaviorist Fred Provenza spent his academic career researching how animals respond to an intricately tuned system of flavor-feedback relationships at the cellular level – a nutritional wisdom that guides them to eat the appropriate combination of foods to meet their dietary needs.

Of course, just like humans, animal's health is directly related to what they eat. Furthermore, the nutrition and health promoting components of what they eat are expressed in the nutritional value of the meat, eggs, or milk they produce (Provenza, 2018).

A recently published study described a US-wide study of 1,163 milk samples collected over three years from cows that were fed a nearly 100% forage-based diet (grass milk) in an organic system compared to those under organic and non-organic management which included a more grain-based ration. The main focus was looking for differences which might help reverse the large changes in fatty acid intakes that have occurred in America over the last century. What is of concern is the omega-6/omega-3 ratios. When that ratio is closer to 1 in any food, such as milk, there is a potential for reducing the risk of cardiovascular and other metabolic diseases. In this study the ratio for omega-6/omega-3 fatty acids dropped from 5.77 in non-organic milk, to 2.28 in organic milk to 0.95 in grass milk. This is a dramatic shift which can lead to better health of grass milk drinkers (Benbrook, 2018).

In CAFOs sometimes the livestock or birds are crowded to the point of creating stress which leads to disease which then requires medication such as antibiotics. Some are given growth hormones to promote growth. Some of these medications and hormones can then end up in the food produced from those operations which is not a positive thing for those eating it. Besides the presence of these negative compounds, there is often a reduction of good nutrition in the food and sometimes an increase of natural compounds which have a negative effect. Antibiotics and hormones in animal products can be avoided by buying certified organic. But if you want to be sure your food is coming from grass-fed organic animals you will have to look for a label on the products making that declaration.

FOUR PILLARS OF GROWING FOOD AS MEDICINE

There are four pillars which must be observed to successfully grow food as medicine. They are: a) starting with good seed, b) putting that seed in healthy soil, c) harvesting at peak nutrient value, and d) processing as minimally and as locally as possible.

Starting with Good Seed

As explained with our example of KAMUT® brand ancient wheat, there is a huge difference in the seeds available for food production. If the seed planted does not have the potential to produce crops with high nutrient levels, then that deficiency cannot be made up by applying the principles that follow. Just as importantly, if those seeds contain elements which cause negative health consequents like bloating, cramps, or problems with the gut microbiota, these cannot be significantly reduced or modified by applying the principles that follow. Often, we hear the adage first coined by R.J. Rodale in the late 1940s, that healthy soil makes healthy plants which makes healthy people. I think now it is time to add one more link to the beginning of that chain of thought: we start with good seed.

Putting Good Seed in Healthy Soil

The importance of healthy soil cannot be underestimated. The best way to achieve this goal is by following the principles which are a part of regenerative organic agriculture. Although a lot of people focus on what organic is not, I prefer to focus on what regenerative organic is. There is a lot of talk about the organic rules of production which generally prohibit chemically compounded inputs. meaning that organic farms are the ones that do not use chemical fertilizers or pesticides. I would rather describe organic agriculture as trying to imitate nature. Organic systems focus on feeding the soil using a myriad cover crop plants which provide through photosynthesis, a rich diet of exudates which feed the microbiome of the soil. In some areas this nutrition is added in the form of compost or other allowed organic inputs. A well-fed, thriving and healthy microbiome defines healthy soil and is in turn responsible for feeding the cash crops which will be harvested for food. Even these cash crops nourish the microbiome while they are growing.

The industrial chemical system so popular and widespread today, focuses on feeding our production crops a predetermined list of additives designed to maximize yield with little or no regard for the importance of soil health. Organic systems depend on diversity and rotations and sometimes biological controls and inputs to break up disease, pest, and weed cycles rather than chemically compounded poisons to address these challenges. In this way organic systems are producing healthy soils full of a great diversity of life, mostly microscopic. These organisms have a symbiotic relationship with the plants. They supply the plants with compounds which they produce which may not exist in the soil or free up other compounds and minerals which may be so tightly bound to soil particles as to be unavailable to the plant. As mentioned above, the cover crops as well as the cash crops supply the microbiology with myriads of compounds, mostly produced in the leaves with the aid of the sun, which are expelled from the roots into the soil to nourish the life in the soil.

In summary, regenerative organic systems do two things, first they increase the nutritional levels of the plants and their fruit or seeds as mentioned earlier in the research review on that subject.

Second, there are no poisons present in the soil or plants from direct application to the field or the crops because that is not allowed. Because there are no poisons applied there are no residues on any part of the pant that will be eaten, including roots, leaves, stems, fruit, or seeds. So, the increase of nutrition on the one hand and the absence of poisonous residues on the other forms an important link in the chain which leads to good health.

Harvesting at the Peak of Nutrient Value

The importance of harvesting at the peak of nutrient value does not apply to all plants equally. For example, grains are generally only harvested when completely ripe so that the combines used to harvest are able to thrash the mature dry kernels out of each head of grain. When the grain is ripe, it is at the peak of its nutritional value. However, when desiccates like glyphosate are used to kill the plants and short-circuit the ripening process, nutrition that might have been gained during the end of the ripening process is lost. When fruits or vegetables are picked green or immature so that they can be shipped great distances without bruising or spoiling, or so they can sit on the store shelves for longer periods of time without spoiling, they are certainly not picked at the peak of their nutrient value. This loss in nutrition is a victim of the industrialization of food and agriculture which relies on consolidation and concentration of processing into larger facilities which may not be even close to the harvest field. The purpose of these large facilities is to increase efficiency in order to reduce cost and guarantee cheap food. So even if we start with the best seeds and put them in healthy soil, nutrition can be lost by picking fruits green before they mature. One solution to this problem is to reduce dependance on long supply lines and encourage more local production of food and also more local processing facilities.

Processing as Minimally and Locally as Possible

Processing is another key component in the chain. It can preserve or even enhance all the effort that went into the previous links in the chain to produce the highest level of nutrition possible. Or it can reduce and even destroy some or much of the nutrition which is in the food when it leaves the harvest field and enters the processing facility. Permit me to give you a couple of examples.

The first is wheat. The crop I know best and love most. I will give you an example of degrading nutrition using processing and then an example of enhancing nutrition using processing. When flour mills create white flour, they remove the bran and the germ which consists of about 25% to 30% of the total. The germ is full of minerals and vitamins like Vitamine E which is sometimes recovered when the germ is pressed to make wheat germ oil but most of the time, it is diverted to animal feed. The bran is a great source of fiber which could be used to nourish our gut microbiome. Unfortunately, it is mostly diverted to animal feed. So, you see why the pigs are eating better that we are. White flour is only a recent development with the advent of the steel roller mill during the late 1800s. The steel roller mills quickly replaced stone mills and white flour became the norm by the early 1900s all over the country. Soon nearly everyone had switched from whole wheat flour to white flour. By the time WWII started, a generation had grown up eating white flour. It was shocking to find that more than 40% of those drafted into the military service could not pass the physical. Fingers were pointed at white flour, stripped of its natural vitamins and minerals as one of the main culprits. By 1942, legislation was introduced in Congress to require fortification of white flour.

It is interesting that in the industrial model of agriculture and medicine that rules us today, what we mostly see is the treating of symptoms rather than causes of disease or agriculture problems. In this case, rather than go to the source of the problem which was the removing of significant nutrition from wheat during processing, the focus was on treating a symptom or trying to modify the end result. So, instead of encouraging everyone to eat whole wheat flour, the push was to add back a few vitamins and minerals so that at least some of the most serious problems of removing them in the first place could be addressed. Of course, fortified white flour does not have all the value of whole wheat flour. For example, no bran is added at all.

But not all processing is bad. Consider the production of whole wheat sourdough bread. We start with whole wheat flour which contains all the nutrition found in the whole kernel. The long fermentation required for the sourdough process has two great benefits: nutrition and digestibility. First, many of the products of fermentation act to reduce inflammation and increase antioxidant capacity. They also produce short chain fatty acids which are important to nutrition. In addition to producing something good, long fermentation reduces a couple things that have the potential for being bad. First, long fermentation decreases the pH by increasing acidity. This deactivates phytate which is naturally found in many grains. Phytate is an antinutrient because it binds to minerals, reducing the body's ability to absorb them. As a result, sourdough bread usually contains less active phytate than other bread, hence more of the nutrition found in the wheat is made available to the body. The second problem that is reduced by long fermentation is the presence of complex starch and gluten. In fact, with fermentations of 48 hours or more nearly 95% of the gluten is broken down. This is a big help for those who have trouble digesting gluten. Making sourdough is like predigesting the grain before it is eaten.

What is true for wheat is also true for so many other foods which are highly processed and highly refined, resulting in the loss of so much nutrition that came out of the field at harvest time. And if that were not enough, take a look at most food labels and consider all the additives which have ended up in our highly processed food. One by one, the most damaging of those to our health are being removed such as some food coloring agents for example. But many still remain which cause more trouble than they are worth as flavors, colors, and shelf-life extenders. It would be interesting to see how much change in shoppers' buying habits would be observed if manufacturers would be required to list what nutrition has been removed from the food rather than just what has been added, as well as listing what is left.

There are many advantages of local processing and selling fresh food grown locally as well as many economic benefits which come to the local community by buying locally produced food and supporting local processing. As mentioned before, this keeps a lot of jobs and money circulating in the community. But an even greater contribution is made by enhancing or preserving nutrition of those food crops which are grown and minimally processed as close to their final point of sale as possible. Another important factor is the proximity of processing plants. The closer the processor is to the field, the shorter the time required to arrive at the processor. This is extremely important for highly perishable foods. Once a fruit or vegetable is picked, the nutritional value begins to decline, so reducing the time between field and processor as much as possible is important. This is even more important for fresh fruit and vegetables headed for the store shelves where they may sit for some time before being sold. To penetrate markets at greater and greater distances with cheap food, often the fruits and vegetables are picked before they are ripe so they can be shipped further before spoiling, as well as to reduce bruising. Picking these crops before their peak can severely reduce their nutritional value compared to vine- or tree-ripened fruit. This is especially true when fresh fruit is picked green and shipped across continents, or even over oceans.

SIMILAR PRINCIPLES APPLY TO ANIMAL PRODUCTS WE USE AS FOOD

A lot of our food comes from animal production and processing. The same principles apply for maximizing the nutrition that comes from animal products as we just discussed with plant-based food. You can start with the importance of the selection of breeds. As with grain and other plants, modern breeding has focused on only yields and achieving them as fast as possible. The meat from modern breeds is mostly less nutritious than heritage breeds, but because modern breeds grow faster and produce more meat, they have replaced heirloom breeds for the most part. The modern breeds are not as hardy or healthy as heirloom breeds, however, and often require more medicating. Probably the difference between modern and heirloom breeds is nowhere more pronounced than in poultry. For example, turkeys have been bred so extremely to produce large breasts that they can no longer mate naturally.

Just as it is important what your plants eat, in other words, what kind of soil they are growing in, it is also important what your animals eat. If they are raised in a free-range situation where they are free to graze on native pastures or restored grasslands planted with a great diversity of plants, so they are able to pick from a diverse plant population to balance their own diet, they will be healthier and hardier. Feed lots which focus only on the fastest weight gain using a ration artificially high in cereal grains and oil seeds produce the most meat in the shortest time for the least cost, but do not produce the best nutrition.

How animals are slaughtered also affects their meat quality. If they are killed under a great deal of stress because of the way they were handled or the amount of time they were in transit to the slaughterhouse, the quality of the meat will be less. So here is another similarity to plant-based foods, the method of meat processing plays an important role in the quality and nutrition of the end product.

So, once again if you are looking for high nutrition which promotes better health, choose heirloom breeds which are grass fed on very diverse grass lands, and have been processed in as low stress an environment as is possible. In summary, for both plants and animals we need to do the same thing. We need to start with heritage varieties and breeds; grow the plants in a regenerative organic way and use free-range systems for the animals with a diverse array of plants for animal feed; let the plants and animals reach their peak nutritional potential; and finally process them all in a way that reduces stress on the animals and maintains the nutritional value of the plants. We need to enhance what is good, reduce or eliminate inputs and additions which are poisonous or bad for us. We need to preserve or increase the nutrition taken from the field through minimal processing and reduce the time spent in transport and sitting on the store shelf before food reaches the buyer's table.

THE FUTURE

So, where do we go from here? There are some good things starting to happen There is more and more talk of Food as Medicine. There is more and more talk of regenerative agriculture. But unfortunately, much of that talk does not seem to include the whole picture. I have attended Food as Medicine conferences where hardly a word was spoken about how the food is grown or processed. I have seen interviews with leading candy companies touting the idea that they can make a more nutritious candy bar. Much of the regenerative talk and promotion is now beginning to come from the largest agrichemical companies. They claim on their websites that they are the leaders in regenerative practices in rural America. They tend to focus on no till as the panacea because there is no soil disturbance such as you have with tillage. However, they completely ignore soil disturbance caused by the poisonous chemicals they sell and recommend as an important necessity in their regenerative program. It is my hope that the future will fill the following four needs: first, research and education which increases the understanding of the link between food and health; second, a good nutrient test for food which could lead to the third, a better labeling system for food; and finally the fourth, making sure that purchasing good food is a possibility for all.

The Need for Research and Education

There is an expanding volume of published research describing the link between food and health, especially chronic disease. The greatest amount of research has focused on what kinds of foods are good or bad for health. The negative effects of too much sugar, any trans-fats, too much soda pop, and too many empty calories are well understood. Some claim we have all the research we need to make the right food choices. It is true that the positive effects of increasing fruit and vegetables and whole grain in the diet is well understood. What is not well understood, however, is the importance of how the food on the good list is produced and processed. What is the real contribution of regenerative organic agriculture in increasing nutrition in the food produced? What is the total effect of common processing systems in preserving, decreasing, or enhancing that nutrition? As mentioned above, there have been some studies comparing crops grown in organic systems with those grown in non-organic systems. Even though what has been published shows a significant advantage in final nutrition in food coming from organic systems, the number of those studies is still small, and great benefit could come from more research. It would be very interesting to see more comparisons of different simplified processing compared to complex systems. Also, it is important to understand better the effect of various processing systems including aids and additives on the nutrition of the processed food.

It would also be interesting to compare monocultures with complex cropping systems which include a multitude of species grown simultaneously. It is often said that the more we can mimic the diversity found in nature, the closer we will come to mimicking the nutrition found in nature, a sustainable system that has been in place for eons. Farmers need help to create more diversity.

Another area of potential research is comparing heirloom and modern varieties of plants and breeds of animals for nutrient density in the end product. We need to focus on nutrition in breeding programs. Finally, it would be important to understand the real health risk of residues of poisons used in industrial agriculture or biological controls which may be used in organic systems.

With the results of this research in hand, the education of our people as to how to make the best choices in a holistic way could begin in earnest. The results of this kind of research and education would make clearer the direction we should go in the future to reduce chronic disease by improving the quality of our food.

The Need for a Good Nutrient Test and Better Labeling

There is one key I think is missing and that prevents us from really making progress with speeding the change from a dominance of cheap food with the focus on high yields and high-speed processing, to a demand for food with high nutrition. That key is a relativity quick, reproducible, inexpensive, and reliable test for nutrition.

There is work now underway to perfect a handheld device, which might end up being your phone, to instantly evaluate nutrition levels of food on the grocery shelf. That would certainly influence purchasing decisions, driving a change in production and processing decisions all the way back to the farm.

From our research in Italy comparing ancient wheat and modern wheat, we found that the greatest difference between ancient and modern wheat was in inflammation and inflammatory markers. The difference ranged generally from about 25–40%. If the production of inflammatory markers or anti-inflammatory markers could be understood as being tied to nutrition levels of the food, this would give us a way to see the end result of efforts to modify nutrition in food without having to understand the mode of action.

The goal would be to produce an assay which could be used to create an inflammatory index from −100 to +100. At zero, there was no effect of inflammation. Negative numbers which were less than zero would indicate that the food causes inflammation. Higher inflammation would be represented by higher negative numbers. While higher positive numbers would indicate an increasing ability of the food to reduce inflammation or be anti-inflammatory. For the past several years, members of our Kamut International research team in Italy have been working on human gut cell cultures to perfect this idea. They have not been completely successful yet, but progress has been made and efforts are continuing.

Imagine the value of an assay such as this to the food system at all levels from breeders to processors. Breeding programs could have an easy way to assess nutrient values along with all the other tests they run such as yield, disease and pest resistance, as well as bread loaf volume in the case of wheat. Selections which do not reach certain goals of nutrition could be discarded immediately, while those reaching or exceeding those goals could move forward. Farmers could then choose varieties with high nutrient potential to plant. This would be of great interest to them if they were given premiums for good results. The premium system would be much easier to establish if there were a simple test for it. Processors would be able to modify their processing procedures based on information telling them if those modifications were increasing or decreasing the nutrient value of the final product. Finally, the main driver in this whole chain, the buyers, would be able to make informed buying decisions based on a label that would tell them the nutrient value of the food they were purchasing. If they could see the value in paying more for higher nutrition, and reducing their health care costs, they would certainly be more likely to do so. In the end, everyone wins.

Ensuring Good Food Is Available for All

One of the most prevalent criticisms of increasing the price of food is based on the fear of putting it out of reach for those with less income. How can people pay for more expensive food? My answer to that is with another question: How can people pay for the more expensive health care required with the ever-increasing rate of chronic disease? I would suggest designing programs to help low-income families afford good food, similar to those which help these same families afford health care. From the experience of the past, I think we can expect the savings in health care realized by increasing the health of our people will more than pay for any new programs to help those in need eat better. Programs such as SNAP (Supplemental Nutrition Assistance Program) are already in place to help provide food. It would not be hard to modify those programs to focus on high nutrition rather than just filling the food basket at the store with the cheapest food available.

In addition to new programs or modification to existing programs, I believe education could play a vital role in long-term change. The areas I see with greatest potential are: reducing food waste, growing a garden, buying less processed food, cooking more at home, buying in season and preserving the excess, and finally, eating more locally produced food.

It is estimated that 40 to 50% of all food produced in the world is wasted and never eaten. In underdeveloped countries this happens on or near the farms because of lack of proper storage to preserve and protect the crops from adverse weather insects, birds, and animals, as well as lack of transportation to get everything to the processor, if available, and/or the market. In developed countries, most of this waste occurs beyond the farm. Think of your own refrigerator and how much uneaten food goes from there directly into the garbage. How much uneaten food from the plates on your table goes into the garbage? Not wasting this food by only putting on your plate what you will eat and then repurposing the leftovers in your refrigerator (my grandmother was a wizard at this) might save enough money in your food costs to pay more for better nutrition.

Another way to save on food costs is to grow some of your food yourself. Communities could provide open space for community gardens in empty lots for those without backyards, or apartment building owners could provide roof tops for gardens. Everyone who desired a garden could have access to one. Education programs should provide instruction on best gardening methods for the region, and help with tools and seeds might be provided for those without resources. To reduce those costs, seed saving could also be taught.

Next, people should be encouraged to buy less of the expensive processed food. Again, education is necessary to help folks learn to cook more at home using food from their garden or unprocessed fruits and vegetables from the store. Perhaps this could be done in school lunch kitchens after school hours or on weekends, or other public kitchens could be created. People should be encouraged to buy in season when bulk food is cheaper. Again, education is necessary to help teach people how to process and preserve this food at harvest time. The more local the high nutrient dense food is purchased, the fresher and more nutritious it is, and the longer shelf life it has. Also, there are the savings from not paying the cost of transportation over long supply lines.

THE CHOICE IS OURS

In the end it will be the buyers, eaters and their friends and family who will decide which path we follow. Will we continue down the path of cheap food, most of which carries a very high cost to the health of our farmers, communities, environment and finally our people? Will the number and vitality of our farms and communities continue to decline? Will we continue to increase chemical pollution on the earth? Will chronic disease continue to grow beyond three-quarters of the population now affected? How bad do all these problems have to get before we decide to take a different path? Will folks adopt the programs mentioned above to reduce the cost of good food? In the end, the cost of producing and processing high nutrient value food that could be considered medicine, could easily be offset by a reduction in health-care costs. Chronic disease and the number of pills taken to treat it, as well as visits to the doctor will decrease with better diets.

So ultimately, the choice is ours. We can vote with our dollars every time we go to the food store. We can vote with our time by devoting even a small portion of it to growing our own food and cooking and preserving more at home. The ultimate driver is a better understanding of the link between food and health and the fact that we can heal the earth by growing and eating food as medicine.

REFERENCES

Abdel-Aal, E. S. M., Young, J. C., Rabalski, I., Hucl, P., & Fregeau-Reid, J. (2007). Identification and quantification of seed carotenoids in selected wheat species. *Journal of Agricultural and Food Chemistry*, 55, 787–794. https://doi.org/10.1021/jf062764p

AFBF. (2025). Fast facts about agriculture & food. https://www.fb.org/newsroom/fast-facts

Baker, D. (2024). In the good old days, one fourth of income went to food. *Center for Economic and Policy Research*. https://cepr.net/publications/in-the-good-old-days-one-fourth-of-income-went-to-food/

Baldi, S., Pagliai, G., & Dinu, M. (2022). Effect of ancient Khorasan wheat on gut microbiota, inflammation, and short-chain fatty acid production in patients with fibromyalgia. *World Journal of Gastroenterology*, 28(18), 1965–1980.

Benbrook, C. M., Davis, D. R., Heins, B. J., Latif, M. A., Leifert, C., Peterman, L., Butler, G., Faergeman, O., Abel-Caines, S., & Baranski, M. (2018). Enhancing the fatty acid profile of milk through forage-based rations, with nutrition modeling of diet outcomes. *Food Science & Nutrition*, 6(3), 681–700.

Benedetti, S., Primiterra, M., Tagliamonte, M. C., Carnevali, A., Gianotti, A., Bordoni, A., & Canestrari, F. (2012). Counteraction of oxidative damage in the rat liver by an ancient grain (Kamut brand Khorasan wheat). *Nutrition*, 28(4), 436–441. https://doi.org/10.1016/j.nut.2011.08.006

Burd, L. A. (2020). EPA finds glyphosate is likely to injure or kill 93% of endangered species. *Center for Biological Diversity Press Release*. https://biologicaldiversity.org/w/news/press-releases/epa-finds-glyphosate-likely-injure-or-kill-93-endangered-species-2020-11-25/#:~:text=EPA%20Finds%20Glyphosate%20Is%20Likely%20to%20Injure%20or%20Kill%2093%25%20of%20Endangered%20Species

Carnevali, A., Gianotti, A., Benedetti, S., Tagliamonte, M. C., Primiterra, M., Laghi, L., Danesi, F., Valli, V., Ndaghijimana, M., Capozzi, F., Canestrari, F., & Bordoni, A. (2014). Role of Kamut® brand Khorasan wheat in the counteraction of non-celiac wheat sensitivity and oxidative damage. *Food Research International*, 63, 218–226. https://doi.org/10.1016/j.foodres.2014.01.065

Chang, F., Simcik, M. F., & Capel, P. D. (2010). Occurrence and fate of the herbicide glyphosate and its degradate aminomethylphosphonic acid in the atmosphere. *Environmental Toxicology and Chemistry*, 30(3), 548–555.

Cicero, A. F. G., Fogacci, F., Veronesi, M., Grandi, E., Dinelli, G., Hrelia, S., & Borghi, C. (2018). Short-term hemodynamic effects of modern wheat products substitution in diet with ancient wheat. *Nutrients*, 10. https://doi.org/10.3390/nu10111666

de Longeril, M., & Salen, P. (2014). Gluten and wheat intolerance today: Are modern wheat strains involved? *International Journal of Food Science and Nutrition*, 65(5), 577–581.

Dinelli, G., Carretero, A. S., Silvestro, R. Di, Marotti, I., Fu, S., Benedettelli, S., Ghiselli, L., & Fernández Gutiérrez, A. (2009). Determination of phenolic compounds in modern and old varieties of durum wheat using liquid chromatography coupled with time-of-flight mass spectrometry. *Journal of Chromatography A*, 1216(43), 7229–7240.

Dinu, M., Whittaker, A., Pagliai, G., Benedettelli, S., & Sofi, F. (2018a). Ancient wheat species and human health: biochemical and clinical implications. *The Journal of Nutritional Biochemistry*, 52, 1–9. https://doi.org/10.1016/J.JNUTBIO.2017.09.001

Dinu, M., Whittaker, A., Pagliai, G., Giangrandi, I., Colombini, B., Gori, A. M., Fiorillo, C., Becatti, M., Casini, A., Benedettelli, S., & Sofi, F. (2018b). A Khorasan wheat-based replacement diet improves risk profile of patients with nonalcoholic fatty liver disease (NAFLD): A randomized clinical trial. *Journal of the American College of Nutrition*, 37(6), 508–514. https://doi.org/10.1080/07315724.2018.1445047

Fan, M. S., Zhao, F. J., Fairweather-Tait, S. J., Poulton, P. R., Dunham, S. J., & McGrath, S. P. (2008). Evidence of decreasing mineral density in wheat grain over the last 160 years. *Journal of Trace Elements in Medicine and Biology*, 22, 315–324. https://doi.org/10.1016/j.jtemb.2008.07.002

Gianotti, A., Danesi, F., Verardo, V., Serrazanetti, D. I., Valli, V., Russo, A., Riciputi, Y., Tossani, N., Caboni, M. F., Guerzoni, M. E., & Bordoni, A. (2011). Role of cereal type and processing in whole grain in vivo protection from oxidative stress. *Frontiers in Bioscience-Landmark*, 16, 1609–1618. https://doi.org/10.2741/3808

Hidalgo, H., Brandolini, A., & Ratti, S. (2009). Influence of genetic and environmental factors on selected nutritional traits of Triticum monococcum. *Journal of Agriculture and Food Chemistry*, 57(14), 6342–6348.

Jessmer, D. (2024). Gulf of Mexico 'dead zone' larger than average, scientists find.https://www.noaa.gov/news-release/gulf-of-mexico-dead-zone-larger-than-average-scientists-find

Jing, F. Y., The Medicine of Buddhism. (2017). Fo Guang Shan international translation center. https://www.fgsitc.org/the-medicine-of-buddhism/

John, C. (2025). 9 confucius quotes on eating well. *Magnifissance Timeless Beauty & Wisdom*. https://magnifissance.com/arts/traditional-chinese-arts/9-confucius-quotes-on-eating-well/

Kopittke, P. M., Dalal, R. C., McKenna, B. A., Smith, P., Wang, P., Weng, Z., van der Bom, F. J. T., & Menzies, N. W. (2024). Soil is a major contributor to global greenhouse gas emissions and climate change, *SOIL*, 10, 873–885, https://doi.org/10.5194/soil-10-873-2024

Lee, J. M., & Johnson, J. A. (2004). An important role of Nrf2–ARE pathway in the cellular defense mechanism. *Journal of Biochemistry and Molecular Biology*, 37, 139–143.

Menegat, S., Ledo, A., & Tirado, R. (2022). Greenhouse gas emissions from global production and use of nitrogen synthetic fertilizers in agriculture *Scientific Reports*, 12, 14490.

Montgomery, D. R. & Biklé, A. (2022). *What Your Food Ate How to Heal our Land & Reclaim our Health.* W. W. Norton & Company Inc.

Murphy, K. M., Reeves, P. G., & Jones, S. S. (2008). Relationship between yield and mineral nutrient concentrations in historical and modern spring wheat cultivars. *Euphytica*, 163, 381–390. https://doi.org/10.1007/s10681-008-9681-x

NFU. (2024). The farmer's share. https://nfu.org/wp-content/uploads/2024/06/06_2024_FarmersShare.pdf

Pagliai, G., Colombini, B., Dinu, M., Whittaker, A., Masoni, A., Danza, G., Amedei, A., Ballerini, G., Benedettelli, S., & Sofi, F. (2020). Effectiveness of a Khorasan Wheat – Based replacement on pain symptoms and quality of life in patients with fibromyalgia. *Pain Medicine*, 21(10), 2366–2372. https://doi.org/10.1093/pm/pnaa134

Pasqualone, A., Piergiovanni, A. R., Caponio, F., Paradiso, V. M., Summo, C., & Simeone, R. (2011). Evaluation of the technological characteristics and bread-making quality of alternative wheat cereals in comparison with common and durum wheat. *Food Science and Technology International*, 17(2), 135–142. https://doi.org/10.1177/1082013210381547

Provenza, F. (2018). *Nurishment What Animals Can Teach Us About Rediscoering Our Nutritional Wisdom.* Chelsea Green Publishing.

Quinn, B., & Carlisle, L. (2019). *Grain by Grain*. Island Press.

Saa, D. T., Turroni, S., Serrazanetti, D. I., Rampelli, S., Maccaferri, S., Candela, M., Severgnini, M., Simonetti, E., Brigidi, P., & Gianotti, A. (2014). Impact of Kamut® Khorasan on gut microbiota and metabolome in healthy volunteers. *Food Research International*, 63, 227–232.

Sereni, A., Cesari, F., Gori, A. M., Maggini, N. O., Marcucci, R., Casini, A., & Sofi, F. (2016). Cardiovascular benefits from ancient grain bread consumption: findings from a double-blinded randomized crossover intervention trial. *International Journal of Food Sciences and Nutrition*, 68(1), 97–103. https://doi.org/10.1080/09637486.2016.1216528

Shewry, P. R., Hassall, K. L., Grausgruber, H., Andersson, A. A. M., Lampi, A. M., Piironen, V., Rakszegi, M., Ward, J. L., & Lovegrove, A. (2020). Do modern types of wheat have lower quality for human health?

In *Nutrition Bulletin* (Vol. 45, Issue 4, pp. 362–373). Blackwell Publishing Ltd. https://doi.org/10.1111/nbu.12461

Simonetti, E., Bosi, S., Negri, L., & Dinelli, G. (2022). Amylase Trypsin Inhibitors (ATIs) in a selection of ancient and modern wheat: Effect of genotype and growing environment on inhibitory activities. *Plants*, 11, 3268.

Sofi, F., Whittaker, A., Cesari, F., Gori, A. M., Fiorillo, C., Becatti, M., Marotti, I., Dinelli, G., Casini, A., Abbate, R., Gensini, G. F., & Benedettelli, S. (2013). Characterization of Khorasan wheat (Kamut) and impact of a replacement diet on cardiovascular risk factors: Cross-over dietary intervention study. *European Journal of Clinical Nutrition*, 67(2), 190–195. https://doi.org/10.1038/ejcn.2012.206

Sofi, F., Whittaker, A., Gori, A. M., Cesari, F., Surrenti, E., Abbate, R., Gensini, G. F., Benedettelli, S., & Casini, A. (2014). Effect of Triticum turgidum subsp. turanicum wheat on irritable bowel syndrome: A double-blinded randomised dietary intervention trial. *British Journal of Nutrition*, 111 (11), 1992–1999. https://doi.org/10.1017/s000711451400018x

Spisni, E., Valerii, M. C., de Fazio, L., Rotondo, E., Di Natale, M., Giovanardi, E., Posabella, G., Bregola, V., Stenico, V., Sferrazza, R. E., Dinelli, G., & Bosi, S. (2020). A Khorasan wheat-based diet improves systemic inflammatory profile in semi-professional basketball players: A randomized crossover pilot study. *Journal of the Science of Food and Agriculture*, 100(11), 4101–4107. https://doi.org/10.1002/jsfa.9947

Trozzi, C., Raffaelli, F., Vignini, A., Nanetti, L., Gesuita, R., & Mazzanti, L. (2019). Evaluation of antioxidative and diabetes-preventive properties of an ancient grain, KAMUT® khorasan wheat, in healthy volunteers. *European Journal of Nutrition*, 58(1), 151–161. https://doi.org/10.1007/s00394-017-1579-8

Truzzi, F., Dinelli, G., Spisni, E., Simonetti, E., Trebbi, G., Bosi, S., & Marotti, I. (2020). Phenolic acids of modern and ancient grains: Effect on in vitro cell model. *Journal of the Science of Food and Agriculture*, 100(11), 4075–4082. https://doi.org/10.1002/jsfa.9796

Truzzi, F., Tibaldi, C., Whittaker, A., Dilloo, S., Spisni, E., & Dinelli, G. (2021). Pro-inflammatory effect of gliadins and glutenins extracted from different wheat cultivars on an in vitro 3d intestinal epithelium model. *International Journal of Molecular Sciences*, 22(1), 1–21. https://doi.org/10.3390/ijms22010172

Valerii, M. C., Ricci, C., Spisni, E., di Silvestro, R., de Fazio, L., Cavazza, E., Lanzini, A., Campieri, M., Dalpiaz, A., Pavan, B., Volta, U., & Dinelli, G. (2015). Responses of peripheral blood mononucleated cells from non-celiac gluten sensitive patients to various cereal sources. *Food Chemistry*, 176, 167–174. https://doi.org/10.1016/j.foodchem.2014.12.061

Valli, V., Danesi, F., Gianotti, A., di Nunzio, M., Taneyo Saa, D. L., & Bordoni, A. (2016). Antioxidative and anti-inflammatory effect of in vitro digested cookies baked using different types of flours and fermentation methods. *Food Research International*, 88, 256–262. https://doi.org/10.1016/j.foodres.2015.12.010

Valli, V., Taccari, A., di Nunzio, M., Danesi, F., & Bordoni, A. (2018). Health benefits of ancient grains. Comparison among bread made with ancient, heritage and modern grain flours in human cultured cells. *Food Research International*, 107, 206–215. https://doi.org/10.1016/j.foodres.2018.02.032

Vauzour, D. (2012). Dietary polyphenols as modulators of brain functions: Biological actions and molecular mechanisms underpinning their beneficial effects. *Oxidative Medicine and Cellular Longevity*, 16(1), 914273.

Watson, K. B., Wiltz, J. L., Nhim, K., Kaufmann, R. B., Thomas, C. W., & Greenlund, K. J. (2025). Trends in multiple chronic conditions among US adults, by life stage, behavioral risk factor surveillance system, 2013–2023. *Preventing Chronic Disease*, 22, 240539. https://doi.org/10.5888/pcd22.240539

Whittaker, A., Dinu, M., Cesari, F., Gori, A. M., Fiorillo, C., Becatti, M., Casini, A., Marcucci, R., Benedettelli, S., & Sofi, F. (2017). A Khorasan wheat-based replacement diet improves risk profile of patients with type 2 diabetes mellitus (T2DM): A randomized crossover trial. *European Journal of Nutrition*, 56(3), 1191–1200. https://doi.org/10.1007/s00394-016-1168-2

Whittaker, A., Sofi, F., Luisi, M. L. E., Rafanelli, E., Fiorillo, C., Becatti, M., Abbate, R., Casini, A., Gensini, G. F., & Benedettelli, S. (2015). An organic khorasan wheat-based replacement diet improves risk profile of patients with acute coronary syndrome: A randomized crossover trial. *Nutrients*, 7(5), 3401–3415. https://doi.org/10.3390/nu7053401

WLN. (2019). Iowa's private wells overrun with agricultural contaminants. https://www.iaenvironment.org/newsroom/water-and-land-news/iowas-private-wells-overrun-with-agricultural-contaminants

Yenagi, N. B., Hanchinal, R. R., Patil, C. S., Koppikar, V., & Halagi, M. (2001). Glycemic and lipidemic response to dicoccum wheat (Triticum dicoccum) in the diet of diabetic patients. *International Journal of Diabetes in Developing Countries*, 21, 153–155.

3 Soil Chemical and Physical Properties, Processes, and Human Health

Eric C. Brevik and Amir Sadeghpour

INTRODUCTION

A relationship between soil and human health has long been recognized. Understanding of the importance of fertile soil to human well-being goes back to at least around 1400 BC as shown in the Biblical description of Moses leading his people into Canaan. Greek writings from about 400 BCE and Roman from around 60 BCE also demonstrate knowledge of a link between soils and human health (Brevik and Sauer, 2015). The origin stories of several cultures and religions recognize links between soil and life. This holds true in narratives from Chinese, Greek, Indian, Native American, Siberian shamanism, and Yoruba mythology, among others (Patzel et al., 2023). If one accepts that a link between soil and life is the ultimate soil–human health relationship, these myths show the depth of that recognition. Despite the long-standing recognition of these links, the scientific investigation of links between soil and human health did not begin until the 20th century (Brevik and Sauer, 2015).

Humans are exposed to soil materials through ingestion, respiration, and skin absorption or penetration (Steffan et al., 2018). Examples of ways soil can impact human health include both nutrient supply and toxic exposures through the food web, medications provided to cure illnesses, and exposure to dust generated from the soil or to soil organisms. Soil affects human health both directly and indirectly. Examples of direct effects include penetration of the skin by a pathogenic organism that lives in soil or inhalation of dust particles that cause respiratory issues. Indirect effects include nutrient supply from foods grown in soil, exposure to heavy metals contained in foods grown in contaminated soil, or taking medication developed from a soil source to cure illness. As can be seen through this discussion, the influence of soils on human health can also be positive or negative, that is, soils may improve or degrade human health depending on the circumstances. Human management of and interactions with the soil have a major impact on whether that soil will support or threaten human health. Pollution or fertilization that causes nutrient imbalances are ways to threaten health while appropriate fertilization and protecting soil from contamination can improve human health (Brevik et al., 2020). In general, management practices that promote soil health are also widely viewed as supporting human health (Pepper, 2013; Lehmann et al., 2020).

Historically, chemical aspects of soil and human health have received the greatest attention (Brevik and Sauer, 2015). This has included topics such as nutrient supply to humans through the soil–plant system, exposure to soil contaminants such as organic chemicals or heavy metals, and the ability of soils to neutralize chemical contaminates through biogeochemical processes. Biological aspects of the soil and human health relationship have also received considerable attention, although this has focused more on the microbiological side and less on macrobiological perspectives (Pereg et al., 2021). Physical soil properties have not received as much attention as biological and chemical properties as regards their influence on human health. Much of the work on physical properties has focused on the filtration capacity of soils and their ability to purify water or function as part of on-site sewage treatment systems (Brevik and Burgess, 2014).

DOI: 10.1201/9781003469742-3

The properties of any given soil are formed through pedogenic (soil forming) processes. Pedogenesis is largely controlled by the five soil forming factors: (1) climate, (2) organisms, (3) parent materials, (4) topography, and (5) time. Early in the pedogenic process parent materials and topography tend to have a large influence on soil properties. The parent material determines the minerals present in the initial soil, significantly influencing the availability of nutrients, soil pH, and the rate at which the mineralogy will change, as well as the predominant texture. Topography influences water movement, solar heating, and erosion rates within the forming soils. As time goes by, climate and organisms begin to have a larger role in determining the properties of a soil. Climate determines the amount of water present, along with average temperatures and the seasonal variations in both water availability and temperature throughout the year. Warmer and wetter climates accelerate the transformations of specific minerals, with some mineral types undergoing more rapid transformations than others. Organisms are important for adding organic matter to a soil and in driving biogeochemical reactions. The end result of these soil-forming factors is the soil at any given place at the current time, with all its corresponding influences on human health. For this reason, pedologists (scientists who study soil formation and change) have an important role to play in future soil and human health research.

While there has been considerable interest in the links between soils and human health for many decades (Brevik and Sauer, 2015), it has been rare for soil–human health links to be implicitly connected to soil properties. Therefore, this chapter seeks to clarify these connections with a focus on soil chemical and physical properties. The biological properties of, and processes in, soil as relates to human health received a fairly recent review in Pereg et al. (2021). It is also important to realize these subdivisions do not exist as separate categories in the soil, so the interactions between them, including soil biology, will also be discussed. The ultimate goal of the chapter is to provide additional insights into the links between soils and human health to facilitate future research and understanding in this area.

SOIL CHEMICAL PROPERTIES AND PROCESSES

Soil, a fundamental component of ecosystems, significantly influences human health through its intricate chemical processes (Duckworth et al., 2022). The regulation of environmental contaminants within the soil matrix dictates the extent and pace at which these substances traverse exposure routes, ultimately impacting human well-being (Elsakhawy et al., 2022). This section introduces soil chemical properties and contaminants and discusses soil as a contaminant regulator, chemical processes for contaminant availability and losses, human health implications, and future research needs in the soil chemistry–environment–human health nexus.

The intricate chemistry of soil serves as a critical determinant in shaping the toxicity of contaminants present within it, thereby influencing the potential health risks associated with human and environmental exposure. Soil chemistry, characterized by factors such as pH, organic matter content, and mineral composition, contributes significantly to the fate and behavior of contaminants, ultimately dictating their impact on ecosystems and human well-being. This section will present those soil chemical factors, indicate how they can affect contaminants availability, and identify practices that could reduce negative effects on human health.

Soil pH

Variations in soil pH can dramatically affect the solubility and mobility of different contaminants (Yu et al., 2023). For instance, certain metals may become more soluble and bioavailable in acidic soils, heightening the risk of uptake by plants and subsequent exposure through the food chain (Brevik et al., 2020). Many heavy metals are much more available for plant uptake in acidic soils and therefore, edible crops (e.g., vegetables, fruits, etc.) can take up large amounts of these metals that could affect the food we eat and thus, human health (Deng et al., 2020). An effective and easy way to fix this issue is by liming and raising the soil pH to neutral (Deng et al., 2021). This not only offers better overall essential nutrient availability but also decreases the availability of metals for luxury consumption by crops. Heavy metals are generally cations. At low soil pH cation exchange capacity (CEC) decreases and there are fewer exchange sites to hold the metal ions. Adding lime to raise the pH increases those exchange sites, and thus reduces the potential for heavy metal uptake by crops.

Organic Matter Content

Soil organic matter (SOM) plays a crucial role in influencing the availability of contaminants, particularly heavy metals, in the soil. The presence and characteristics of SOM impact various physicochemical and biological processes that govern the fate and behavior of contaminants (Yu et al., 2023). Soil organic matter contains functional groups such as carboxyl, hydroxyl, and amino groups that can act as binding sites for heavy metal ions (Brady and Weil, 2016). These functional groups facilitate adsorption and complexation, reducing the mobility and bioavailability of heavy metals. The adsorption and complexation of heavy metals by SOM decreases their concentration in the soil solution, limiting their availability for plant uptake and potential leaching into groundwater.

Soil organic matter contributes significantly to the CEC of soil (Brady and Weil, 2016). The CEC is the soil's ability to retain and exchange cations (Figure 3.1), including heavy metal ions. Higher SOM content generally results in increased CEC. An elevated CEC enhances the retention of heavy metals in the soil matrix, reducing their availability for uptake by plants. This is particularly important for mitigating the risks associated with heavy metal contamination. SOM also serves as a nutrient source for soil microorganisms (Rieke et al., 2022b). Microbial activity influenced by SOM can indirectly affect the mobility and availability of heavy metals through processes like microbial degradation or reduction (Zhen et al., 2019). Adding organic amendments to soils, especially those with low SOM content, can be an effective way to facilitate retaining heavy metals and other contaminants in the soil. Organic materials such as compost or manure contribute to SOM, enhancing its content and functionality. Adding cover crops and ensuring soil cover through no-till practices can also contribute to maintaining and then increasing SOM over time (Weidhuner et al., 2021).

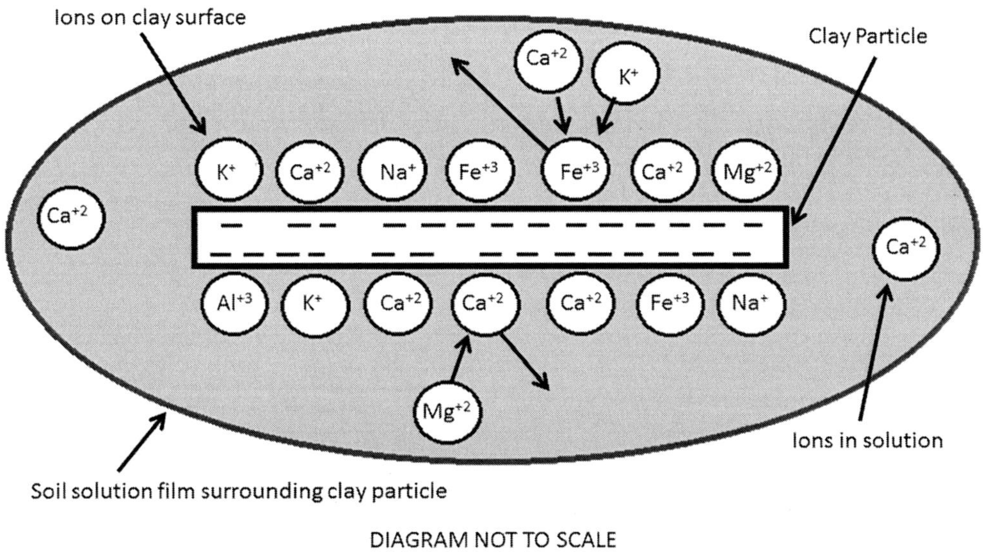

FIGURE 3.1 Negatively charged particles in the soil, such as soil organic matter and clay minerals, can attract cations to them and exchange them for other cations in charge-for-charge exchange reactions.

Pathogens (e.g., bacteria, viruses, parasites) can contaminate the soil through the application of animal waste and improper handling of manure. Improper management practices and inadequate handling of manure introduce these microorganisms into agricultural soils, thereby creating a risk for both farmers and consumers. The presence of pathogens in the soil can pose a significant threat to food safety, as contaminated produce may serve as a vehicle for the transmission of harmful microorganisms to consumers (Bicudo and Goyal, 2003).

Redox Potential

Oxidation-reduction (redox) reactions are chemical processes that involve the transfer of electrons between substances. In the context of soil, redox reactions play a crucial role in influencing the chemical forms of contaminants, particularly heavy metals. The redox potential, a measure of a system's ability to undergo oxidation or reduction, determines the direction and extent of these reactions (Yu et al., 2023). Changes in redox potential can result in the transformation of contaminants into forms that vary in toxicity, ultimately shaping their impact on ecosystems and human health. Soil conditions can exhibit variability in redox potential due to factors such as moisture content, organic matter decomposition, and microbial activity (Fiedler et al., 2007). Soils may experience periods of oxidation or reduction depending on environmental conditions. In oxidized conditions, iron oxides in the soil act as electron acceptors, promoting the oxidation of organic matter. During reduction, microorganisms use iron oxides as electron donors, leading to the release of electrons (Yu et al., 2023). This process can influence the solubility and mobility of heavy metals. Chromium exists in various oxidation states, with hexavalent chromium (Cr(VI)) being more toxic than trivalent chromium (Cr(III)) (DesMarias and Costa, 2019). Under oxidizing conditions, Cr(VI) may predominate, but under reducing conditions, it can be transformed into less toxic Cr(III). This transformation is significant in influencing the environmental fate of chromium. Arsenic (As) exists in multiple forms, and its mobility is influenced by redox conditions (Yu et al., 2023). Under reducing conditions, As may be mobilized in the more toxic form of arsenite (As(III)), while oxidizing conditions can lead to the immobilization of As as arsenate (As (V)), reducing its mobility (Abbas et al., 2018). Microbial activity in soils is often linked to redox reactions. For instance, in the process of nitrate reduction, microorganisms use nitrate as an electron acceptor, leading to the formation of nitrogen (N) gas. This process affects the overall redox potential and can influence the fate of other contaminants in the soil (Kamp et al., 2015). In certain anaerobic environments, such as wetlands, redox conditions can contribute to the methylation of mercury (Hg). Methylmercury, the transformed species, is highly toxic and can biomagnify in aquatic food chains, posing risks to human health through fish consumption (Helmrich et al., 2022).

SOIL PHYSICAL PROPERTIES AND PROCESSES

Soil physical properties control the movement of air, water, heat, and chemicals within the soil (Jat et al., 2018). Because of this they have profound impacts on soil biological and chemical processes, and through that on human health. However, the effects of soil physical properties on human health, with a few exceptions, tend to be more indirect than the effects of biological and chemical properties. This is probably why soil physical properties have received less attention historically in discussions of the links between soil and human health (Brevik and Sauer, 2015). Important soil physical properties include texture, structure, aggregate stability, porosity and permeability, bulk density, water holding capacity, and color (Brady and Weil, 2016). These properties do not exist in isolation. There are several connections between the physical properties, as well as many connections between biological, chemical, and physical properties. Some of these links will be highlighted in the discussion below.

TEXTURE

Soil texture is determined by the relative distribution of sand, silt, and clay-sized particles in the soil. It is one of the most commonly performed analyses in soil science because many of the physical and chemical properties of the soil are dependent on texture. Clay is particularly important to soil chemical properties because of its high surface area per unit mass (Brady and Weil, 2016). A significant portion of a soil's CEC comes from the clay content (Helling et al., 1964), with both the percentage and type of clay being important. The CEC by type of clay is kaolinite < chlorite = illite < smectite < vermiculite. The net negative charge on clay particles comes from isomorphic substitutions in the clay crystal structure and hydroxyl groups found at the crystal edges. The charges from isomorphic substitutions are termed permanent charges and are not pH dependent. The charges from the hydroxyl groups are known as variable charges and increase with increasing soil pH as hydrogen ions are released from the hydroxyls (Schaetzl and Thompson, 2015). Soil CEC, which is partially due to clay minerals (SOM is also an important source of soil CEC), is an important soil property for human health.

Many essential plant nutrients, such as calcium (Ca), magnesium (Mg), ammonium-N, potassium (K), iron (Fe), copper (Cu), manganese (Mn), and zinc (Zn) are available to plants in cation form. These nutrients can be held on the CEC sites in soil and made available to plants through exchange reactions (Figure 3.1). Therefore, soils with high CEC will also often be highly fertile. This does two things to support human health, it (1) allows high crop yields that provide adequate food amounts, and (2) helps ensure an abundance of nutrients so the food products from those crops are nutritious.

Most heavy metals, such as cadmium (Cd), Cu, lead (Pb), Mn, Hg, uranium (U), vanadium (V), and Zn are found in cation form in soils. Soil pH can be effectively adjusted with a range of amendments that either raise or lower the pH. It is common to manage heavy metal-contaminated soils with lime to raise the pH, therefore increasing the variable-charge CEC on clay minerals (and SOM) and providing exchange sites that adsorb the metals (Houben et al., 2013; Perez-Esteban et al., 2014). However, it is important to note that other soil properties come into play as well. For example, the adsorption of Cd is influenced by Ca and Zn in solution, because these metals can desorb Cd from soil particles and compete for sorption sites (Rodriguez-Martin et al., 2006). Calcium in solution also competes for sorption sites with Zn (Bradl, 2004), and the type and content of SOM can also make a difference in heavy metal adsorption (Vamerali et al., 2010). Therefore, while a high soil CEC can help reduce the hazardous cations available for plants to take up from the soil and pass up the food chain, the overall picture is more complex than just the CEC.

Clay content and CEC have also been shown to have a relationship to the adsorption of viruses that pass through soil with infiltrating water. In particular, soils with high CEC have been shown to remove poliovirus 1 and reovirus 3 from the soil solution. Montmorillonite has also been shown to absorb poliovirus 1. The sorption of these viruses to soil surfaces likely reduces the chance they will end up in groundwater resources and cause human disease. However, not all viruses are readily absorbed to soil particles. Differences in absorption are probably related to differences in the isoelectric points of various viruses (Schijven and Hassanizadeh, 2000).

There are some instances when the physical properties of soil, such as texture, can directly impact human health. One is podoconiosis, which occurs when clay is absorbed through the feet and blocks lymph node macrophages in the legs. In its advanced stage this causes swelling of the legs known as elephantiasis (Figure 3.2) (Thiele-Bruhn, 2021).

STRUCTURE

Soil structure refers to the arrangement of sand, silt, clay, and organic matter into aggregates (Figure 3.3). This is accomplished through the flocculation of primary particles (sand, silt, and clay) due to electrostatic charges followed by stabilization through organic and mineral binding

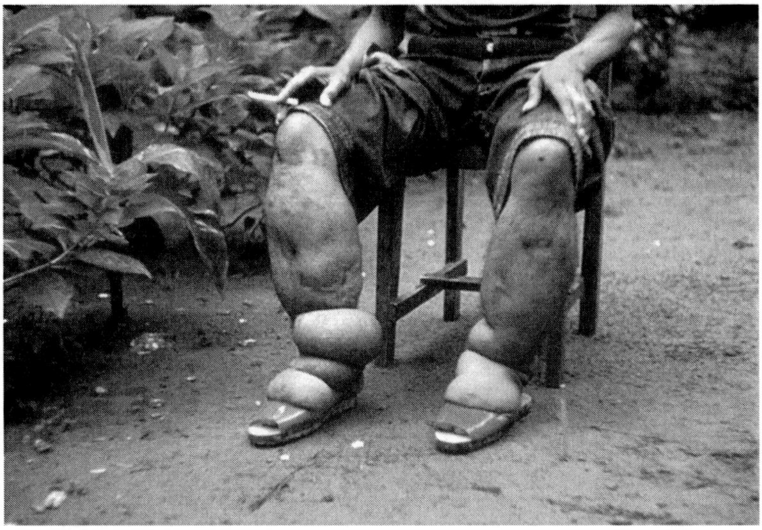

FIGURE 3.2 Elephantiasis as shown by swelling in the lower legs.
Image courtesy of the Centers for Disease Control and Prevention Public Health Image Library. Photo in the public domain.

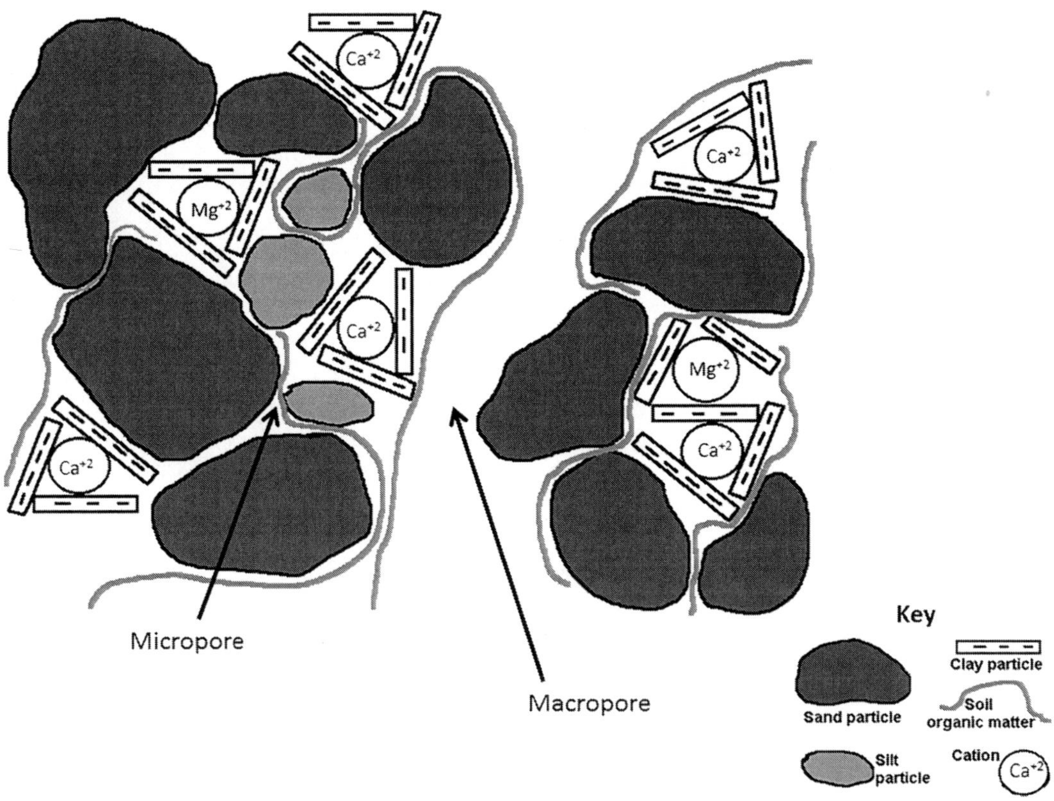

FIGURE 3.3 Soil aggregates are formed by the linking together of sand, silt, and clay particles. This happens through flocculation (as shown by the cations attracting the clay particles) and stabilization (as shown by the soil organic matter binders). The various particles shown in this image are not to scale.

agents (Sharma and Kumar, 2023). One of the most important aspects of forming stable aggregates is having adequate levels of SOM (Six et al., 2004). These aggregates create micropores within the aggregates that provide water-holding capacity and macropores between the aggregates that serve as corridors for soil organisms, roots, water, and air to move through (Brady and Weil, 2016). Mesopores, which are between micropores and macropores in size and are important for holding plant available water, are also created by the structure. Because of their importance in aeration, water movement and storage, soil temperature, and root penetration, well-aggregated soils are important for crop production.

The most direct effect of structure on crop production is through root growth. A well-structured soil provides abundant macropores that allow easy penetration of roots through the soil. This in turn allows an extensive root system that can access water and nutrients throughout a large soil volume. Roots may send signals that restrict shoot growth when growing in soils that are difficult to penetrate, so a well-aggregated soil can also facilitate above-ground plant growth (Sharma and Kumar, 2023). This in turn promotes food security, nutrient-dense foods, and through that human health.

Well-aggregated soils provide a range of pore sizes within the soil matrix. This range of pore sizes provides the full suite of services needed from aggregates. Water can both move through the soil and be stored for crop use. Soil organisms have suitable habitat and can help create an ecosystem that is conducive to plant growth (Sharma and Kumar, 2023). Air, which is essential for nutrient uptake by roots, can move into the soil macropores (Havlin et al., 2013). In short, well-aggregated soils create an environment that supports strong plant growth and crop yields, and, therefore, human health.

Soil organic matter contributes to soil structure and aggregation. Well-aggregated soils with adequate SOM content create microenvironments where heavy metals are sequestered, reducing their availability. Soil aggregation promoted by SOM helps in creating stable aggregates (Six et al., 2004) that can encapsulate heavy metals, minimizing their exposure and potential movement (Shen et al., 2022) and therefore protecting human health.

Aggregates in the soil serve as reservoirs of SOM that is protected from ready decomposition (Sharma and Kumar, 2023). Given the importance of SOM to crop production, this is another place where aggregates are important for food security. In addition, soil is an important part of the global carbon cycle, which in turn is important in climate change. By helping hold carbon-based SOM in the soil, aggregates also help keep carbon-based greenhouse gases out of the atmosphere (Brevik, 2012; Lal et al., 2021). Climate change is identified as a major factor shaping future human health (Barrett et al., 2015). In this context, soils become influential in human health through their impact on climate change.

Aggregate Stability

Soil structure is a dynamic property; it can quickly change given changes to the soil system. Therefore, the resistance of structural aggregates to being destroyed by forces such as water, wind, or tillage is an important physical property of soil. Aggregate stability can be measured using wet sieving, simulated raindrop impact, sedimentation techniques, or dry sieving, depending on the information desired (Rieke et al., 2022a). For example, dry sieving would be used when evaluating aggregate stability for a wind erosion study, while one or more of the wet methods would be used for a water erosion study.

Stable aggregates minimize soil erosion, as the aggregates are more difficult for wind or water to move than the smaller individual sand, silt, or clay particles that compose the aggregates (Sharma and Kumar, 2023). Erosion is the leading global cause of soil degradation, which in turn decreases crop yields (Shaheen and Bukhari, 2018). Zornoza et al. (2015) also identified aggregate stability as an important sensitive indicator of soil quality or health that could then be linked back to human health. Therefore, the formation of stable aggregates is important to human health.

Porosity and Permeability

Porosity and permeability are separate but related properties of the soil. Porosity refers to the amount of open, or void, space within the soil. Pores are created by aggregation (see discussion above) as well as the movement of earthworms, ants, termites, plant roots, and other organisms through the soil. Permeability refers to the degree of interconnectedness between pores. In some cases there is a high level of connectivity between the pores in a soil, while at other times the pores are largely isolated. The former is referred to as high permeability, while the latter is low permeability.

Soil pores are important for multiple soil functions. The importance of micropores and mesopores for water holding capacity and macropores for air and water movement and root penetration were already discussed in the section on structure. Larger pores also serve as the habitat of many soil microorganisms (Smiles, 1988), while cryptopores are those that are too small for soil microorganisms to move through (Sharma and Kumar, 2023). Microorganisms can be trapped by filtration as soil water moves through small pores; some bacterial varieties that have shown low movement through soil included *Bacillus* strain CB3, *Flavobacterium* strain DF3, *Pseudomonas* strain CB1, and *Enterobacter* strain Tol1. A common theme to these bacteria was a long cell length, and they were likely filtered out of the moving water by soil pores too small for the bacteria to pass through (Gannon et al., 1991). This can have both positive and negative effects on human health. For example, on the positive side pathogenic microorganisms that are filtered out cannot get to groundwater, potentially infecting water used for human consumption. On the negative side, bacteria being used as part of bioremediation may not be able to move through the soil to the contaminant of interest if they are filtered out or have their movement restricted by the soil.

Both high and low permeability can promote human health depending on the circumstances. High permeability is desirable for growing crops because it allows for root penetration along with air and water movement throughout the soil profile. This increases crop yields (Vinueza, 2018). However, high permeability may also allow contaminants to move through the soil and into groundwater (la Cecilia et al., 2020), potentially threatening human health. Therefore, high permeability into the subsoil layers may be undesirable when soil contamination is an issue, particularly contaminants that are highly mobile such as nitrates and herbicides like atrazine and glyphosate. Such contaminants can move into groundwater supplies, causing human health problems.

Permeability can also influence the movement of soil organisms, which can impact human health. Soils with high permeability allow predatory organisms to move through the soil, seeking out and consuming organisms that may be plant or human parasites. Soils with low permeability are more favorable to parasites, as they are not as reliant on the free movement required by predators to conduct life functions (Wolf and Snyder, 2003). Lower parasite populations are beneficial for crop production, food security, and human health.

Bulk Density

Bulk density is the measure of the mass of dry material within a given volume of soil. A soil with low bulk density will have high porosity, while a soil with high bulk density will have low porosity. For that reason, bulk density is a frequently used measure of soil compaction (Brady and Weil, 2016). Compacted soils (high bulk density) exhibit many of the same problems discussed for soil with poor aggregation and low permeability, including poor aeration and being difficult for roots to penetrate (Table 3.1) and soil organisms to navigate. Studies have also shown that compaction limits additions of fresh organic matter to soil (Brevik et al., 2002). This decreases the ability of the soil to maintain/increase SOM content and thus improve soil resiliency against abnormal weather conditions for plant production, among other problems. Lack of root penetration also is directly linked to poor nutrient uptake and, therefore, negatively influences not only plant production (Figure 3.4) but also the production of nutritious food and thus negatively influences human health.

TABLE 3.1
Typical Root Limiting Bulk Densities by Soil Texture

Texture	Root-Limiting Bulk Density (g/cm^3)
Sand	1.80
Fine sand	1.75
Sandy loam	1.70
Fine sandy loam	1.65
Loam	1.55
Silt loam	1.45
Clay loam	1.50
Clay	1.40

Source: Data from Coder (1998).
Note: Root growth is limited when bulk density values are at or above the values given. Coarser soils tend to have higher root-limiting bulk densities because in poorly structured soils the coarse particles create more macropores than do the smaller particles of fine-textured soils.

FIGURE 3.4 The dark gray strip through the center of this photograph shows forages growing in an area of compacted soil through a pasture in Iowa. Note the poor forage growth as shown by the plants that are much shorter and less mature than those to either side.
Photo by E.C. Brevik.

Water Holding Capacity

Water holding capacity is the ability of a soil to store water within its pores. This stored water can be very important to keep plants alive during dry periods within a growing season. Water is stored within a soil's micropores and mesopores, so texture and aggregation are important properties that determine water holding capacity. For crop production, the most important measure is plant

TABLE 3.2
Typical Available Water Holding Capacities for Soils of Various Textures

Soil Texture	Plant-Available Water (cm of Water per 30 cm of Soil Depth)
Sand	0.6–2.5
Loamy sand	2.8–3.0
Sandy loam	3.2–3.6
Silt loam	5.1–6.4
Silty clay loam	4.6–5.1
Silty clay	3.8–4.3
Clay	3.0–3.8

Source: Data from Ball (2001).

available water, which is a portion of the overall water holding capacity (Table 3.2). Soils with adequate water holding capacity tend to produce larger crop yields in average-to-dry years, but soils with low water holding capacity may perform better during wet years as they can move excess water away from the crop root zone. Soils with good water holding capacity also support more abundant microbial communities (Garaycochea, 2020). The effects of water holding capacity on human health come primarily through its impact on crop yields and overall soil ecology.

Color

Soil color is an important physical property with implications for understanding the chemical and physical conditions during soil formation. While the color alone does not directly influence human health, it acts as an informative indicator. As discussed in the sections above, these chemical and physical conditions, in turn, affect human health. Examples of what color can communicate about the soil environment are given in Table 3.3.

INTERACTIONS BETWEEN SOIL PROCESSES

Soil serves as a dynamic matrix intricately involved in dictating the fate and transport of environmental contaminants (Brevik and Sauer, 2015). The multifaceted interplay of soil processes exerts a profound influence on how air, chemicals, heat energy, organisms, and water move within the soil profile, ultimately shaping the destiny of various factors that influence human health and the potential impact of contaminants on the environment. Biological, chemical, and physical soil properties all combine to determine the soil processes occurring at a given location and time. This section will give some examples of important ways that soil properties interact to regulate the soil environment and influence human health.

Adsorption and Desorption

Soil particles possess charged sites capable of attracting and retaining contaminants and some microorganisms (Steffan et al., 2020). This process, known as adsorption, immobilizes charged particles on soil surfaces, reducing their mobility. Conversely, contaminants or microorganisms can detach from soil particles through desorption, becoming mobile again. This dynamic equilibrium between adsorption and desorption significantly influences the fate of contaminants (Yu et al., 2023) and is an excellent example of the interplay between soil biological, chemical, and physical properties. Clay minerals and SOM exhibit significant CEC, with negative charges that enable the absorption of both positive and negative ions through ion exchange mechanisms (Yu et al., 2023). The CEC allows clay particles and SOM to absorb positively charged metal cations or microorganisms

TABLE 3.3
Some Common Soil Colors, What Causes Them, and What They Tell Us about the Soil

Iron Oxide Pigmenting Agents

Color	Cause (Mineral Name)	Soil Environment
Yellow	Goethite	Goethite forms in cool temperatures, or in moist soils in warm settings that are not water saturated. Goethite often occurs in association with hematite.
Red	Hematite	Hematite forms in high temperature soils that are well oxygenated.
Brown	Maghemite	Maghemite is formed when other iron oxides are transformed by heating in association with organic matter, i.e., in forest fires.
Orange	Lepidocrocite	Lepidocrocite forms in a local zone of oxidation in soils that are otherwise reduced, i.e., around the roots of wetland plants.
Gray-green	Hydromagnetite	Hydromagnetite forms in soils that are frequently saturated, creating reducing conditions.

Other Pigmenting Agents

Color	Cause	Indicates
Black or dark colors near the surface	Coatings of organic matter on soil particles	Conditions are favorable for the accumulation of organic matter in surface horizons. This may indicate wet conditions or high levels of organic matter additions.
Black in the subsurface	Organic matter or manganese oxides	Organic matter – intense leaching is moving organic matter from surface horizons into the subsurface. Manganese oxides – form in wet soils that undergo frequent alterations between oxidizing and reducing.
White	Carbonates or other salts	The soils form in an environment that is so drythat there is not enough water to leach the salts from the soil.

from the soil solution (Figure 3.1). Under low pH conditions (e.g., pH 3–6), clay minerals initially undergo ion exchange with cationic heavy metals, while at higher pH levels they form surface complexes. The relatively low content of aluminum oxide in the soil at higher pH values diminishes the contribution of aluminum oxides to the adsorption and fixation of heavy metals compared to iron oxides. Generally, swelling clays exhibit the highest absorption rate for positively charged particles, followed by mixed clay structures containing swelling components (Sipos et al., 2008). Some heavy metals have a higher affinity for humus (the stable fraction of SOM) as compared to clay minerals and iron (hydroxide) oxides (Yu et al., 2023). Interestingly, clay types have preferential absorption for certain heavy metals. For example, kaolin preferentially adsorbs chromium (Cr), vermiculite preferentially adsorbs Cu and Zn, while mica absorbs Cr. Also, kaolin was found not to effectively fix Cu (Covelo et al., 2007).

Chemical Transformations

Soil chemical processes encompass a spectrum of reactions that can transform the chemical structure of contaminants. These transformations may render contaminants less toxic or, in some cases, convert them into more harmful substances. For example, redox reactions, influenced by soil microbial activity and influenced by the soil aeration status, can either oxidize or reduce contaminants, altering their solubility and mobility in the soil (Kelderman and Osman, 2007). Other reactions that influence the transformation of contaminants include acid-base reaction, hydrolysis, complexation, precipitation, ion exchange, and chelation (Hashim et al., 2011; Yu et al., 2023). Soil is inherently buffered, containing minerals with varying acid-neutralizing capacities. Contaminants exhibiting acidic or basic properties may undergo reactions that neutralize their pH, reducing their potential for harm. Some contaminants react with water molecules in hydrolysis reactions, leading to the breakdown of complex compounds into simpler, less harmful forms. Contaminants can form complexes with minerals and organic matter in the soil (Hashim et al., 2011). This process can reduce the bioavailability of certain elements, mitigating their potential impact, while other contaminants can precipitate into solid forms in certain soil conditions, minimizing their solubility and availability for plant uptake or leaching (Yu et al., 2023). Soil particles have charged sites capable of exchanging ions with the soil solution. This ion exchange process can alter the availability and mobility of contaminants, influencing their impact on human health. Also, SOM and certain microbial byproducts can act as chelating agents, forming stable complexes with metal contaminants (Sears, 2013). This process can decrease the bioavailability of metals and mitigate their toxic effects. All these chemical reactions are influenced by physical properties that control air and water movement through the soil as well as microbial community structures.

Volatilization and Leaching

Certain contaminants can undergo volatilization, transitioning from a solid or liquid state to a gaseous form (Zhao et al., 2019). Soil conditions, such as temperature, moisture, and permeability influence the rate at which volatilization occurs (Morra et al., 2011). Leaching, the downward movement of contaminants through the soil, is a critical aspect of transport. Soil chemical properties such as pH, SOM, and cation/anion exchange capacities (CEC and AEC) play an important role in the availability of nutrients and their loss through the leaching pathway. Soil physical properties such as aggregation and permeability help control nutrient loss by leaching in their control of water movement through the soil profile.

Soil Structure

Structure is an important property of soil that is formed through a combination of biological, chemical, and physical processes. The spaces between soil structural units are an important source of macropores that allow air, organisms, and water to pass through the soil, while the spaces within the structural units are an important location of micropores that store water within the soil. Mesopores may form between or within structural units. This makes structure important to crop growth, filtration, provision of soil habitat, and other soil ecosystem services, and through that to human health. Because it takes biological, chemical, and physical processes to create structure, it is the perfect example of interactions between the soil processes.

CONTAMINANTS IN THE SOIL

Agricultural soils can be contaminated with various substances that influence the food we eat and thus pose risks to human health (Gu et al., 2021). The type and extent of contamination depend on factors such as agricultural practices, pesticide and fertilizer use, industrial activities, soil chemical and physical processes, and the environmental context (Brevik et al., 2021). Several common contaminants may be found in agricultural soils, end up in our foods, and ultimately be harmful to human health.

Pesticides and Herbicides

Organophosphates, organochlorines, and glyphosates are common pesticides/herbicides used in agricultural practices that can cause human health risks (Kim et al., 2017). Organophosphates are a class of pesticides widely used in agriculture to control insects, pests, and diseases affecting crops. While they are effective in pest management, these chemicals pose significant risks to human health. In particular, residues of organophosphates can persist on treated crops, posing a potential risk to human health through the consumption of contaminated food.

Organochlorines are a class of synthetic pesticides that contain carbon (C), chlorine (Cl), and hydrogen (H) atoms in their molecular structure. These pesticides were once extensively used in agriculture for their effectiveness against a wide range of pests (Jayaraj et al., 2016). However, due to environmental concerns related to their persistence, bioaccumulation, and potential health risks, many organochlorines have been phased out or had their use restricted through the Stockholm Convention (Brevik et al., 2020). These pesticides have a long environmental half-life and are therefore very persistent in soils (Jayaraj et al., 2016). Organochlorines can be transported from agricultural fields to water bodies through runoff and leaching. This can contribute to the contamination of surface water and groundwater, posing risks to aquatic ecosystems, our drinking water, and potentially the aquatic food we eat. Although it is now less common to use these pesticides, their residues may still be present in some agricultural soils long after their last use. These residues can be taken up by crops, posing a risk to food safety and thus human health.

Glyphosate is a broad-spectrum herbicide widely used in agriculture and non-agricultural settings to effectively control weeds and unwanted vegetation (Bai and Ogbourne, 2016). However, the extensive use of glyphosate can influence human health through different pathways. Since glyphosate has a low soil binding capacity, its residues can persist in the soil. Additionally, the potential for glyphosate runoff into water bodies may contribute to environmental contamination and thus affect aquatic organisms. The presence of glyphosate in soil and water could result in the accumulation of this chemical in foods that humans consume, leading to severe health problems. The International Agency for Research on Cancer (IARC), a branch of the World Health Organization (WHO), classified glyphosate as a probable human carcinogen in 2015. However, other regulatory agencies, such as the US Environmental Protection Agency (EPA), have concluded that glyphosate is not likely to be carcinogenic to humans when used according to label instructions.

Persistent Organic Pollutants (POPs)

Persistent organic pollutants include substances like polychlorinated biphenyls (PCBs), dioxins, and certain pesticides. They can accumulate in the environment and the food chain. Some POPs can disrupt the endocrine system, leading to reproductive and developmental issues. Certain POPs, such as dioxins, are known carcinogens. Also, POPs can impair the immune system, making individuals more susceptible to infections (Alharbi et al., 2018). POPs are evaluated under the Stockholm Convention, with regular consideration of organic chemicals for addition to the list of banned chemicals (Brevik et al., 2020).

Several of the POPs are organic pesticides. Pesticide exposure can occur through residues in food, occupational exposure in agriculture, and residential use. Organophosphate pesticides can affect the nervous system, leading to symptoms such as dizziness, headaches, and in severe cases, seizures. They also can cause respiratory problems, especially in cases of acute poisoning. Some pesticides are associated with reproductive issues, including fertility problems and birth defects.

Fertilizers

Nitrogen and phosphorus (P) fertilizers, when used excessively, can affect both soil and groundwater and thus food safety and human health (Steffan et al., 2018). Potassium (K) deficiency also causes health problems in humans (Sun and Weaver, 2020). Nitrogen fertilizers, particularly ammonium-based, can decrease soil pH when over-applied and thus increase the availability of certain elements such as heavy metals (Deng et al., 2023) and induce aluminum toxicity in crops (Zhao and Shen, 2018). Over-use of N fertilizers also results in excess N in the soil that could be lost to groundwater (Sadeghpour et al., 2017a) or be lost to air through denitrification processes (Sadeghpour et al., 2017b) and result in contaminated drinking water and reduced air quality. The WHO has established a maximum standard of 10-ppm nitrate-N for safe drinking water. However, this standard is frequently surpassed globally (Townsend et al., 2003). In the United States, where the Safe Drinking Water Act enforces this standard, regional studies indicate that 10–20% of groundwater sources may exceed the 10-ppm limit (https://www.epa.gov/nutrientpollution/estimated-nitrate-concentrations-groundwater-used-drinking). Instances of severe groundwater contamination are often associated with concentrated animal feeding operations (CAFOs), particularly those involving swine and poultry (Mallin, 2000). High nitrate levels have various potential health effects, including reproductive problems (Kramer et al., 1996), methemoglobinemia, and cancer. Infants, in particular, are vulnerable to methemoglobinemia, commonly known as "blue-baby" syndrome. While conclusive evidence for this disorder occurring at nitrate levels below 10-ppm is limited, higher values have been found to significantly increase the risk (Townsend et al., 2003).

Phosphorus fertilizers also pose threats to human health by creating imbalances in the nutrient concentrations of crops. While P itself is not typically a health concern in food, imbalances in nutrient content due to fertilizer use can impact the overall nutritional value of crops (Calvo et al., 2014). Proper nutrient management is essential to ensure a balanced and healthy diet for consumers. In some cases, improper application or runoff of P may lead to the contamination not only of water but also of food and forage crops (Sharpley et al., 2001). While direct health effects from P ingestion are unlikely, it underscores the importance of responsible fertilizer application practices to minimize environmental impacts.

Potassium, due to its reactive nature, is considered one of the most easily leachable and erodible nutrient elements in soil (Alfaro et al., 2004). Its monovalent and electropositive properties make it highly soluble in water, facilitated by the polar nature of water molecules. In regions characterized by substantial weathering, particularly those with elevated precipitation, temperature, and bioactivity, there is a natural depletion of soil K over time (Sardans and Peñuelas, 2015). This phenomenon is noticeable in the southeastern US, which experiences the highest levels of precipitation, temperature, and bioactivity in the continental US, resulting in the lowest K levels in the topsoil (Sun and Weaver, 2020). Although this is a broad estimate, the topsoil K content in around 63% of counties across the 48 contiguous US states was lower in 2010 than in 1976, as indicated by the interpolation of spot soil K data by the US Geological Survey (USGS). In agricultural settings, the loss of K is further exacerbated not only by leaching from precipitation runoff but also by irrigation water runoff and the removal of crop biomass. The loss of K in farm soils persists unless there is compensation through K gain from rock weathering and/or the application of K fertilizer and manure. Without such replacement, there is a continual net loss of plant-available soil K over time. Reduction in available K for crop uptake along with K dilution due to increases in yields results in K reduction in most crops including forages and grain, and thus, meat (because of grazing low K forages).

Radionuclides

Naturally occurring radionuclides, as well as those related to human activities, have the capacity to contaminate agricultural soils, posing potential risks to both the environment and human health (Steffan et al., 2018). These radionuclides can originate from sources such as geological formations, cosmic radiation, and mineral decay. Additionally, human activities, including industrial processes, nuclear power generation, and the use of certain technologies contribute to the release of artificial radionuclides into the environment. Once introduced into the soil, radionuclides have the propensity to accumulate and become incorporated into the soil matrix. This accumulation raises concerns as agricultural crops, being directly in contact with the soil, may absorb these radionuclides during their growth and development. The uptake of radionuclides by crops can vary depending on factors such as soil type, plant species, soil amendments, and environmental conditions. The potential consequences of radionuclide accumulation in crops are twofold. First, there are direct implications for human health if contaminated crops enter the food chain. Ingestion of food items containing elevated levels of radionuclides can lead to internal radiation exposure, posing risks such as increased cancer incidence and other radiation-related health issues. Second, the contamination of agricultural soils with radionuclides has broader environmental implications. It can affect soil microbial communities, disrupt nutrient cycling processes, and impact overall ecosystem health. Furthermore, the potential for radionuclides to leach into groundwater presents an additional concern for both environmental and human health (Olagbaju et al., 2021).

Plastic Residues

Plastics, frequently derived from improperly disposed consumer items in the environment, gradually deteriorate due to photo and thermooxidative processes (Prata et al., 2020). Biodegradation contributes to this process to a lesser extent, causing a reduction in material integrity and resulting in fragmentation into particles smaller than 5 mm, commonly referred to as secondary microplastics (Andrady, 2011). Microplastics have been found in soil (Watteau et al., 2018), sea and freshwater (Rodrigues et al., 2018), sediment (Reed et al., 2018), and even the air we breathe (Abbasi et al., 2019). Recent reports indicate microplastics are present in consumer products like beer, sea salt, and tap water (Kosuth et al., 2018). Given their pervasive presence across various environmental compartments and consumer goods, the exposure of humans to microplastics has become inevitable. Inhaled or ingested microplastics may trigger inflammatory responses in the body. Since microplastics are a recently discovered issue, long-term health effects are not fully understood, and research is needed on the long-term health issues related to microplastic exposure. However, microplastics may potentially cause digestive problems, respiratory disease, decreased growth rate and reproduction when ingested, and inflammatory and immune reactions (Jung et al., 2022; Prata et al., 2020; Wang et al., 2021).

Microplastic impacts on the soil include reduced bulk density and aggregate stability, altered geochemical cycling, and negative effects on soil invertebrates (Qiu et al., 2022). They may also reduce plant biomass, shoot growth, and seed germination (Chia et al., 2021) and have been found in plant tissues (Jung et al., 2022). All these issues can reduce food security.

Heavy Metals

Heavy metals in the soil may have the greatest impact of any contaminant on foods and human health (Deng et al., 2020). Among heavy metals, several such as As, lead (Pb), cadmium (Cd), and Hg are categorized as not essential for metabolic and various other biological functions (Rai et al., 2019). In general, the primary factors contributing to increases in heavy metals are attributed to swift urbanization, alterations in land use, and industrialization, particularly in densely populated developing nations like India and China (Rai et al., 2019). Such activities can also introduce these metals to agricultural soils leading to accumulation in crops (Deng et al., 2021) and, therefore, pose human health risks, especially if these crops are part of the human diet. They are introduced into the soil environment through industrial activities, waste disposal, fossil fuel combustion, as impurities in fertilizers, and heavy metal compounds including As, Cu, and Pb have been used as inorganic pesticides (Table 3.4) (Brevik et al., 2020).

Lead (Pb)

Lead exposure often occurs through lead-based paints, contaminated water, industrial sites, leaded gasoline, and disposal of wastes, leading to several human health issues. Lead is particularly harmful to the developing nervous system in children. It can lead to intellectual disabilities, learning deficits, and behavioral problems. Moreover, Pb exposure has been associated with increased blood pressure and cardiovascular diseases in adults. The literature also agrees that chronic exposure to lead can affect kidney function (Boskabady et al., 2018).

Mercury (Hg)

Mercury can be found in soils impacted by gold mining, coal combustion, and the production of chlorine. Soil organisms can methylate Hg, which can lead to plant uptake or water contamination (Xu et al., 2015). Subsequent exposure to Hg may occur through the consumption of contaminated crops, water, or seafood, particularly those at higher trophic levels. Methylmercury can cause neurological damage, impacting cognitive function and motor skills. Pregnant women exposed to high levels of Hg may pass it on to their unborn children, leading to developmental issues. Also, high levels of Hg exposure have been linked to cardiovascular diseases (Ha et al., 2017).

TABLE 3.4
Common Sources of and Health Problems Associated with Selected Heavy Metals

Heavy Metal	Anthropogenic Sources	Health Problems
Hg	Electrical switches, fluorescent light bulbs, mercury lamps, batteries, thermometers, dental fillings, burning of coal and fuel oil, medical wastes	Central nervous system damage, coordination difficulties, eyesight problems, problems with the sense of touch, liver, heart, and kidney damage
Pb	Batteries, solder, ammunition, pigments, ceramic glaze, hair coloring, fishing equipment, leaded gasoline	Neurological impacts, lowers IQ and attention spans, impaired hand-eye coordination, encephalopathy, deterioration of bones, hypertension
Cd	Zinc smelting, burning coal or Cd-containing garbage, rechargeable batteries, pigments, TVs, solar cells, steel, phosphorus fertilizer	Liver and kidney damage, carcinogenic, low bone density
As	Pesticides, mining and smelting of gold, lead, copper, and nickel, iron, and steel production, burning of coal, wood preservatives	Gastrointestinal damage, skin damage, carcinogenic, heart, neurologic, and liver damage
Cr	Electroplating, corrosion protection, leather tanning, wood preservative	Carcinogenic, gastrointestinal disorders, hemorrhagic diathesis, convulsions

Source: Table based on information from Baird and Cann (2005) and http://www.cdc.gov/.

Arsenic (As)

Arsenic contamination has been identified in soils near treated wood structures (Gardner et al., 2013) and in old orchards where As compounds were employed as pesticides (Cabibbo et al., 2014). Additionally, elevated levels of As can naturally occur in soils with high organic matter content in the parent material (Steffan et al., 2018). Arsenic can contaminate drinking water, especially in areas with natural deposits or from industrial discharges, and contaminate crops when present in irrigation water. Long-term exposure to high levels of arsenic is associated with an increased risk of skin, lung, bladder, and other cancers. Chronic exposure can cause skin lesions, leading to discoloration and thickening of the skin. Arsenic exposure has been linked to cardiovascular diseases, including hypertension and atherosclerosis (Abdul et al., 2015).

Cadmium (Cd)

Cadmium is a heavy metal that can have adverse effects on human health, primarily through exposure to contaminated air, water, and food. Cadmium exposure is a concern because it is a cumulative toxin, meaning it tends to accumulate in the body over time. Cadmium can contaminate certain foods, especially shellfish, liver, and select grains and vegetables if grown in Cd-contaminated soil. Some industrial activities, such as mining, smelting, and manufacturing can release cadmium into the air and water, potentially exposing nearby populations. Cadmium has a strong affinity for the kidneys and can accumulate there over time. Chronic exposure to high levels of Cd is associated with kidney damage, which can progress to kidney disease. Moreover, Cd can replace calcium in bones, leading to skeletal damage. Long-term exposure may contribute to osteoporosis and fractures (Suhani et al., 2021).

Zinc (Zn)

Zinc is an essential trace element that plays a crucial role in both crop production and in various physiological processes within the human body. Zinc is often deficient in soils that are sandy, have pH values over 7, high calcium carbonate content, are highly leached, or frequently flood. It is often present at toxic levels in soils with a pH value of less than 4 (Oliver and Gregory, 2015). While Zn is an essential nutrient for the human body, excessive intake through food sources can lead to adverse effects. Excessive Zn intake can cause gastrointestinal issues such as nausea, vomiting, and abdominal pain. High Zn levels may interfere with the absorption of other essential minerals like Cu and Fe, leading to nutritional imbalances. Prolonged exposure to high levels of Zn may suppress the immune system and impair immune function. Elevated Zn levels have been associated with neurological symptoms such as headache and dizziness (Plum et al., 2010). Zinc deficiency has negative effects on human growth, immune system function, reproductive health, and neurological function (Oliver and Gregory, 2015).

THE IMPORTANCE OF AN INTERDISCIPLINARY APPROACH

Recognizing the interconnectedness of a wide range of environmental properties and processes and anthropogenic activities is important for addressing complex issues related to soils and human health (Brevik et al., 2020). For example, soil chemistry is influenced by microbial activities, such as nutrient cycling and organic matter decomposition. Interactions between microorganisms and soil chemistry impact the availability of nutrients and the transformation of contaminants (Brady and Weil, 2016). Microbes play a crucial role in the biodegradation of organic pollutants, contributing to soil remediation efforts. Similarly, redox processes in soil can affect the mobility and speciation of elements, influencing their bioavailability and environmental fate (Yu et al., 2023). As an example, iron and manganese redox cycling can influence the transformation of contaminants like As, affecting its mobility and toxicity (Fan et al., 2023). At the same time, soil structure influences redox status through its impact on gas exchange within the soil (Mattila, 2023). This example shows the blending of interdisciplinary concepts from the biological, chemical, and physical sciences that are necessary to understand in order to fully grasp redox reactions in the soil and how they influence human health.

Environmental processes, including water movement and precipitation, can influence the transport of contaminants from soil to water bodies, impacting water quality. For instance, runoff from agricultural fields can carry pesticides and fertilizers into water sources, posing risks to aquatic ecosystems and potentially impacting human drinking water. There is a need to develop sustainable soil-management practices and remediation technologies to improve human health (Steffan et al., 2018). As an example, designing phytoremediation strategies that use plants to extract and accumulate contaminants from soil and investigate the links between environmental exposures and human health outcomes (Deng et al., 2020). Overall, connecting disciplines should allow for a comprehensive assessment of soil, environment, and human health to ensure food security, safety, and human health. Therefore, we propose that future research should connect soil science, hydrology, food contamination, and human health. This would involve specialists from disciplines such as agronomy, atmospheric science, biology, food science, geology, human health and nutrition, medical science, and soil science. Several research themes that could increase our understanding of soil contaminants and their effect on human health are discussed in the following sections.

Climate Change Effect on Soil Contamination and Human Health

Climate change has influenced factors such as temperature and moisture through altering precipitation patterns and creating extreme weather events. This affects the transport, fate, and bioavailability of contaminants in soil, as well as the delivery of contaminants to soil and the movement of those contaminants beyond the soil. There is a need to explore the potential for increased mobility and leaching of contaminants, leading to changes in food safety and human exposure.

Long-Term Microplastic Effects on Soils, the Food Chain, and Human Health

Research is still lacking on the occurrence, distribution, and fate of microplastics in soil, their potential to transfer into the food chain, and their health effects. Particularly, the health implications of microplastic ingestion through contaminated food and assessment of their long-term effects on human health is a research gap identified in the literature.

Combined Contaminant Effects on Human Health

While previous studies have demonstrated that combined pollution enhances the bioavailability of heavy metals (An et al., 2023), the simplicity of the types of HMs, microplastics, and land uses considered necessitates a broader scope of research in this area. To date, research on the impact of heavy metals bioavailability has predominantly focused on isolated factors, such as specific characteristics of microplastics, or physicochemical soil indexes. Insufficient attention has been given to exploring the interactive effects of these factors on heavy metals bioavailability, posing potential ecological risks. Urgent efforts are required to delve into multi-factor interactions, including the role of soil microorganisms in these complex processes.

Considering the increased bioavailability of heavy metals and the bioaccumulation and biomagnification tendencies associated with them, there is a heightened risk of toxicity through the food chain. In alignment with the life cycle assessment concept, it is imperative to undertake research on the varying periods of impact intensity of microplastics pollution on heavy metals bioavailability. This research is crucial for conducting comprehensive health risk assessments and gaining insights into the broader implications of these environmental interactions.

Similar needs exist with organic chemicals in the soil system. Most human health research involving organic chemicals entails these chemicals in their manufactured form. However, this is not the state of these chemicals in the natural or agricultural environment. For example, agricultural management in a given field during a single growing season typically involves combinations of organic chemicals, fertilizers, and possibly other treatments. How these combinations of chemicals interact with each other and with the very biogeochemically active soil environment is largely unknown, which also means their impact on human health when exposures are through the soil system is unknown (Burgess, 2013).

Precision Agriculture for Reducing Nutrient Loss and Its Effect on Human Health

Developing innovative nutrient management (N and P for example) techniques using variable rate technologies has been studied for a while; however, study of their effect on nutrient loss reduction, potential for reduced contamination of water, and its effects on human nutrition and health is lacking. This should be evaluated to better understand the potential of these technologies to improve human health.

Advanced Models and Tools for Assessing the Human Health Risks Associated with Contaminants

Improved models and tools to assess cumulative risks associated with multiple contaminants in the soil, considering exposure pathways, environmental factors, and human behaviors are needed to effectively analyze complex datasets, convey meaningful information to the public, and suggest policy changes to improve human health (Brevik et al., 2021).

CONCLUDING STATEMENTS

We have long known that soils exert an influence on human health. However, there are still many things we do not know about the soil–human health nexus. In particular, little attention has been paid to the relationships between soil physical properties and human health. While the literature is replete with discussions of how these properties impact crop production or contaminant movement, human health connections are largely left to be extrapolated. The same is true of how soil physical properties affect water, nutrient, and contaminant movement through the soil and into groundwater, surface water, or other environmental media. While we need to know more about the links between soil biological and chemical properties and human health as well, carefully designed studies to address physical properties are crucial.

Several classic soil and human health issues were discussed, including nutrient supply, heavy metals, organic chemicals, and fertilizer issues. But there are also emerging areas that need to be researched, such as microplastics in soil, precision agriculture and its effect on human health, the soil-climate change-human health nexus, and the risk posed by combinations of chemicals found in the soil environment. Each of these needs to be addressed from the lens of soil chemical and physical properties as well as from other perspectives. Each represents an area of pressing need.

In all the cases noted above, an interdisciplinary approach is needed. Soil scientists need to reach out and work with experts from many other disciplines to tease out the complex relationships that exist in the soil–human health nexus. The problem is larger than any one discipline; the solution will be, too.

REFERENCES

Abbas, G., Murtaza, B., Bibi, I., Shahid, M., Niazi, N.K., Khan, M.I., Amjad, M., Hussain, M., and Natasha. 2018. Arsenic uptake, toxicity, detoxification, and speciation in plants: Physiological, biochemical, and molecular aspects. *International Journal of Environmental Research and Public Health* 15(1), 59. https://doi.org/10.3390/ijerph15010059

Abbasi, S., Keshavarzi, B., Moore, F., Turner, A., Kelly, F.J., Dominguez, A.O., and Jaafarzadeh, N. 2019. Distribution and potential health impacts of microplastics and microrubbers in air and street dusts from Asaluyeh County, Iran. *Environmental Pollution* 244, 153–164. https://doi.org/10.1016/j.envpol.2018.10.039

Abdul, K.S.M., Jayasinghe, S.S., Chandana, E.P.S., Jayasumana, C., and De Silva, P.M.C.S. 2015. Arsenic and human health effects: A review. *Environmental Toxicology and Pharmacology* 40, 828–846. https://doi.org/10.1016/j.etap.2015.09.016

Alfaro, M.A., Jarvis, S.C., and Gregory, P.J. 2004. Factors affecting potassium leaching in different soils. *Soil Use and Management* 20(2), 182–189.

Alharbi, O.M.L., Basheer, A.A., Khattab, R.A., and Ali, I. 2018. Health and environmental effects of persistent organic pollutants. *Journal of Molecular Liquids* 263, 442–453. https://doi.org/10.1016/j.molliq.2018.05.029

An, Q., Zhou, T., Wen, C., and Yan, C. 2023. The effects of microplastic on heavy metals bioavailability in soils: a meta-analysis. *Journal of Hazardous Materials* 460, 132369. https://doi.org/10.1016/j.jhazmat.2023.132369

Andrady, A.L. 2011. Microplastics in the marine environment. *Marine Pollution Bulletin* 62(8), 1596–1605. https://doi.org/10.1016/j.marpolbul.2011.05.030

Bai, S.H., and Ogbourne, S.M. 2016. Glyphosate: Environmental contamination, toxicity and potential risks to human health via food contamination. *Environmental Science and Pollution Research* 23, 18988–19001.

Baird, C., and Cann, M. 2005. *Environmental Chemistry*, 3rd Ed. W.H. Freeman and Company, New York.

Ball, J. 2001. Soil and water relationships. *Ag News and Views*. https://kbsgk12project.kbs.msu.edu/wp-content/uploads/2011/09/Soil-and-Water-Relationships.pdf. Accessed 12 December 2023.

Barrett, B., Charles, J.W., and Temte, J.L. 2015. Climate change, human health, and epidemiological transition. *Preventive Medicine* 70, 69–75. https://doi.org/10.1016/j.ypmed.2014.11.013

Bicudo, J.R., and Goyal, S.M. 2003. Pathogens and manure management systems: A review. *Environmental Technology* 24(1), 115–130.

Boskabady, M., Marefati, N., Farkhondeh, T., Shakeri, F., Farshbaf, A., and Boskabady, M.H. 2018. The effect of environmental lead exposure on human health and the contribution of inflammatory mechanisms, a review. *Environment International* 120, 404–420. https://doi.org/10.1016/j.envint.2018.08.013

Bradl, H.B. 2004. Adsorption of heavy metal ions on soils and soil constituents. *Journal of Colloid and Interface Science* 277, 1–18. https://doi.org/10.1016/j.jcis.2004.04.005

Brady, N.C., and Weil, R.R. 2016. *Nature and Properties of Soils*, 15th Ed. Pearson, Upper Saddle River, NJ.

Brevik, E., Fenton, T., and Moran, L. 2002. Effect of soil compaction on organic carbon amounts and distribution, South-Central Iowa. *Environmental Pollution* 116, S137–S141.

Brevik, E.C. 2012. Soils and climate change: Gas fluxes and soil processes. *Soil Horizons* 53(4), 12–23. https://doi.org/10.2136/sh12-04-0012

Brevik, E.C., Khaledian, Y., and El-Ramady, H. 2021. Assessing the complex links between soils and human health: An area of pressing need. *Frontiers in Soil Science* 1, 731085. https://doi.org/10.3389/fsoil.2021.731085

Brevik, E.C., and Sauer, T.J. 2015. The past, present, and future of soils and human health studies. *SOIL* 1, 35–46. www.soil-journal.net/1/35/2015/

Brevik, E.C., Slaughter, L., Singh, B.R., Steffan, J.J., Collier, D., Barnhart, P., and Pereira, P. 2020. Soil and human health: Current status and future needs. *Air, Soil and Water Research* 13, 1–23. https://doi.org/10.1177/1178622120934441

Brevik, Eric C., and Lynn C. Burgess. 2014. The influence of soils on human health. *Nature Education Knowledge* 5(12), 1. http://www.nature.com/scitable/knowledge/library/the-influence-of-soils-on-human-health-127878980. Accessed 2 January 2024.

Burgess, L.C. 2013. Organic pollutants in soil. In: Brevik, E.C., and Burgess, L.C. (Eds.), *Soils and Human Health*, pp. 83–106. CRC Press, Boca Raton, FL.

Cabibbo, A.A., Hall, C.A., and Helmke, M.F. 2014. Evaluation of soil arsenic contamination at an abandoned apple orchard by portable X-Ray spectrometry. *Soil Horizons* 55. https://doi.org/10.2136/sh14-06-0006

Calvo, M.S., Moshfegh, A.J., and Tucker, K.L. 2014. Assessing the health impact of phosphorus in the food supply: issues and considerations. *Advances in Nutrition* 5(1), 104–113.

Chia, R.W., Lee, J.-Y., Kim, H., and Jang, J. 2021. Microplastic pollution in soil and groundwater: A review. *Environmental Chemistry Letters* 19, 4211–4224.

Coder, K.D. 1998. *Soil Constraints on Root Growth*. University of Georgia Cooperative Extension Service Forest Resources Publication FOR98-10, Athens, GA, USA.

Covelo, E.F., Vega, F.A., and Andrade, M.L. 2007. Competitive sorption and desorption of heavy metals by individual soil components. *Journal of Hazardous Materials* 140, 308–315. https://doi.org/10.1016/j.jhazmat.2006.09.018

Deng, M., Ma, H., Sadeghpour, A., Yang, G., Hu, Y., and Yang, D. 2023. Responses of crop production and soil health to chemical nitrogen fertilization in a maize-wheat rotation system. *Frontiers in Environmental Science* 11. https://doi.org/10.3389/fenvs.2023.1108288

Deng, M., Malik, A., Zhang, Q., Sadeghpour, A., Zhu, Y., and Li, Q. 2021. Improving Cd risk managements of rice cropping system by integrating source-soil-rice-human chain for a typical intensive industrial and agricultural region. *Journal of Cleaner Production*, 313, 127883. https://doi.org/10.1016/j.jclepro.2021.127883

Deng, M., Yang, X., Dai, X., Zhang, Q., Malik, A., and Sadeghpour, A. 2020. Heavy metal pollution risk assessments and their transportation in sediment and overlay water for the typical Chinese reservoirs. *Ecological Indicators*, 112, 106166. https://doi.org/10.1016/j.ecolind.2020.106166

DesMarias, T.L., and Costa, M. 2019. Mechanisms of chromium-induced toxicity. *Current Opinion in Toxicology* 14, 1–7. https://doi.org/10.1016/j.cotox.2019.05.003

Duckworth, O.W., Polizzotto, M.L., and Thompson, A. 2022. Bringing soil chemistry to environmental health science to tackle soil contaminants. *Frontiers in Environmental Science* 10. https://doi.org/10.3389/fenvs.2022.981607

Elsakhawy, T.A., Omara, A.E., Amer, M.M., El-Ramady, H.R., Prokisch, J., and Brevik, E. 2022. A diagrammatic mini-review on soil–human health health-nexus: With focus on soil nano-pollution. *Environment, Biodiversity and Soil Security* 6, 327–338.

Fan, Q., Wang, L., Fu, Y., Li, Q., Liu, Y., Wang, Z., and Zhu, H. 2023. Iron redox cycling in layered clay minerals and its impact on contaminant dynamics: A review. *Science of the Total Environment* 855, 159003. https://doi.org/10.1016/j.scitotenv.2022.159003

Fiedler, S., Vepraskas, M.J., and Richardson, J.L. 2007. Soil redox potential: Importance, field measurements, and observations. *Advances in Agronomy* 94, 1–54. https://doi.org/10.1016/S0065-2113(06)94001-2

Gannon, J.T., Manilal, V.B., and Alexander, M. 1991. Relationship between cell surface properties and transport of bacteria through soil. *Applied and Environmental Microbiology* 57, 190–193. https://doi.org/10.1128/aem.57.1.190-193.1991

Garaycochea, S., Romero, H., Beyhaut, E., Neal, A.L., and Altier, N. 2020. Soil structure, nutrient status and water holding capacity shape Uruguayan grassland prokaryotic communities. *FEMS Microbiology Ecology* 96, fiaa207. https://doi.org/10.1093/femsec/fiaa207

Gardner, D., Weindorf, D.C., and Flynn, M. 2013. Presence of chromium, copper, and arsenic in schoolyard soils. *Soil Horizons*, 54. https://doi.org/10.2136/sh12-12-0032

Gu, B., Chen, D., Yang, Y., Vitousek, P., and Zhu, Y.G. 2021. Soil-food-environment-health nexus for sustainable development. *Research*, 9804807. https://doi.org/10.34133/2021/9804807

Ha, E., Basu, N., Bose-O'Reilly, S., Dórea, J.G., McSorley, E., Sakamoto, M., and Chan, H.M. 2017. Current progress on understanding the impact of mercury on human health. *Environmental Research* 152, 419–433. https://doi.org/10.1016/j.envres.2016.06.042

Hashim, M.A., Mukhopadhyay, S., Sahu, J.N., and Sengupta, B. 2011. Remediation technologies for heavy metal contaminated groundwater. *Journal of Environmental Management* 92, 2355–2388. https://doi.org/10.1016/j.jenvman.2011.06.009

Havlin, J.L., Tisdale, S.L., Nelson, W.L., and Beaton, J.D.. 2013. *Soil Fertility and Fertilizers*, 8th Ed. Pearson-Prentice Hall, Upper Saddle River, NJ.

Helling, C.S., Chesters, G., and Corey, R.B. 1964. Contribution of organic matter and clay to soil cation-exchange capacity as affected by the pH of the saturating solution. *Soil Science Society of America Journal* 28, 457–584. https://doi.org/10.2136/sssaj1964.03615995002800040020x

Helmrich, S., Vlassopoulos, D., Alpers, C.N., and O'Day, P.A. 2022. Critical review of mercury methylation and methylmercury demethylation rate constants in aquatic sediments for biogeochemical modeling. *Critical Reviews in Environmental Science and Technology* 52, 4353–4378. https://doi.org/10.1080/10643389.2021.2013073

Houben, D., Evrard, L., and Sonnet, P. 2013. Mobility, bioavailability and pH dependent leaching of cadmium, zinc and lead in a contaminated soil amended with biochar. *Chemosphere* 92, 1450–1457. https://doi.org/10.1016/j.chemosphere.2013.03.055

Jat, M.L., Singh, B., Stirling, C.M., Jat, H.S., Tetarwal, J.P., Jat, R.K., Singh, R., Lopez-Ridaura, S., and Shirsath, P.B. 2018. Soil processes and wheat cropping under emerging climate change scenarios in South Asia. *Advances in Agronomy* 148, 111–171. https://doi.org/10.1016/bs.agron.2017.11.006

Jayaraj, R., Megha, P., and Sreedev, P. 2016. Review article. Organochlorine pesticides, their toxic effects on living organisms and their fate in the environment. *Interdisciplinary Toxicology* 9(3-4), 90–100. https://doi.org/10.1515/intox-2016-0012

Jung, Y.S., Sampath, V., Prunicki, M., Aguilera, J., Allen, H., LaBeaud, D., Veidis, E., Barry, M., Erny, B., Patel, L., Akdis, C., Akdis, M., and Nadeau, K. 2022. Characterization and regulation of microplastic pollution for protecting planetary and human health. *Environmental Pollution* 315, 120442.

Kamp, A., Høgslund, S., Risgaard-Petersen, N., and Stief, P. 2015. Nitrate storage and dissimilatory nitrate reduction by eukaryotic microbes. *Front. Microbiol.* 6. https://doi.org/10.3389/fmicb.2015.01492

Kelderman, P., and Osman, A.A. 2007. Effect of redox potential on heavy metal binding forms in polluted canal sediments in Delft (The Netherlands). *Water Research* 41, 4251–4261. https://doi.org/10.1016/j.watres.2007.05.058

Kim, K.H., Kabir, E., and Jahan, S.A. 2017. Exposure to pesticides and the associated human health effects. *Science of The Total Environment* 575, 525–535. https://doi.org/10.1016/j.scitotenv.2016.09.009

Kosuth, M., Mason, S.A., and Wattenberg, E.V. 2018. Anthropogenic contamination of tap water, beer, and sea salt. *Plos One*. https://doi.org/10.1371/journal.pone.0194970

Kramer, M.H., Herwaldt, B.L., Craun, G.F., Calderon, R.L., and Juranek, D.D. 1996. Surveillance of water-borne-disease outbreaks – United States, 1993–1994. *CDCP Surveillance Summaries, MMWR* 45, 1–33.

la Cecilia, D., Porta, G.M., Tang, F.H.M., Riva, M., and Maggi, F. 2020. Probabilistic indicators for soil and groundwater contamination risk assessment. *Ecological Indicators* 115, 106424. https://doi.org/10.1016/j.ecolind.2020.106424

Lal, R., Monger, C., Nave, L., and Smith, P. 2021. The role of soil in regulation of climate. *Philosophical Transactions of the Royal Society B* 376, 20210084. http://doi.org/10.1098/rstb.2021.0084

Lehmann, J., Bossio, D.A., Kögel-Knabner, I., and Rillig, M.C. 2020. The concept and future prospects of soil health. *Nature Reviews Earth & Environment* 1, 544–553. https://doi.org/10.1038/s43017-020-0080-8

Mallin, MA. 2000. Impacts of industrial animal production on rivers and estuaries. *American Scientist* 88, 2–13.

Mattila, T.J. 2023. Redox potential as a soil health indicator – how does it compare to microbial activity and soil structure? *Plant and Soil*. https://doi.org/10.1007/s11104-023-06305-y

Morra, P., Leonardelli, L., and Spadoni, G. 2011. The volatilization of pollutants from soil and groundwater: Its importance in assessing risk for human health for a real contaminated site. *Journal of Environmental Protection* 2, 1192–1206. https://doi.org/10.4236/jep.2011.29137

Olagbaju, P.O., Wojuola, O.B., and Tshivhase, V. 2021. Radionuclides contamination in soil: Effects, sources and spatial distribution. *EPJ Web of Conferences* 253, 09006. https://doi.org/10.1051/epjconf/202125309006

Oliver, M.A., and Gregory, P.J. 2015. Soil, food security and human health: A review. *European Journal of Soil Science* 66, 257–276. https://doi.org/10.1111/ejss.12216

Patzel, N., Grunwald, S., Brevik, E.C., and Feller, C. 2023. Summary and conclusions: Understanding soil-cultural relations. In: Patzel, N., Grunwald, S., Brevik, E.C., and Feller, C. (Eds.), *Cultural Understanding of Soils: The Importance of Cultural Diversity and of the Inner World*, pp. 521–540. Springer, Cham. https://doi.org/10.1007/978-3-031-13169-1_26

Pepper, I.L. 2013. The soil health-human health nexus. *Critical Reviews in Environmental Science and Technology* 43(24), 2617–2652. https://doi.org/10.1080/10643389.2012.694330

Pereg, L., Steffan, J.J., Gedeon, C., Thomas, P., and Brevik, E.C. 2021. Medical geology of soil ecology. In: Siegel, M., Selinus, O., and Finkelman, R. (Eds.). *Practical Applications of Medical Geology*, pp. 343–401. Springer, Cham.

Perez-Esteban, J., Ecolastico, C., Masaguer, A., Vargas, C. and Moliner, A. 2014. Soluble organic carbon and pH of organic amendments affect metal mobility and chemical speciation in mine soils. *Chemosphere* 103, 167–171. https://doi.org/10.1016/j.chemosphere.2013.11.055

Plum, L.M., Rink, L., and Haase, H. 2010. The essential toxin: Impact of zinc on human health. *International Journal of Environmental Research and Public Health* 7, 1342–1365. https://doi.org/10.3390/ijerph7041342

Prata, J.C., da Costa, J.P., Lopes, I., Duarte, A.C., and Rocha-Santos, T. 2020. Environmental exposure to microplastics: An overview on possible human health effects. *Science of the Total Environment* 702, 134455. https://doi.org/10.1016/j.scitotenv.2019.134455

Qiu, Y., Zhou, S., Zhang, C., Zhou, Y., and Qin, W. 2022. Soil microplastic characteristics and the effects on soil properties and biota: A systematic review and meta-analysis. *Environmental Pollution* 313, 120183.

Rai, P.K., Lee, S.S., Zhang, M., Tsang, Y.F. and Kim, K.H., 2019. Heavy metals in food crops: Health risks, fate, mechanisms, and management. *Environment international*, 125, 365–385. https://doi.org/10.1016/j.envint.2019.01.067

Reed, S., Clark, M., Thompson, R., and Hughes, K.A. 2018. Microplastics in marine sediments near Rothera Research Station, Antarctica. *Marine Pollution Bulletin* 133, 460–463. https://doi.org/10.1016/j.marpolbul.2018.05.068

Rieke, E.L., Bagnall, D.K., Morgan, C.L., Flynn, K.D., Howe, J.A., Greub, K.L., Mac Bean, G., Cappellazzi, S.B., Cope, M., Liptzin, D., Norris, C.E., Tracy, P.W., Aberle, E., Ashworth, A., Tavarez, O.B., Bary, A.I., Baumhardt, R.L., Gracia, A.B., Brainard, D.C., Brennan, J.R., Reyes, D.B., Bruhjell, D., Carlyle, C.N., Crawford, J.J.W., Creech, C.F., Culman, S.W., Deen, B., Dell, C.J., Derner, J.D., Ducey, T.F., Duiker, S.W., Dyck, M.F., Ellert, B.H., Entz, M.H., Solorio, A.E., Fonte, S.J., Fonteyne, S., Fortuna, A.M., Foster, J.L., Fultz, L.M., Gamble, A.V., Geddes, C.M., Griffin-LaHue, D., Grove, J.H., Hamilton, S.K., Hao, X., Hayden, Z.D., Honsdorf, N., Ippolito, J.A., Johnson, G.A., Kautz, M.A., Kitchen, N.R., Kumar, S., Kurtz, K.S.M., Larney, F.J., Lewis, K.L., Liebman, M., Ramirez, A.L., Machado, S., Maharjan, B., Gamiño, M.A.M., May, W.E., Mc Claran, M.P., McDaniel, M.D., Millar, N., Mitchell, J.P., Moore, A.D., Moore Jr, P.A., Gutiérrez, M.M., Nelson, K.A., Omondi, E.C., Osborne, S.L., Alcalá, L.O., Owens, P., Pena-Yewtukhiw, E.M., Poffenbarger, H.J., Lira, B.P., Reeve, J.R., Reinbott, T.M., Reiter, M.S., Ritchey, E.L., Roozeboom, K.L., Rui, Y.H., Sadeghpour, A., Sainju, U.M., Sanford, G.R., Schillinger, W.F., Schindelbeck, R.R., Schipanski, M.E., Schlegel, A.J., Scow, K.M., Sherrod, L.A., Shober, A.L., Sidhu, S.S., Moya, E.S., St. Luce, M., Strock, J.S., Suyker, A.E., Sykes, V.R., Tao, H., Campos, A.T., Van Eerd, L.L., Van Es, H.M., Verhulst, N., Vyn, T.J., Wang, Y., Watts, D.B., Wright, D.L., Zhang, T., and Honeycutt, C.W. 2022a. Evaluation of aggregate stability methods for soil health. *Geoderma* 428, 116156. https://doi.org/10.1016/j.geoderma.2022.116156

Rieke, E.L., Cappellazzi, S.B., Cope, M., Liptzin, D., Bean, G.M., Greub, K.L.H., Norris, C.E., Tracy, P.W., Aberle, E., Ashworth, A., Tavarez, O.B., Bary, A.I., Baumhardt, R.L., Gracia, A.B., Brainard, D.C., Brennan, J.R., Reyes, D.B., Bruhjell, D., Carlyle, C.N., Crawford, J.J.W., Creech, C.F., Culman, S.W., Deen, B., Dell, C.J., Derner, J.D., Ducey, T.F., Duiker, S.W., Dyck, M.F., Ellert, B.H., Solorio, A.E., Fonte, S.J., Fonteyne, S., Fortuna, A., Foster, J.L., Fultz, L.M., Gamble, A.V., Geddes, C.M., Griffin-LaHue, D., Grove, J.H., Hamilton, S.K., Hao, X., Hayden, Z.D., Honsdorf, N., Howe, J.A., Ippolito, J.A., Johnson, G.A., Kautz, M.A., Kitchen, N.R., Kumar, S., Kurtz, K.S.M., Larney, F.J., Lewis, K.L., Liebman, M., Ramirez, A.L., Machado, S., Maharjan, B., Gamiño, M.A.M., May, W.E., McClaran, M.P., McDaniel, M.D., Millar, N., Mitchell, J.P., Moore, A.D., Moore Jr., P.A., Gutiérrez, M.M., Nelson, K.A., Omondi, E.C., Osborne, S.L., Alcalá, L.O., Owens, P., Pena-Yewtukhiw, E.M., Poffenbarger, H.J., Lira, B.P., Reeve, J.R., Reinbott, T.M., Reiter, M.S., Ritchey, E.L., Roozeboom, K.L., Rui, Y., Sadeghpour, A., Sainju, U.M., Sanford, G.R., Schillinger, W.F., Schindelbeck, R.R., Schipanski, M.E., Schlegel, A.J., Scow, K.M., Sherrod, L.A., Shober, A.L., Sidhu, S.S., Moya, E.S., St Luce, M., Strock, J.S., Suyker, A.E., Sykes, V.R., Tao, H., Campos, A.T., Van Eerd, L.L., Verhulst, N., Vyn, T.J., Wang, Y., Watts, D.B., William, B.B., Wright, D.L., Zhang, T., Morgan, C.L.S., and Honeycutt, C.W. 2022b. Linking soil microbial community structure to potential carbon mineralization: A continental scale assessment of reduced tillage. *Soil Biology and Biochemistry* 168, 108618. https://doi.org/10.1016/j.soilbio.2022.108618

Rodrigues, M.O., Abrantes, N., Gonçalves, F.J.M., Nogueira, H., Marques, J.C., and Gonçalves, A.M.M. 2018. Spatial and temporal distribution of microplastics in water and sediments of a freshwater system (Antuã River, Portugal). *Science of the Total Environment* 633, 1549–1559. https://doi.org/10.1016/j.scitotenv.2018.03.233

Rodriguez-Martin, J.A., Arias, M.L., and GrauCorbi, J.M. 2006. Heavy metal contents in agricultural topsoils in the Ebro basin (Spain). Application of multivariate geostatistical methods to study spatial variations. *Environmental Pollution* 144, 1001–1012. https://doi.org/10.1016/j.envpol.2006.01.045

Sadeghpour, A., Ketterings, Q.M., Godwin, G.S., and Czymmek, K.J. 2017a. Under- or over-application of nitrogen impact corn yield, quality, soil, and environment. *Agronomy Journal* 109, 343–353. https://doi.org/10.2134/agronj2016.06.0355

Sadeghpour, A., Ketterings, Q.M., Godwin, G.S., Czymmek, K.J., and Vermeylen, F. 2017b. Agro-environmental consequences of shifting from nitrogen- to phosphorus-based manure management of corn. *Soil Science Society of America Journal* 81, 1127–1138. https://doi.org/10.2136/sssaj2016.03.0417

Sardans, J., and Peñuelas, J. 2015. Potassium: A neglected nutrient in global change. *Global Ecology and Biogeography* 24(3), 261–275. https://doi.org/10.1111/geb.12259

Schaetzl, R.J., and Thompson, M.L. 2015. *Soils: Genesis and Geomorphology*, 2nd Ed. Cambridge University Press, New York, NY.

Schijven, J.F., and Hassanizadeh, S.M. 2000. Removal of viruses by soil passage: Overview of modeling, processes, and parameters. *Critical Reviews in Environmental Science and Technology* 30, 49–127. https://doi.org/10.1080/10643380091184174

Sears, M.E. 2013. Chelation: Harnessing and enhancing heavy metal detoxification—A review. *The Scientific World Journal* 2013, 219840. https://doi.org/10.1155/2013/219840

Shaheen, A., and Bukhari, S.T.A. 2018. Potential of sawdust and corn cobs derived biochar to improve soil aggregate stability, water retention, and crop yield of degraded sandy loam soil. *Journal of Plant Nutrition* 41, 2673–2682. https://doi.org/10.1080/01904167.2018.1509092

Sharma, P.K., and Kumar, S. 2023. *Soil Physical Environment and Plant Growth*. Springer, Cham.

Sharpley, A.N., McDowell, R.W., and Kleinman, P.J.A. 2001. Phosphorus loss from land to water: integrating agricultural and environmental management. *Plant and Soil* 237, 287–307.

Shen, Q., Wu, M., and Zhang, M. 2022. Accumulation and relationship of metals in different soil aggregate fractions along soil profiles. *Journal of Environmental Sciences* 115, 47–54. https://doi.org/10.1016/j.jes.2021.07.007

Sipos, P., Németh, T., Kis, V.K., and Mohai, I. 2008. Sorption of copper, zinc and lead on soil mineral phases. *Chemosphere* 73, 461–469. https://doi.org/10.1016/j.chemosphere.2008.06.046

Six, J., Bossuyt, H., Degryze, S., and Denef, K. 2004. A history of research on the link between (micro) aggregates, soil biota, and soil organic matter dynamics. *Soil and Tillage Research* 79, 7–31. https://doi.org/10.1016/j.still.2004.03.008

Smiles, D.E. 1988. Aspects of the physical environment of soil organisms. *Biology and Fertility of Soils* 6, 204–215. https://doi.org/10.1007/BF00260817

Steffan, J.J., Brevik, E.C., Burgess, L.C., and Cerdà, A. 2018. The effect of soil on human health: An overview. *European Journal of Soil Science* 69, 159–171. https://doi.org/10.1111/ejss.12451

Steffan, J.J., Derby, J.A., and Brevik, E.C. 2020. Soil pathogens that may potentially cause pandemics, including severe acute respiratory syndrome (SARS) coronaviruses. *Current Opinion in Environmental Science & Health* 17, 35–40. https://doi.org/10.1016/j.coesh.2020.08.005

Suhani, I., Sahab, S., Srivastava, V., and Singh, R.P. 2021. Impact of cadmium pollution on food safety and human health. *Current Opinion in Toxicology* 27, 1–7. https://doi.org/10.1016/j.cotox.2021.04.004

Sun, H., and Weaver, C.M. 2020. Rise in potassium deficiency in the US population linked to agriculture practices and dietary potassium deficits. *Journal of Agricultural and Food Chemistry* 68(40), 11121–11127. https://doi.org/10.1021/acs.jafc.0c05139

Thiele-Bruhn, S. 2021. The role of soils in provision of genetic, medicinal and biochemical resources. *Philosophical Transactions of the Royal Society, Series B* 376, 20200183.

Townsend, A.R., Howarth, R.W., Bazzaz, F.A., Booth, M.S., Cleveland, C.C., Collinge, S.K., Dobson, A.P., Epstein, P.R., Holland, E.A., Keeney, D.R., and Mallin, M.A. 2003. Human health effects of a changing global nitrogen cycle. *Frontiers in Ecology and the Environment* 1(5), 240–246.

Vamerali, T., Bandiera, M., and Mosca, G. 2010. Field crops for phytoremediation of metal-contaminated land. A review. *Environmental Chemistry Letters* 8, 1–17. https://doi.org/10.1007/s10311-009-0268-0

Vinueza, R.E.A. 2018. *Subsurface and Surface Tillage Practices and Their Effects on Structure, Permeability and Crop Yield in a Chernozemic and Solonetzic Soil.* Unpublished Masters Thesis, University of Saskatchewan, Saskatoon.

Wang, C., Zhao, J., and Xing, B. 2021. Environmental source, fate, and toxicity of microplastics. *Journal of Hazardous Materials* 407, 124357. https://doi.org/10.1016/j.jhazmat.2020.124357

Watteau, F., Dignac, M.F., Bouchard, A., Revallier, A., and Houot, S. 2018. Microplastic detection in soil amended with municipal solid waste composts as revealed by transmission electronic microscopy and pyrolysis/GC/MS. *Frontiers in Sustainable Food Systems* 2. https://doi.org/10.3389/fsufs.2018.00081

Weidhuner, A., Hanauer, A., Krausz, R., Crittenden, S.J., Gage, K., and Sadeghpour, A. 2021. Tillage impacts on soil aggregation and aggregate-associated carbon and nitrogen after 49 years. *Soil and Tillage Research* 208, 104878. https://doi.org/10.1016/j.still.2020.104878

Wolf, B., and G.H. Snyder. 2003. *Sustainable Soils: The Place of Organic Matter in Sustaining Soils and their Productivity.* Haworth Press, New York.

Xu, J., Bravo, A.G., Lagerkvist, A., Bertilsson, S., Sjöblom, R., and Kumpiene, J. 2015. Sources and remediation techniques for mercury contaminated soil. *Environment International* 74, 42–53.

Yu, H., Li, C., Yan, J., Ma, Y., Zhou, X., Yu, W., Kan, H., Meng, Q., Xie, R., and Dong, P. 2023. A review on adsorption characteristics and influencing mechanism of heavy metals in farmland soil. *RSC Advances* 13, 3505–3519. https://doi.org/10.1039/D2RA07095B

Zhao, C., Dong, Y., Feng, Y., Li, Y., and Dong, Y. 2019. Thermal desorption for remediation of contaminated soil: A review. *Chemosphere* 221, 841–855. https://doi.org/10.1016/j.chemosphere.2019.01.079

Zhao, X.Q., and Shen, R.F. 2018. Aluminum–nitrogen interactions in the soil–plant system. *Frontiers in Plant Science* 9. https://doi.org/10.3389/fpls.2018.00807

Zhen, Z., Wang, S., Luo, S., Ren, L., Liang, Y., Yang, R., Li, Y., Zhang, Y., Deng, S., Zou, L., Lin, Z., and Zhang, D. 2019. Significant impacts of both total amount and availability of heavy metals on the functions and assembly of soil microbial communities in different land use patterns. *Frontiers in Microbiology* 10. https://doi.org/10.3389/fmicb.2019.02293

Zornoza, R., Acosta, J.A., Bastida, F., Domínguez, S.G., Toledo, D.M., and Faz, A. 2015. Identification of sensitive indicators to assess the interrelationship between soil quality, management practices and human health. *SOIL* 1, 173–185. https://doi.org/10.5194/soil-1-173-2015

4 Soil Health and Its Impact on Human Health in Pakistan

Noor Us Sabah, Rattan Lal, Mukkram Ali Tahir, and Rafia Mubaraka

INTRODUCTION

The rapid increase in world population is a cause of concern for agricultural scientists in fulfilling the demand for food, feed, fiber, fuel, and shelter. While increase in population is a concern for many countries, the problem is worst in developing countries because of the negative relation between availability of resources and population (Alonso et al., 2018). Feeding an estimated world population of 9.6 billion by 2050 (Gu et al., 2021) is a major concern. In contrast to the growing population, the arable land area is either stagnant or decreasing because of soil degradation and soil-specific constraints that limit crop yields and their nutritional quality. Therefore, it is critically important to increase and sustain the average agronomic yield per unit area available for cultivation (Hu et al., 2023).

Efforts are being made to improve crop yield in order to cope with the demand of the growing population. Since 95% of food supplies for supporting human health and well-being come from soils (Oliver and Gregory, 2015; Ville et al., 2019), only healthy soils can support this extended supply of food, feed, fuel, and fiber in order to achieve the daunting task of food security (i.e., the provision of ample supplies of good quality food on a continuous basis to every person). Food security is compromised by soil contamination and pollution which not only reduces crop yields, but also decreases the quality of produce, making it unsafe and unfit for human consumption (Hegland, 2017).

The "One Health" concept acknowledges the close association between human, animal, and environmental health, with specific focus on the association between soil and human health (Zinsstag et al., 2015). The One Health concept recognizes that the health of humans, animals, plants, and the environment / ecosystems are closely associated and interconnected. Soil health ultimately contributes to healthy life, since human beings are at the top of food chain. Soil health is of foremost importance to achieving food security). Soil health can affect human health positively and negatively in many ways whether directly or indirectly. For example, soil is the primary source for the provision of a number of essential elements and medicines necessary for human growth and well-being. On the other hand, imbalances in nutrients (deficiency, or toxicity, or both) and the existence of disease-causing agents in soil can negatively affect human health (Oliver & Gregory, 2015). Healthy soils produce nutritious food for humans and feed for animals, thus contributing to healthy human life (Nieder et al., 2018). Contamination of the soil with heavy metals is a global issue that has a significant impact on ecosystems and human health. Soil contaminated with heavy metals presents a serious risk to human health and ecosystems. Not only does heavy metal pollution have an impact on crop yield and quality, but it also degrades the atmosphere and waterways, and endangers the lives and health of both humans and animals (Brevik et al., 2017).

Soil and human and environmental health are well connected (Nieder et al., 2018). The human body needs 25 essential elements for normal growth and well-being. The demand for these elements is met either by direct consumption of plants, or indirectly from animals feeding on these plants (Abrahams, 2002). Soil is the source of about 14 elements that are considered essential for human health (Combs, 2005). With the exception of carbon, hydrogen and oxygen, 22 elements are mainly

DOI: 10.1201/9781003469742-4

provided by soil. About 99.9% of the atoms in human body are made up of 11 elements, out of which the four elements: hydrogen (H), oxygen (O), carbon (C), and nitrogen (N), classified as major elements, account for 99%, and the remaining seven elements: phosphorus (P), potassium (K), calcium (Ca), magnesium (Mg), sulfur (S), sodium (Na), and chlorine (Cl) are minor elements and account for 0.9% (Combs, 2005). With the exception of C, H, and O, the rest of these are provided by soil (Brevik, 2013). Soil is also the source of about 18 additional essential elements (micronutrients) that account for 0.1% of the human body (Combs, 2005). Thus, only healthy soils can grow nutrient-enriched plants that ultimately lead to healthy human life.

SOIL HEALTH: A FUNDAMENTAL CONCEPT

Soil health refers to the capacity of soil to function as a living system that sustains biological productivity, maintains environmental quality, and promotes plant, animal, and human health (Doran and Zeiss, 2000). It encompasses the physical, chemical, and biological properties of soil, which collectively determine its ability to support sustainable agricultural practices and ecosystem stability. Healthy soil contributes to clean water, carbon sequestration, and nutrient cycling, making it essential for environmental and human well-being (Brevik, 2013; Brevik et al., 2017).

Components of Soil Health

The key components of soil health include:

- **Physical Properties**: Soil texture, structure, porosity, and water retention capacity that influence plant root growth and water availability.
- **Chemical Properties**: Soil pH, salinity, and the availability of macronutrients like nitrogen (N), phosphorus (P), potassium (K), and micronutrients such as zinc (Zn), iron (Fe), copper (Cu) that are crucial for crop growth and quality (Broadley et al., 2012).
- **Biological Properties**: Soil microbial diversity and organic matter content that are vital for nutrient cycling and maintaining soil fertility (Lal, 2018).

Role of Soil Organic Matter (SOM)

Soil organic matter (SOM) is a cornerstone of soil health, contributing to improved soil structure, water retention, and nutrient availability. In Pakistan, however, more than 90% of soils have low SOM content ($<1\%$), resulting in poor fertility and limited productivity (Rafique et al., 2023). SOM also plays a crucial role in mitigating climate change by storing carbon in the soil and reducing greenhouse gas emissions.

Soil as a Provider of Essential Nutrients

Soil is the primary source of 22 out of the 25 essential elements required by humans, plants, and animals (Combs, 2005). Macronutrients like N, P, and K are needed in large quantities for plant growth, while micronutrients like Zn and Fe are critical for plant physiological functions and human nutrition (Abrahams, 2002; Oliver & Gregory, 2015). Without healthy soils, crops fail to meet the nutritional needs of humans and animals, creating cascading effects on health and food security.

Ecosystem Services Provided by Soils

Healthy soils support numerous ecosystem services:

- **Nutrient Cycling**: Healthy soils ensure the availability of nutrients for plants through natural cycles.

- **Water Filtration**: Soil acts as a natural filter, reducing pollutants in water.
- **Carbon Sequestration**: By storing carbon, soils play a pivotal role in mitigating climate change (Lal, 2004).
- **Habitat for Microorganisms**: Soil is home to billions of microorganisms that drive nutrient cycling and support plant health (Brevik, 2013).

Threats to Soil Health

Several threats undermine soil health, including:

- Nutrient imbalances caused by unbalanced fertilizer application (Imtiaz et al., 2006).
- Soil contamination from heavy metals, pesticides, and untreated wastewater (Pečiulytė et al., 2006).
- Physical degradation, such as erosion and compaction, exacerbated by poor agricultural practices and deforestation (Lal, 2018).

CHALLENGES TO SOIL HEALTH IN PAKISTAN

Soils in Pakistan face numerous challenges that undermine their health and capacity to support sustainable agricultural productivity. These challenges stem from a combination of climatic, geographic, and anthropogenic factors, which contribute to soil degradation, nutrient deficiencies, and contamination. Addressing these challenges is essential to improving agricultural output, ensuring food security, and safeguarding environmental sustainability.

Climatic and Geographic Factors

Pakistan's predominantly arid to semi-arid climate poses inherent challenges to soil health. Approximately 80% of the country experiences low rainfall, high temperatures, and high evaporation rates, which exacerbate soil-degradation processes like salinization, sodicity, and desertification (Azam and Shafique, 2017). These conditions reduce soil organic matter (SOM) content and fertility, making the soils less productive.

The country's geography further compounds the problem. Soils derived from alluvium and loess parent materials are often low in SOM content and prone to nutrient imbalances (Wakeel et al., 2022). Accelerated erosion due to deforestation and improper land management exacerbates the loss of fertile topsoil, leaving large tracts of land unsuitable for cultivation.

Nutrient Deficiencies and Imbalances

Nutrient deficiencies are a major constraint to Pakistan's agricultural productivity. Over 90% of soils in the country are deficient in N and P, while K deficiency is also widespread due to its neglect in fertilizer applications (Rafique et al., 2023). Micronutrient deficiencies are similarly prevalent, with 70% of soils deficient in Zn, 25% deficient in Fe, and site-specific deficiencies in Cu and manganese (Mn) (Imtiaz et al., 2010; Rashid et al., 1997).

The imbalance in fertilizer use, particularly the overapplication of N and P without K, reduces nutrient-use efficiency and exacerbates soil fertility issues. High fertilizer costs, limited availability of fertilizers at critical times, and a lack of awareness among farmers further aggravate these problems (Imtiaz et al., 2006).

Soil Contamination

Soil contamination is another significant threat to soil health in Pakistan. The use of untreated wastewater for irrigation, especially in urban and peri-urban areas, introduces heavy metals like cadmium (Cd), lead (Pb), and nickel (Ni) into the soils. These contaminants accumulate in crops, entering the food chain and posing severe risks to human and environmental health (Ahmed et al., 2020; Pečiulytė et al., 2006).

Pesticides and chemical fertilizers contribute to soil pollution by introducing harmful residues that disrupt microbial activity and soil fertility. Industrial effluents and urban waste, often disposed of improperly, further contribute to the contamination of agricultural soil. For example, a survey in Punjab found that vegetables irrigated with wastewater contained toxic levels of Cd and Pb, exceeding permissible limits (Shakir et al., 2017; Sharif et al., 2010).

Physical Degradation

Physical degradation, including soil erosion, compaction, and crusting, is prevalent in Pakistan. Improper irrigation practices, overgrazing, excessive plowing, and deforestation accelerate these processes, reducing soil water-holding capacity and fertility (Lal, 2018). Salinization and sodicity affect an estimated 6.7 million hectares of cultivated land, rendering large areas unproductive (Azam and Shafique, 2017).

The lack of proper drainage systems further exacerbates waterlogging and salinity problems, particularly in irrigated areas. These issues are major contributors to the moderate productivity of crops and the vulnerability of the agricultural sector to climate change.

LINK BETWEEN SOIL HEALTH AND HUMAN HEALTH

Soil health and human health are intricately connected through food production, nutrient cycling, and the prevention or propagation of contaminants. Healthy soils are fundamental to human well-being, as they ensure the production of nutrient-rich crops, provide clean water, and mitigate exposure to harmful substances. Conversely, degraded or contaminated soils can lead to poor nutrition and increased health risks.

Soil as a Source of Essential Nutrients

Soil is the primary source of essential nutrients for humans, as it supplies 22 out of the 25 elements required by the human body. These nutrients are absorbed by plants and passed on to humans and animals through the food chain (Combs, 2005). Macronutrients such as N, P and K are critical for plant growth and development, while micronutrients like Zn, Fe and Cu are essential for maintaining human health (Brevik, 2013).

In Pakistan, nutrient-deficient soils result in crops that lack the necessary vitamins and minerals to meet dietary needs. For instance, Zn deficiency in soils contributes to widespread zinc deficiency in humans, leading to weakened immune systems and developmental issues, particularly among children (Ahmed et al., 2017). Similarly, Fe-deficient soils contribute to anemia, which affects 62% of children under five in Pakistan (Purohit et al., 2024).

Pathways of Exposure to Soil Contaminants

While healthy soils enhance human health, contaminated soils pose significant risks. Humans are exposed to soil contaminants through three primary pathways: ingestion, inhalation, and dermal contact. For example:

- **Ingestion**: Consuming crops grown in contaminated soils introduces heavy metals, pesticide residues, and pathogens into the body. This is particularly common in Pakistan, where untreated wastewater is frequently used for irrigation (Ahmed et al., 2020; Pečiulytė et al., 2006; von Lindern et al., 2016).

- **Inhalation**: Dust from polluted soils can carry harmful particles, including asbestos and pathogenic fungi, which may cause respiratory conditions such as bronchitis and asthma (Stockamp & Thompson, 2016; Zosky et al., 2014).
- **Dermal Contact**: Prolonged exposure to contaminated soils, particularly among barefoot farmers, can lead to conditions such as podoconiosis, a type of non-filarial elephantiasis caused by volcanic clay soils (Deribe et al., 2017).

Health Risks from Soil Contamination

Soil contamination with heavy metals such as Cd, Pb, and arsenic (As) is a global issue, and its impacts are particularly severe in Pakistan. These contaminants enter the food chain through crops and water, leading to serious health conditions. For instance:

- **Lead (Pb)**: Chronic exposure to Pb-contaminated soils and crops can cause brain and nerve disorders, kidney damage, and developmental delays, especially in children (Ara & Usmani, 2015; Henry & Cring, 2013).
- **Cadmium (Cd)**: Exposure to Cd through contaminated food can result in kidney damage, bone disorders, and increased cancer risk (Rafique et al., 2023).
- **Arsenic (As)**: Prolonged exposure to As, particularly through groundwater used for irrigation, can cause skin lesions, cardiovascular disease, and cancer (Ahmed et al., 2017).

The widespread use of untreated wastewater for irrigation in Pakistan exacerbates these risks, introducing toxic substances into the soil and crops. A survey in Punjab revealed that vegetables irrigated with wastewater contained Cd and Pb levels exceeding permissible limits (Shakir et al., 2017).

Nutritional and Public Health Implications

Degradation of soil health leads to the production of nutrient-deficient crops, contributing to malnutrition and "hidden hunger" in Pakistan. Deficiencies in micronutrients like Zn, Fe, and iodine (I) result in stunted growth, weakened immunity, and cognitive impairments, particularly in children and pregnant women (Khalid et al., 2014; Khattak et al., 2017; Mahmood et al., 2023).

Moreover, the consumption of contaminated food exacerbates health issues, including gastrointestinal diseases, respiratory problems, and chronic conditions. The burden of these health challenges is disproportionately borne by vulnerable populations, such as rural communities and low-income households.

FOOD SECURITY IN PAKISTAN

Food security is a critical concern in Pakistan, where the interplay of soil degradation, malnutrition, and population growth creates significant challenges. Defined as the availability, accessibility, and utilization of sufficient, safe, and nutritious food, food security is deeply tied to soil health. Degraded soils with low fertility and nutrient deficiencies compromise the quality and quantity of food, exacerbating malnutrition and limiting Pakistan's progress toward global development goals.

Current Status and Challenges

Pakistan is ranked 164th out of 188 countries in achieving the United Nations Sustainable Development Goals (Mahmood et al., 2023), reflecting the severity of its food insecurity. Despite being an agricultural economy, a substantial portion of the population faces food shortages due to low crop yields and poor nutritional content. The Global Hunger Index ranks Pakistan 102nd out of 125 countries, highlighting the persistent issue of hunger and malnutrition (Purohit et al., 2024).

The country's reliance on nutrient-deficient soils reduces crop productivity, while the use of contaminated water for irrigation introduces toxins into the food supply. Climatic factors, such as arid and semi-arid conditions, exacerbate soil degradation (Niazi, 2003), further straining agricultural production. High population growth and diminishing arable land contribute to an ongoing cycle of food insecurity.

Malnutrition and Hidden Hunger

Malnutrition is one of the most pressing outcomes of food insecurity in Pakistan. An estimated 40.2% of children under five are stunted due to inadequate nutrition, while over half of this age group suffers from deficiencies in essential vitamins and minerals (UNICEF). Key nutrient deficiencies include:

- **Zinc (Zn)**: Affects immunity and cognitive development. Zn deficiency is common in populations reliant on cereals grown in Zn-deficient soils (Akhtar, 2013; Ahmed et al., 2017; Khera et al., 2020).
- **Iron (Fe)**: Leads to anemia, which affects 62% of children under five, causing fatigue, cognitive delays, and weakened immunity (Purohit et al., 2024).
- **Vitamin A and D**: Deficiencies in these vitamins result in vision problems, weakened bones, and impaired growth (Mora et al., 2008).

The lack of nutrient-dense crops grown in degraded soils exacerbates "hidden hunger," where calorie intake is sufficient, but critical micronutrients are missing. Women and children, particularly in rural areas, are the most affected by these deficiencies, leading to intergenerational health impacts (Mahmood et al., 2023).

Impacts on Vulnerable Populations

Rural populations, especially farmers and low-income households, are disproportionately affected by food insecurity and malnutrition. Limited access to balanced diets, dependency on low-quality crops and exposure to soil contamination contribute to higher rates of health issues in these communities (Imtiaz et al., 2010). Vulnerable groups, including pregnant women and children (West, 2002), are more likely to suffer from deficiencies in Zn, Fe, and vitamin D, resulting in developmental delays, complications during childbirth, and increased mortality rates.

SUSTAINABLE SOLUTIONS FOR SOIL AND FOOD SECURITY

Addressing soil degradation and its impact on food security in Pakistan requires the adoption of sustainable solutions tailored to the country's unique challenges. These solutions should focus on improving soil fertility, minimizing contamination, and enhancing agricultural productivity while ensuring environmental sustainability. Implementing these measures is critical for achieving food security and better nutrition for the population.

Soil Fertility Management

Sustainable soil fertility management is essential for reversing nutrient deficiencies and enhancing crop yields. Key strategies include:

- **Balanced Fertilizer Application**: Using the right proportions of N, P and K along with micronutrients like Zn and Fe can address nutrient imbalances and improve soil productivity (Imtiaz et al., 2006). Farmers must be educated on applying fertilizers appropriately to enhance nutrient-use efficiency and reduce wastage.

- **Organic Amendments**: Adding organic matter through compost, manure, or crop residues can restore SOM content, improve soil structure, and boost nutrient availability. This is particularly critical in Pakistan, where more than 90% of soil is deficient in SOM (Rafique et al., 2023).
- **Conservation Practices**: Techniques such as crop rotation, cover cropping, and reduced tillage can prevent erosion, enhance water retention, and maintain soil fertility (Lal, 2018).

Technological Interventions

Modern agricultural technologies offer promising solutions for improving soil management and increasing crop productivity:

- **Precision Agriculture**: Precision farming tools (digital agriculture), such as soil sensors and GPS mapping, enable site-specific nutrient management, optimizing fertilizer use and minimizing environmental impacts (Ahmed et al., 2020).
- **Soil Fertility Mapping**: Detailed soil maps help identify nutrient deficiencies and salinity issues at the local level, allowing for targeted interventions to improve soil health (Ahmed et al., 2017).
- **Remote Sensing**: Satellite imagery and drones can monitor soil conditions, crop health, and land use changes, providing valuable data for decision-making.

These technologies can help Pakistan's farmers achieve higher productivity, while reducing input costs and minimizing resource wastage.

Policy Recommendations

Effective policy frameworks are critical for promoting sustainable soil management practices. Key policy interventions include:

- **Payments for Ecosystem Services**: Providing financial incentives through payments for ecosystem services can enhance use of organic fertilizers, soil-testing kits, and precision agriculture technologies can encourage farmers to adopt sustainable practices (Amanullah & Irfanullah, 2016).
- **Regulating Contamination Sources**: Policies to control the use of untreated wastewater reduce industrial effluent discharge, and limit pesticide overuse are essential for minimizing soil contamination (Pečiulytė et al., 2006).
- **Farmer Education and Training**: Awareness campaigns and training programs can help farmers understand the importance of soil health and adopt practices like balanced fertilization and conservation tillage.

Collaboration between government agencies, research institutions, and international organizations is crucial for implementing these policies effectively.

Community and Farmer Engagement

Engaging farmers and local communities is key to ensuring the success of sustainable solutions:

- **Capacity Building**: Providing farmers with knowledge and resources for soil testing, nutrient management, and sustainable practices can empower them to restore soil health.
- **Local Demonstrations**: On-ground demonstrations of successful soil restoration techniques can encourage adoption among farmers.
- **Community-Based Programs**: Initiatives that involve local communities in soil conservation projects can create a sense of ownership and ensure long-term sustainability.

HEAVY METAL POLLUTION: CAUSES AND EFFECTS

Heavy metal pollution is a significant environmental challenge that poses severe risks to soil health, ecosystems, and human health. In Pakistan, industrialization, urbanization, and unsustainable agricultural practices contribute to the accumulation of toxic metals in the soil, affecting crop quality and threatening food security. Understanding the causes and impacts of heavy metal pollution is essential for developing mitigation strategies and safeguarding public health.

Sources of Heavy Metal Pollution

Heavy metals, including Cd, Pb, As, chromium (Cr), and mercury (Hg), originate from both natural and anthropogenic sources. In Pakistan, the primary contributors to heavy metal pollution include:

- **Industrial Effluents and Urban Waste**: Discharge from industries such as textiles, tanneries, and metal processing introduces significant amounts of heavy metals into the soil. Urban waste, including untreated sewage, is another major source of contamination (Ahmed et al., 2017; Pečiulytė et al., 2006).
- **Agricultural Practices**: The extensive use of chemical fertilizers, especially phosphate-based ones, and pesticides adds toxic metals to the soil. Additionally, organic fertilizers and irrigation with untreated wastewater contribute to soil contamination (Maric et al., 2013).
- **Mining and Smelting Activities**: Mining operations release heavy metals into the environment, contaminating nearby soil and water sources (Shahid et al., 2014).
- **Atmospheric Deposition**: Heavy metals are deposited in soil through air pollution caused by vehicle emissions, coal combustion, and industrial activities (Nagajyoti et al., 2010).

In Punjab, surveys revealed high concentrations of Cd (71.4%) and Pb (70.7%) in vegetables irrigated with wastewater, exceeding permissible limits (Ayub Research, 2019–2020).

Impacts on Ecosystems and Crop Production

Heavy metal contamination significantly affects soil ecosystems and agricultural productivity. Toxic metals disrupt soil microbial activity, impairing nutrient cycling and reducing soil fertility. Common impacts on crops include:

- **Reduced Yield**: Heavy metals like Cd and Pb interfere with nutrient uptake, leading to stunted growth and reduced crop yields (Rizwan et al., 2016).
- **Physiological and Structural Damage**: Heavy metal toxicity causes oxidative stress in plants, leading to chlorosis, necrosis, and reduced photosynthetic efficiency (Frayssinet et al., 2023).
- **Bioaccumulation**: Metals absorbed by plant roots accumulate in edible parts such as fruits and grains, posing risks to human health through the food chain.

For example, rice grown in AS-contaminated soils has been identified as a significant source of As exposure in humans (Brammer, 2009).

Human Health Implications

The entry of heavy metals into the food chain has dire consequences for human health. Table 4.1 summarizes the major health impacts associated with common heavy metals, highlighting their toxicity and the risks they pose to human populations.

TABLE 4.1
Impacts of Heavy Metals Toxicity on Human Health

Heavy Metal	Impact on Human Health	References
Cr	Skin allergies, dermatitis, skin diseases, cancers, irritation and diseases related to digestive, excretory, respiratory, and reproductive systems	Scragg (2006)
Cu	Eye problems, cataract, headache, chills	Ahmed et al. (2014)
	Skin irritation and skin and hair discoloration	Eck and Wilson (1989)
As	Arsenicosis, "pins and needles" sensation in hands and feet, gastrointestinal irritation, decrease in white and red blood cell production, disrupts the heart rhythms and damages blood vessels	Abernathy et al. (2003)
	Long term exposure to high concentrations causes cancer, cardiovascular disease, melanosis, leukomelanosis, hyperkeratosis, black foot disease, and neuropathy	Caussy (2005); Gary (2012)
Cd	Kidney damage	Barbier et al. (2005)
	Reproductive problems	Johnson et al. (2003); Piasek and Laskey (1999)
	Cancer	Waalkes et al. (1988)
	Bone damage	Kazantzis (1979)
	Vomiting and diarrhea nausea, muscle cramps, salvation, sensory disturbances, liver injury, convulsions shocks, and renal failure	Nordberg (2004)
	Cellular damage of intestine and liver	Aziz et al. (2014)
	Renal dysfunction, lung cancer and respiratory problems	Hussain et al. (2014)
Hg	Affects nervous system, disrupts neurotransmitter production, and decreases the production of hormones like thyroid and testosterone	Fatoki and Awofolu (2003)
	Pneumonitis and bronchial irritation	Gary (2012)
Pb	Brain and nerve disorders, kidney damage, digestive problems, cardiovascular problems, and hypertension	Gidlow (2004); Riess and Halm (2007);
	Subencephalopathic, neurological and behavioral effects	Lehloesa and Muyima (2000)

CASE STUDIES AND REGIONAL INSIGHTS

Understanding the regional dimensions of soil health challenges in Pakistan provides valuable context for targeted interventions. Case studies and regional insights illustrate the specific issues faced by different areas of the country and highlight opportunities for improvement through sustainable practices.

Soil Degradation and Salinity in Punjab

Punjab, the agricultural heartland of Pakistan, faces severe challenges due to soil salinity and degradation. With 6.7 million hectares (Mha) of land affected by salinity, large tracts of fertile land are rendered unproductive (Azam and Shafique, 2017). This issue is exacerbated by improper irrigation practices, leading to water logging and further salinization.

A notable case involves the use of untreated wastewater for irrigation in areas like Faisalabad and Gujranwala. Surveys in these regions revealed that over 70% of vegetables irrigated with wastewater contained Cd and Pb concentrations exceeding permissible limits (Ayub Research, 2019–2020). These practices not only degrade soil quality but also pose significant health risks to consumers.

Efforts to address salinity through gypsum application and improved drainage systems have shown promise. However, large-scale implementation of these solutions requires coordinated efforts between local governments and farming communities.

Heavy Metal Pollution in Urban Agriculture

In urban areas, the extensive use of untreated municipal wastewater for irrigation is a major contributor to heavy metal contamination. A study in Punjab reported toxic levels of Cd, Pb and Ni in soils and crops grown in wastewater-irrigated fields. The accumulation of these metals in edible parts of crops like wheat, rice, and vegetables poses significant health risks, including kidney damage and developmental issues in children (Anwar, 2005; Rafique et al., 2023).

For instance, in Lahore, the use of contaminated water has led to widespread Pb pollution in crops, which has been linked to neurological disorders among residents. The local government has implemented pilot projects for wastewater treatment and awareness campaigns among farmers to mitigate the problem.

Arsenic Contamination in Sindh

Sindh faces unique challenges with As contamination, particularly in groundwater used for irrigation. Regions near the Indus River have reported As concentrations in water and soil that exceed safe limits, impacting both crop quality and human health (Ahmed et al., 2017).

Rice grown in As-contaminated soils is a primary concern, as As accumulates in grains and enters the food chain. This has led to increased cases of arsenicosis, a chronic health condition characterized by skin lesions, cardiovascular issues, and cancer (Brammer, 2009). Efforts to promote As-tolerant crop varieties and alternative irrigation sources are underway, but progress remains limited due to resource constraints.

Soil Erosion in Balochistan

Balochistan, with its arid climate and mountainous terrain, experiences significant soil erosion due to overgrazing, deforestation, and poor land management. The loss of topsoil has reduced agricultural productivity and increased desertification in the region (Lal, 2018).

Small-scale initiatives, such as reforestation projects and terracing techniques, have shown success in reducing soil erosion. For example, community-led projects in Ziarat and Quetta have improved soil stability and water retention, thus supporting local agriculture and reducing vulnerability to climate change.

Success Stories: Integrated Soil Management

Despite these challenges, there are examples of successful soil restoration efforts in Pakistan. In the Indus Basin, precision agriculture techniques and balanced fertilizer applications have significantly improved crop yields and soil health. These practices have been particularly effective in addressing micronutrient deficiencies, such as Zn and Fe, in staple crops like wheat and rice (Ahmed et al., 2017).

In Punjab, farmers adopting conservation agriculture (CA) practices, including no-till farming and crop rotation, have reported reduced soil degradation and increased water use efficiency. These successes highlight the potential of integrated soil management to transform agricultural landscapes and enhance food security.

THE ROLE OF RESEARCH AND COLLABORATION

Research and collaboration are critical for addressing the multifaceted challenges of soil degradation, food insecurity, and public health risks in Pakistan. By fostering partnerships between research institutions, government agencies, and international organizations, Pakistan can develop innovative, evidence-based solutions to restore soil health and improve agricultural sustainability.

This section explores the vital contributions of research and collaborative efforts in advancing soil management and addressing the interconnected issues of food security and human health.

Contributions of Research Institutions

Research institutions play a central role in identifying soil health issues and developing strategies to mitigate them. In Pakistan, universities and research centers, such as the University of Agriculture Faisalabad and the Soil Salinity Research Institute, focus on soil fertility studies, nutrient management, and remediation techniques.

Key research contributions include:

- **Soil Testing and Nutrient Mapping**: Research-driven initiatives have developed soil fertility maps that highlight macronutrient and micronutrient deficiencies across Pakistan. These maps provide critical data for implementing site-specific soil management practices (Ahmed et al., 2017).
- **Development of Sustainable Practices**: Institutions have proposed solutions such as balanced fertilizer applications, organic amendments, and conservation tillage to address nutrient imbalances and soil erosion (Lal, 2018; Rafique et al., 2023).
- **Innovative Technologies**: Research has introduced precision agriculture tools, including remote sensing and GIS-based soil monitoring, which enable farmers to optimize resource use and enhance productivity (Ahmed et al., 2020).

International Collaboration

International partnerships bring valuable expertise, funding, and technological support to soil health initiatives in Pakistan. Collaborations with global organizations such as the Food and Agriculture Organization (FAO) and the International Atomic Energy Agency (IAEA) have supported capacity-building programs and advanced research in soil conservation and nutrient management.

Significant collaborative efforts include:

- **FAO's Role in Sustainable Agriculture**: The FAO has provided technical assistance for projects aimed at combating soil salinity, promoting sustainable land management, and improving irrigation practices (FAO, 2017).
- **Carbon Sequestration Projects**: Partnerships with international research centers, such as the CFAES Rattan Lal Center for Carbon Management and Sequestration, have introduced strategies for enhancing SOM content and mitigating climate change impacts (Lal, 2018).

Government and Policy Support

Collaboration between research institutions and government agencies is essential for translating research findings into actionable policies. In Pakistan, policies promoting sustainable agriculture and soil health have emerged from collaborative research efforts. Key initiatives include:

- **Payments for Ecosystem Services by Saving in Fertilizers and Soil Testing**: Policies incentivizing the use of micronutrient-enriched fertilizers and subsidizing soil-testing kits aim to encourage farmers to adopt nutrient-balanced practices (Amanullah & Irfanullah, 2016).
- **Wastewater Management Programs**: Research-backed policies addressing the use of untreated wastewater for irrigation have highlighted the need for stricter regulations to prevent soil contamination (Pečiulytė et al., 2006).

Farmer and Community Engagement

Effective collaboration extends beyond research and policy development to include farmers and local communities, who are the primary stewards of soil resources. Engaging farmers through training programs, demonstration plots, and knowledge-sharing initiatives ensures the successful implementation of research findings.

Successful examples of farmer engagement include:

- **Training in Conservation Practices**: Programs teaching farmers about CA, crop rotation, and organic amendments have demonstrated tangible improvements in soil health and crop yields (Ahmed et al., 2017).
- **Community-Led Soil Restoration**: In regions like Balochistan, community-driven reforestation and terracing projects have reduced soil erosion and improved water retention, showcasing the power of grassroots collaboration.

FUTURE DIRECTIONS AND CONCLUSION

The challenges of soil degradation, food insecurity, and malnutrition in Pakistan require a forward-looking approach that integrates sustainable soil management, innovative technologies, and effective policies. Addressing these issues is not only vital for improving agricultural productivity but also for safeguarding human health and achieving long-term environmental sustainability. This section outlines key future directions and concludes with actionable insights.

Future Directions

To overcome the interconnected challenges of soil health and food security, Pakistan must focus on the following strategies:

1. Enhancing Soil Fertility:
    a. Promote the use of balanced fertilizers enriched with macronutrients and micronutrients to restore soil fertility and improve crop quality (Imtiaz et al., 2006; Rafique et al., 2023).
    b. Encourage organic amendments, such as compost and biochar, to restore SOM content and enhance nutrient cycling (Lal, 2018).
2. Implementing Sustainable Agricultural Practices:
    a. Adopt CA techniques, including no-till farming, crop rotation, and cover cropping, to minimize soil erosion and improve water retention (Ahmed et al., 2017).
    b. Address salinity and sodicity through gypsum application, improved drainage systems, and salt-tolerant crop varieties.
3. Leveraging Technology for Precision Agriculture:
    a. Invest in soil fertility mapping and remote sensing technologies to identify site-specific soil health issues and optimize resource use (Ahmed et al., 2020).
    b. Provide access to precision farming tools for smallholder farmers to reduce input costs and improve efficiency.
4. Strengthening Policy and Institutional Support:
    a. Develop and enforce regulations to control soil contamination from untreated wastewater, industrial effluents, and excessive pesticide use (Pečiulytė et al., 2006).
    b. Implement subsidy programs for sustainable inputs like organic fertilizers and soil-testing kits to encourage adoption of best practices (Amanullah & Irfanullah, 2016).
5. Building Capacity through Research and Collaboration:
    a. Enhance research funding for soil science and agriculture to develop innovative solutions for Pakistan's unique challenges.
    b. Foster international collaborations to access advanced technologies and expertise in soil health management (FAO, 2017; Lal, 2018).

6. Engaging Communities for Long-Term Impact:
   a. Conduct farmer-training programs on sustainable practices, nutrient management, and the importance of soil testing.
   b. Involve local communities in reforestation, soil restoration, and conservation projects to ensure grassroots participation and ownership.

Soil health is the foundation of agricultural sustainability, food security, and human well-being. In Pakistan, degraded soils, nutrient deficiencies, and contamination pose significant threats to these goals. However, through a combination of sustainable practices, advanced technologies, effective policies, and collaborative efforts, these challenges can be addressed.

Restoring soil health is not just an agricultural imperative – it is a societal priority that directly impacts public health and economic stability. By focusing on integrated solutions and fostering partnerships across all levels of stakeholders, Pakistan can pave the way for a resilient agricultural system and a healthier, more food-secure population. The time to act is now, and the efforts invested today will determine the country's future for generations to come.

REFERENCES

Abernathy, C. O., Thomas, D. J., & Calderon, R. L. (2003). Health effects and risk assessment of arsenic. *Journal of Nutrition*, *133*, 1536–1538. https://doi.org/10.1093/jn/133.5.1536s

Abrahams, P. W. (2002). Soils: their implications to human health. *Science of the Total Environment*, *291*(1–3), 1–32. https://doi.org/10.1016/S0048-9697(01)01102-0

Ahmed, A., Ahmad, A., Khalid, N., David, A., Sandhu, M. A., Randhawa, M. A., & Suleria, H. A. R. (2014). A question mark on iron deficiency in 185 million people of Pakistan: its outcomes and prevention. *Critical Reviews in Food Science and Nutrition*, *54*(12), 1617–1635. https://doi.org/10.1080/10408398.2011.645087

Ahmed, S., Akhtar, E., Roy, A., von Ehrenstein, O. S., Vahter, M., Wagatsuma, Y., & Raqib, R. (2017). Arsenic exposure alters lung function and airway inflammation in children: A cohort study in rural Bangladesh. *Environment International*, *101*, 108–116. https://www.sciencedirect.com/science/article/pii/S0160412016304263

Ahmed, W., Tan, Q., Shaikh, G. M., Waqas, H., Kanasro, N. A., Ali, S., & Solangi, Y. A. (2020). Assessing and prioritizing the climate change policy objectives for sustainable development in Pakistan. *Symmetry*, *12*(8), 1203. https://doi.org/10.3390/sym12081203

Akhtar, S. (2013). Zinc status in South Asian populations-an update. *Journal of Health, Population, and Nutrition*, *31*(2), 139–149. https://doi.org/10.3329/jhpn.v31i2.16378

Alonso, J., Liu, Z., Evans-Lacko, S., Sadikova, E., Sampson, N., Chatterji, S., Jibril, A., Sergio, A., Ali, A., Laura, H. A., Ronny, B., Graça, C., Alfredo, C., Silvia, F., de Giovanni, G., Oye, G., Josep, M. H., Yanling He, M. D., Peter, d. J., Elie, G. K., Norito, K., Viviane, K.M., Sing, L., Daphna, L., Maria, E. M. M., Fernando, N.-M., Beth, E. P., Marina, P., José, P.-V., Margreet, Zahari, Z., Ronald, C. K., Graham, T., & WHO World Mental Health Survey Collaborators. (2018). Treatment gap for anxiety disorders is global: Results of the World Mental Health Surveys in 21 countries. *Depress Anxiety*, *35*(3), 195–208. https://doi.org/10.1002/da.22711

Amanullah, I. A., & Irfanullah, H. Z. (2016). Potassium management for improving growth and grain yield of maize (*Zea mays* L.) under moisture stress condition. *Scientific Reports*, *6*, 34627. https://doi.org/10.1038/srep34627

Anwar, M. (2005). Arsenic, cadmium and lead levels in hair and toenail samples in Pakistan. *Environmental Sciences*, *12*, 71–86. https://pubmed.ncbi.nlm.nih.gov/15915148/

Ara, A., & Usmani, J. A. (2015). Lead toxicity: A review. *Interdisciplinary Toxicology*, *8*(2), 55–64. https://doi.org/10.1515/intox-2015-0009

Ayub Agricultural Research Institute. (2019–2020). Annual detailed report 2019–2020. Retrieved from https://aari.punjab.gov.pk/system/files/Annual%20Report%20ISCES%2C%20KSK%202019-20%20.pdf

Azam, A., & Shafique, M. (2017). Agriculture in Pakistan and its impact on economy. A review. *International Journal of Advanced Science and Technology*, *103*, 47–60. http://dx.doi.org/10.14257/ijast.2017.103.05

Aziz, R., Rafiq, M. T., Yang, J., Liu, D., Lu, L., He, Z., Daud, M. K., Li, T., & Yang, X. (2014). Impact assessment of cadmium toxicity and Its bioavailability in human cell lines (Caco-2 and HL-7702). *BioMed Research International*, *2014* (10), 839538. https://doi.org/10.1155/2014/839538

Barbier, O., Jacquillet, G., Tauc, M., Cougnon, M., & Poujeol, P. (2005). Effect of heavy metals on, andhandling by, the kidney. *Nephron Physiology*, *99*, 105–110. https://pubmed.ncbi.nlm.nih.gov/15722646/

Brammer, H. (2009). Mitigation of arsenic contamination in irrigated paddy soils in South and South-east Asia. *Environment International*, *35*(6), 856–863. https://www.sciencedirect.com/science/article/pii/S0160412009000609

Brevik, E. C. (2013). Soils and human health: An overview. In Brevik, E. C. & Burgess, L. C. (eds), *Soils and human health* (pp. 29–58). Boca Raton: CRC Press.

Brevik, E. C., Steffan, J. J., Burgess, L. C., Cerdà, A. (2017). Links between soil security and the influence of soil on human health. In Field, D. J., Morgan, C. L. S., & McBratney, A. B. (eds), *Global soil security. Progress in soil science*. Cham: Springer. https://doi.org/10.1007/978-3-319-43394-3_24

Broadley, M., Brown, P., Cakmak, I., Ma, J. F., Rengel, Z., & Zhao, F. (2012). Beneficial elements. In *Marschner's mineral nutrition of higher plants* (pp. 249–269). Elsevier Ltd. https://doi.org/10.1016/B978-0-12-384905-2.00008-X

Caussy, D. M. (2005). *A field guide for detection, management and surveillance of arsenicosis cases. Regional office for South-East Asia*. New Delhi: World Health Organization.

Combs. (2005). *Committee on minerals and toxic substances in diets and water for animals*. National Research Council.

Deribe, K., Cano, J., Giorgi, E., Pigott, D. M., Golding, N., Pullan, R. L., ... & Davey, G. (2017). Estimating the number of cases of podoconiosis in Ethiopia using geostatistical methods. *Wellcome Open Research*, *2*, 78. https://doi.org/10.12688/wellcomeopenres.12483.2. PMID: 29152596; PMCID: PMC5668927

Doran, J. W., and Zeiss, M. R. (2000). Soil health and sustainability: Managing the biotic component of soil quality. *Applied Soil Ecology*, *15*, 3–11. https://doi.org/10.1016/S0929-1393(00)00067-6

Eck, P., & Wilson, L. (1989). *Toxic metals in human health and disease*. Phoenix: Eck Institute of Applied Nutrition and Bioenergetics, Ltd.

FAO. (2017). *Soil fertility atlas of Pakistan* (p. 1). The Punjab Province

Fatoki, O. S., & Awofolu, R. (2003). Levels of Cd, Hg and Zn in some surface waters from the Eastern Cape Province, South Africa. *Water SA*, *29*, 375–380. https://doi.org/10.4314/wsa.v29i4.5042

Frayssinet, C., Osterrieth, M. L., Fernández Honaine, M., Marcovecchio, J. E., La Colla, N., & Villagran, D. (2023). Relationship between silica and silicophytoliths with heavy metals in a wheat (Triticum aestivum) planted with silicon fertilization. *Sustainable Water Resources Management*, *9*(3), 84. https://doi.org/10.1007/s40899-023-00861-y

Gary, F. (2012). Common heavy metals: sources and specific effects. http://www.becomehealthynow.com/article/productshair/122/3/

Gidlow, D. A. (2004). Lead toxicity. *Occupational Medicine*, 54, 76–81. https://doi.org/10.1093/occmed/kqh019

Gu, D., Andreev, K., & Dupre, M. E. (2021). Major trends in population growth around the world. *China CDC Weekly*, *3*(28), 604.

Hegland, S. J. (2017). *FAO-IPCC expert meeting on climate change*, Land Use and Food Security: Final Meeting Report.

Henry, J. M., & Cring, D. (2013). Geophagy: An anthropological perspective. In E. C. Brevik, L. C. Burgess (eds), *Soils and human health, edition* (pp. 179–198). CRC Press Taylor & Francis Group. https://doi.org/10.1201/b13683-12

Hu, T., Ju, Z., & Liu, X. (2023). Towards sustainable food security through regional grain supply and demand analysis in China. *International Journal of Environmental Research and Public Health*, *20*(4), 3434. https://doi.org/10.3390/ijerph20043434

Hussain, R., Ali, L., Hussain, I., & Khattak, S. A. (2014). Source identification and assessment of physico-chemical parameters and heavy metals in drinking water of Islampur area, Swat, Pakistan. *Journal of Himalayan Earth Science*, *47*, 99–106. https://www.researchgate.net/publication/282184667

Imtiaz, M., Alloway, B. J., Memon, M. Y., Khan, P., Siddiqui, S. U. H., Aslam, M., & Shah, S. K. H. (2006). Zinc tolerance in wheat cultivars as affected by varying levels of phosphorus. *Communications in Soil Science and Plant Analysis*, *37*(11–12), 1689–1702. http://dx.doi.org/10.1080/00103620600710363

Imtiaz, M., Rashid, A., Khan, P., Memon, M. Y., & Aslam, M. (2010). The role of micronutrients in crop production and human health. *Pakistan Journal of Botany*, *42*(4), 2565–2578. http://mail.pakbs.org/pjbot/PDFs/42(4)/PJB42(4)2565.pdf

Johnson, M. D., Kenney, N., Stoica, A., Hilakivi-Clarke, L., Singh, B., Chepko, G., Clarke, R., Sholler, P. F., Lirio, A., Foss, C., Reiter, R., Trock, B., Paik, S., & Martin, M. B. (2003). Cadmium mimics the in vivo effects of estrogen in the uterus and mammary gland. *Nature Medicine*, *9*, 1081–1084. https://doi.org/10.1038/nm902

Kazantzis, G. (1979). Renal tubular dysfunction and abnormalities of calcium metabolism in cadmium workers. *Environmental Health Perspectives*, *28*, 155–159. https://doi.org/10.1289/ehp.7928155

Khalid, N., Ahmed, A., Bhatti, M. S., Randhawa, M. A., Ahmad, A., & Rafaqat, R. (2014). A question mark on zinc deficiency in 185 million people in Pakistan-possible way out. *Critical Reviews in Food Science and Nutrition*, *54*(9), 1222–1240. https://pubmed.ncbi.nlm.nih.gov/24499152/

Khattak, R. M., Khattak, M. N. K., Ittermann, T., & Völzke, H. (2017). Factors affecting sustainable iodine deficiency elimination in Pakistan: A global perspective. *Journal of Epidemiology*, *27*(6), 249–257. https://pmc.ncbi.nlm.nih.gov/articles/PMC5463024/

Khera, D., Singh, S., Purohit, P., Sharma, P., & Singh, K. (2020). Prevalence of zinc deficiency and the effect of zinc supplementation on the prevention of acute respiratory infections. *Turkish Thoracic Journal*, *21* (6), 371. https://pmc.ncbi.nlm.nih.gov/articles/PMC7752109/

Lal, R. (2004). Soil carbon sequestration impacts on global climate change and food security. *Science*, *304* (5677), 1623–1627. https://doi.org/10.1126/science.1097396

Lal, R. 2018. Managing agricultural soils of Pakistan for food and climate. *Soil Environment*, *37*(1), 1–10.

Lehloesa, L. J., & Muyima, N. Y. O. (2000). Evaluation of the impact of household treatment procedures on the quality of groundwater supplies in the rural community of Victoria district, Eastern Cape: technical note. *Water SA*, *26*, 285–290. https://www.researchgate.net/publication/228600283

Mahmood, T., Kumar, R., Ali, T. M., Naeem, N., & Pongpanich, S. (2023). Determinants of the food insecurity at household level in Pakistan: A multilevel model approach. *PLoS One*, *18*(10), e0291343. https://doi.org/10.1371/journal.pone.0291343

Maric, M., Antonijevic, M., & Alagic, S. (2013). The investigation of the possibility for using some wild and cultivated plants as hyperaccumulators of heavy metals from contaminated soil. *Environmental Science and Pollution Research*, *20*, 1181–1188. https://pubmed.ncbi.nlm.nih.gov/22661260/

Mora, J. R., Iwata, M., & Von Andrian, U. H. (2008). Vitamin effects on the immune system: vitamins A and D take centre stage. *Nature Reviews Immunology*, *8*(9), 685–698. https://doi.org/10.1038/nri2378

Nagajyoti, P. C., Lee, K. D., & Sreekanth, T. V. M. (2010). Heavy metals, occurrence and toxicity for plants: A review. *Environmental Chemistry Letters*, *8*, 199–216. https://doi.org/10.1007/s10311-010-0297-8

Niazi, T. (2003). Land tenure, land use, and land degradation: A case for sustainable development in Pakistan. *The Journal of Environment & Development*, *12*(3), 275–294. https://doi.org/10.1177/1070496503255485

Nieder, R., Benbi, D. K., & Reichl, F. X. (2018). *Soil components and human health. 978-94* (223–225). Dordrecht, The Netherlands: Springer Publisher. https://doi.org/10.1007/978-94-024-1222-2

Nordberg, G. F. (2004). Cadmium and health in the 21st century-historical remarks and trends for the future. *Biometals*, *17*, 485–489. https://pubmed.ncbi.nlm.nih.gov/15688851/

Oliver, M. A., & Gregory, P. J. (2015). Soil, food security and human health: A review: Soil, food security and human health. *European Journal of Soil Science*, *66*(2), 257–276. https://doi.org/10.1111/ejss.12216

Pečiulytė, D., Repečkienė, J., Levinskaitė, L., Lugauskas, A., Motuzas, A., & Prosyčevas, I. (2006). Growth and metal accumulation ability of plants in soil polluted with Cu, Zn and Pb. *Ekologija*, *1*, 48–52.

Piasek, M., & Laskey, J. W. (1999). Effects of in vitro cadmium exposure on ovarian steroidogenesis in rats. *Journal of Applied Toxicology*, *19*, 211–217. https://pubmed.ncbi.nlm.nih.gov/10362273/

Purohit, N., Kaur, N., Zaidi, S. R., Rajapaksa, L., Sarker, M., Adhikari, S. R., & Prinja, S. (2024). Assessing the WHO-UNICEF primary health-care measurement framework; Bangladesh, India, Nepal, Pakistan and Sri Lanka. *Bulletin of the World Health Organization*, *102*(7), 476. https://pmc.ncbi.nlm.nih.gov/articles/PMC11197645/

Rafique, E., Gill, S. M., Ahmed, H., Hussain, G. S., & Raza, M. M. (2023). Mapping nitrogen status in soil and foliage of cotton producing area of southern punjab, pakistan. *Pakistan Journal of Botany*, *55*(6), 2275–2281. https://www.pakbs.org/pjbot/papers/1689661894.pdf

Rashid, A., Rafique, E., Bughio, N., & Yasin, M. (1997). Micronutrient deficiencies in rainfed calcareous soils of Pakistan. IV. Zinc nutrition of sorghum. *Communications in Soil Science and Plant Analysis*, *28*(6-8), 455–467. https://doi.org/10.1080/00103629709369803

Riess, M. L., & Halm, J. K. (2007). Lead poisoning in an adult: lead mobilization by pregnancy. *Journal of General Internal Medicine*, *22*, 1212–1215. https://doi.org/10.1007/s11606-007-0253-x

Rizwan, M., Ali, S., Rizvi, H., Rinklebe, J., Tsang, D. C., Meers, E., . . . & Ishaque, W. (2016). Phytomanagement of heavy metals in contaminated soils using sunflower: A review. *Critical Reviews in Environmental Science and Technology*, *46*(18), 1498–1528. https://biblio.ugent.be/publication/8514495

Scragg, A. (2006). *Environmental biotechnology*(2nd Edition, p. 456). Oxford, UK: Oxford University Press.

Shahid, M., Pourrut, B., Dumat, C., Nadeem, M., Aslam, M., & Pinelli, E. (2014). Heavy-metal-induced reactive oxygen species: phytotoxicity and physicochemical changes in plants. *Reviews of Environmental Contamination and Toxicology*, *232*, 1–44. http://dx.doi.org/10.1007/978-3-319-06746-9_1

Shakir, S. K., Azizullah, A., Murad, W., Daud, M. K., Nabeela, F., Rahman, H., Shafiq, U. R., & Häder, D. P. (2017). Toxic metal pollution in Pakistan and its possible risks to public health. *Reviews of Environmental Contamination and Toxicology*, *242*, 1–60. https://doi.org/10.1007/398_2016_9

Sharif, M., Nergis, Y., & Farooq, M. A. (2010). Soil contamination of toxic elements irrigated with mixed industrial effluent and its environmental impact on urban area of Karachi, Pakistan. *American Eurasian Journal of Agriculture and Environmental Sciences*, *9*, 584–591. https://www.researchgate.net/publication/226894179

Stockamp, N. W., & Thompson, G. R. (2016). Coccidioidomycosis. *Infectious Disease Clinics*, *30*(1), 229–246. https://pubmed.ncbi.nlm.nih.gov/26739609/

Ville, S. A., Po, J. Y. T., Sen, A., Bui, A., & Melgar-Quiñonez, H. (2019). Food security and the Food Insecurity Experience Scale (FIES): ensuring progress by 2030. *Food Security*, *11*, 483–491. https://doi.org/10.1007/s12571-019-00936-9

von Lindern, I., Spalinger, S., Stifelman, M. L., Stanek, L. W., & Bartrem, C. (2016). Estimating children's soil/dust ingestion rates through retrospective analyses of blood lead biomonitoring from the Bunker Hill Superfund Site in Idaho. *Environmental Health Perspectives*, *124*(9), 1462–1470. https://pubmed.ncbi.nlm.nih.gov/26745545/

Waalkes, M. P., Rehm, S., Riggs, C. W., Bare, R. M., Devor, D. E., & Poirier, L. A. (1988). Cadmium carcinogenesis in male Wistar [Crl:(WI) BR] rats: dose-response analysis of tumor induction in the prostate and testes and at the injection site. *Cancer Research*, *48*, 4656–4663. https://pubmed.ncbi.nlm.nih.gov/3396014/

Wakeel, A., Ishfaq, M., Wakeel, A., & Ishfaq, M. (2022). Use of Potash in Pakistan. In *Book: Potash use and dynamics in agriculture* (1st Edition, Chapter 9, pp. 87–97). Singapore: Springer. https://doi.org/10.1007/978-981-16-6883-8_9

West Jr, K. P. (2002). Extent of vitamin A deficiency among preschool children and women of reproductive age. *The Journal of nutrition*, *132*(9), 2857S–2866S. https://doi.org/10.1093/jn/132.9.2857S

Zinsstag, J., Schelling, E., Waltner-Toews, D., & Tanner, M., eds. (2015). One health, the theory and practice of integrated health approaches. *CABI*. https://doi.org/10.1079/9781780643410.0000

Zosky, G. R., Boylen, C. E., Wong, R. S., Smirk, M. N., Gutiérrez, L., Woodward, R. C., Siah, W. S., Devine, B., Maley, F., & Cook, A. (2014). Variability and consistency in lung inflammatory responses to particles with a geogenic origin. *Respirology*, 19(1), 58–66. https://doi.org/10.1111/resp.12150

5 Zinc and Its Role in Soil, Plant, and Human Health

A Fundamental Yet Crucial Nexus

Waqar Ahmad, Muhammad Ansar Farooq, Munir Zia, and Rattan Lal

INTRODUCTION

Zinc (Zn) – an essential micronutrient plays a pivotal role in soil, plant as well as human health, forming a crucial nexus between agricultural productivity and human well-being. The essentiality of Zn was established by Sommer and Lipman in 1926. As an essential micronutrient, Zn is required for plant growth, contributing to chlorophyll synthesis and protein metabolism, and serving as a cofactor to regulate multiple plant enzymes (Asadi and Akhlaghi, 2020; Rudani et al., 2018). However, Zn deficiency is prevalent in agricultural soils on a global scale, affecting over half of these lands and subsequently impacts nutritional quality of food crops (Martínez-Ríos et al., 2024). This deficiency hampers plant health and has profound implications for human health, as Zn is crucial for immune function, cell division, and DNA synthesis (Chasapis et al., 2020; Lin et al., 2017). In many developing countries, where soils are inherently deficient in Zn or have low levels due to its fixation, and where Zn-based fertilizers are applied inadequately relative to plant uptake, Zn deficiency can result in severe health consequences to humans. These impacts include weakened immune systems and increased susceptibility to infections (Alloway, 2008; Podewils et al., 2004). Therefore, a comprehensive understanding of Zn bioavailability and its significance at soil–plant–human continuum is essential to effectively address these deficiencies and ensure food and nutritional security, particularly in vulnerable populations. At the same time, climate change has caused serious damage to crop production and food security around the globe. For example, extreme heat, droughts and floods have seriously impacted the quality of staple food crops as well as crop productivity, nutrition and food security (Ray et al., 2015). At global scale, the climatic variability has caused ~5.0 and 1.6 million tons decline in wheat and rice yields per annum, respectively (Arora, 2019; Ray et al., 2015). The consequent reduction in crop yields has intensified the global challenge of meeting daily nutrients and calories demands of the growing world population. The UN Sustainable Development Goal (SDG) for zero hunger (SDG# 2) and good health and well-being (SDG# 3) for all cannot be met without alleviating Zn malnutrition (Gupta et al., 2020). This scenario further warrants investigating the role of Zn in governing soil, plant, and human health under the changing climate.

EXTENT OF ZN DEFICIENCY IN SOILS: CURRENT STATE AND CHALLENGES (GLOBAL AND REGIONAL SCALES)

Zinc (Zn) is bio-geochemically distributed across various pools in the soil *viz.* water-soluble, chelated, complexed, exchangeable, and adsorbed pools. The potential of these pools to provide phyto-available, leachable and extractable Zn varies significantly (Alloway, 2008). The relative Zn distribution in each pool is dependent on various aspects such as parent material, pH, soil texture,

DOI: 10.1201/9781003469742-5

the presence of competing ions and soil Zn levels (Ahmad et al., 2012; Alloway, 2008; Natasha et al., 2022). These factors affect Zn deficiency patterns at global, regional, and national scales.

The presence of diverse soil types in India contributes to varying Zn deficiency. Alluvial soils (~40% of India's total area), typically found in river valleys and deltas, are fertile but prone to Zn deficiency due to intensive farming practices. Black cotton soils (~15% of India's total area), rich in clay, are often Zn deficient due to high calcium carbonate ($CaCO_3$) content as well as high pH. Red soils (~10% of India's total area), common in southern India, are generally acidic and may also face Zn deficiency issues. Laterite soils (~5% of India's total area), found in tropical regions, are often nutrient-poor, including low Zn levels. Additionally, saline and alkaline soils in coastal and arid regions can have low Zn levels due to their high pH and salt content. Table 5.1 data demonstrates that in India, particularly Alfisols, Entisols, and Vertisols, Zn concentrations

TABLE 5.1
Zn Content in Major Cropland Soils across the World (Alfisols, Entisols and Vertisols)

Soil Order	Total Zn (mg kg^{-1})	Exch. Zn (mg kg^{-1})	Region/Country	Remarks	References
Alfisols	~90	~0.15	Global/Various	Higher total Zn than Entisols but lower mobility than Vertisols.	Kabata-Pendias and Pendias (2001)
Entisols	~50	~0.05	-do-	Generally, Zn-deficient due to sandy texture and low OM.	Kabata-Pendias and Pendias (2001)
Vertisols	~70–100	~0.2	-do-	Have moderate Zn with higher mobility than Entisols and Alfisols.	Kabata-Pendias and Pendias (2001)
Alfisols	44	0.55	South Asia/India	Highest Exchangeable Zn but lowest total Zn amongst three soil orders.	Katyal and Sharma (1991)
Entisols	47	0.44	-do-	Medium values observed for total and Exchangeable Zn.	Katyal and Sharma (1991)
Vertisols	63	0.41	-do-	Vertisols showed the highest total Zn content but lowest Exchangeable Zn across three soil orders.	Katyal and Sharma (1991)
Alfisols	1.40 ± 1.60	0.11–3.88	South Asia/Pakistan	Zn deficiency is widespread in calcareous soils.	Alloway (2008)
Entisols	NS	NS	-do-	Zn-deficient due to low OM and high pH in calcareous soils.	Alloway (2008)
Vertisols	NS	NS	-do-	Zn deficiency affects yield and nutrition.	Alloway (2008)
Alfisols	13-108	0.1-9.1	Sub-Saharan Africa (Zambia)	Zn deficiency is common in West African Alfisols, particularly in semi-arid and subhumid zones.	Yerokun and Chirwa (2014)
Entisols	0.4-4.9	0.01-0.11 (E – Zn)	North Africa (Algeria)	Very low total Zn (median ~2.21 mg kg^{-1}) and extremely limited exchangeable Zn	Degui et al. (2024)
Vertisols	60–120	0.2–0.6	Sub-Saharan Africa/Malawi, Tanzania	Maize reported to contain ~30% higher grain Zn.	Botoman et al., (2020, 2022)

Note: **NS**: Not specified.

(total and exchangeable) varied distinctly by soil order, as reported by Katyal and Sharma (1991). Alfisols exhibited the lowest total Zn at 44 mg kg^{-1} but the highest exchangeable Zn was reported at 0.55 mg kg^{-1}. Entisols showed medium values, with total Zn at 47 mg kg^{-1} and exchangeable Zn was reported to be 0.44 mg kg^{-1}. Vertisols had the highest total Zn at 63 mg kg^{-1} but the lowest exchangeable Zn was 0.41 mg kg^{-1} amongst the three orders of fifty-seven benchmark soils.

Overall, approximately 39% to 50% of soils in India are Zn-deficient, with projections indicating that this could rise to 63% by 2025 (Khokhar et al., 2024). Given the current state of Zn fertilizer application on benchmark soils across various cropping systems in India, Zn could be depleted in as little as 165 to 320 years (Takkar, 1996).

In Pakistan, soils are predominantly low in organic matter (OM) with high alkalinity and calcareousness. According to the geo-spatial map developed by the Fauji Fertilizer Company Limited (FFC) in Pakistan, about 85% of the soils (n: 1,500) collected during the year 2014–2017 were found to be Zn deficient (<1.0 mg kg^{-1} soil; DTPA-extractable) (Figure 5.1). The FFC routinely collects thousands of soil samples annually and provides on-farm guidance through its farmer advisory and extension services, aimed at optimizing soil fertility management and improving field crop productivity (Zia et al., 2021). Similar to India, Entisols in Pakistan also suffer from Zn deficiency probably due to low OM contents, and high pH (Table 5.1). While Alfisols have total Zn in the range of 1.40 ± 1.60 mg kg^{-1}, and the exchangeable Zn content ranging between 0.11–3.88 mg kg^{-1} (Alloway, 2008). On the other hand, Vertisols are reported to demonstrate the highest Zn mobility compared to Entisols and Alfisols (Table 5.1). Zn deficiency in Pakistan is exacerbated by intensive cropping and insufficient application of Zn-based fertilizers. The first recognized case of Zn deficiency in soils of Pakistan was reported by Yoshida and Tanaka (1969) of

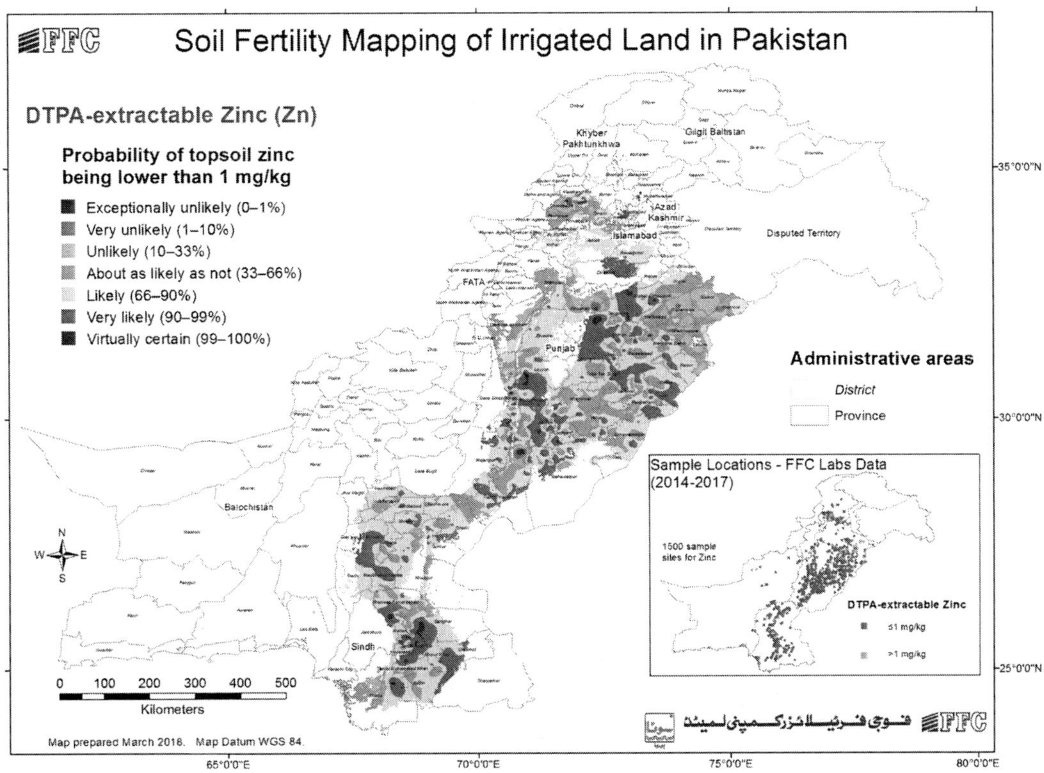

FIGURE 5.1 Probability of topsoil Zn being lower than the threshold limit of 1 mg Zn kg^{-1} soil. **Published with the permission of FFC, Pakistan.**

the IRRI (International Rice Research Institute), who reported '*Hadda*' disease in rice to be caused by Zn deficiency. Simultaneously, Zn deficiency-induced disease in rice was referred to as '*Kharia*' in India (Tandon, 2013). Fertilization with Zn in deficient soils has proven economically beneficial because of significant improvement in crop yields. In Pakistan, increase in Zn fertilizers' use could reduce the inadequate intake of dietary Zn by almost half (Joy et al., 2017). In Sindh and Punjab provinces, the use of Zn fertilizers could generate an estimated US $800 million in increased yields, with a benefit-to-cost ratio of approximately 15 (Rashid et al., 2022). Additionally, full utilization of Zn fertilizer for wheat could provide an additional revenue of US $ 405–1,216 million in returns in the Punjab province alone. Thus, enhancing Zn fertilizer use could increase farm productivity, profitability as well as public human outcomes (Joy et al., 2015, 2017; Rashid et al., 2022).

Despite strong evidence of Zn deficiencies in crop plants, and the economics of Zn fertilizer use, currently the application rate of the micronutrient fertilizers in Pakistan is far below the required levels, with use 5–23 times lower than the actual requirement (Rashid et al., 2022). The UN Food & Agriculture Organization (UN-FAO), Pakistan in collaboration with respective Provincial Agriculture Extension Departments, conducted Inputs [fertilizer] Use Assessments under diverse agro-ecoregions across the country during 2015 (Punjab and Sindh) and 2018 (Khyber Pakhtunkhwa and Balochistan) (Figure 5.2). More than 6,500 farmers were interviewed from all the provinces, maintaining the suitable proportion/weights of the farmers based on the fertilizer consumption statistics populated by the National Fertilizer Development Centre (NFDC). Importantly, the variations in sampling size or selection of different numbers of farmers were made to have uniformity in the information obtained by the respondents as well. The selection of farmers reveals

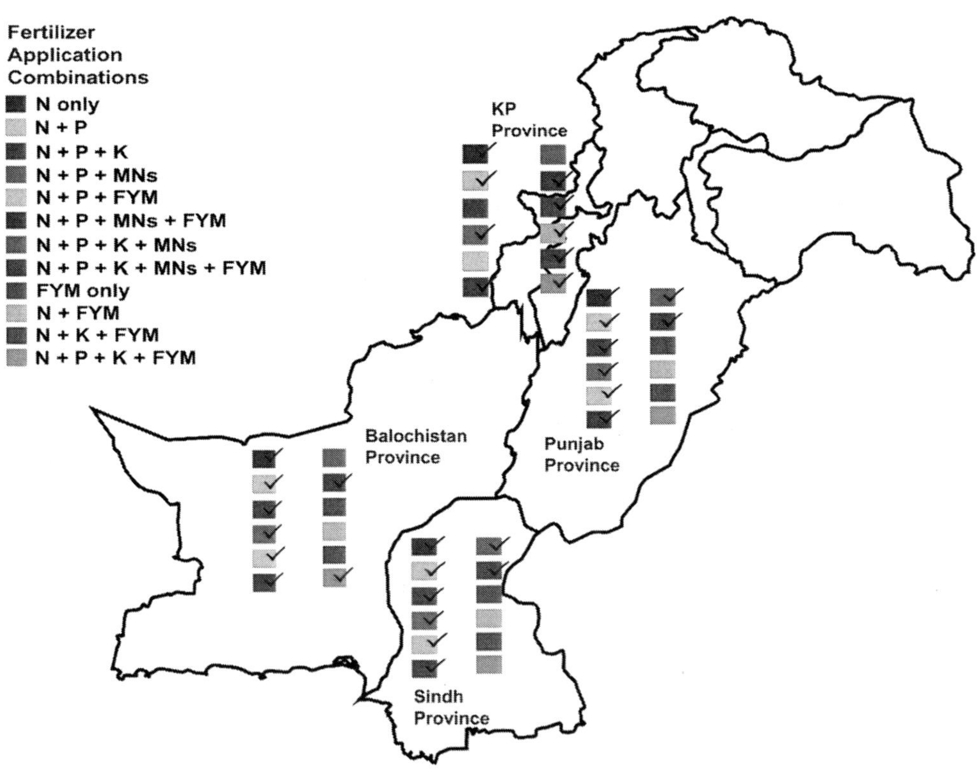

FIGURE 5.2 Fertilizer application combinations in the provinces of Pakistan. A national trend shows predominant use of nitrogen-based (urea) fertilizers with limited application of micronutrients such as zinc (Zn). Micronutrient use, particularly Zn, is more common in rice–wheat systems. Farmers also apply farmyard manure (FYM), usually without chemical analysis.

Figure courtesy of Hassan Waqar.

that the sample size was skewed toward medium-to-progressive farmers with whom agriculture extension workers frequently interact. Moreover, the secondary data were also collected and maintained through the online published fertilizer industries annual reports (financial and sustainability); national (NFDC, Pakistan Bureau of Statistics and Pakistan Economic Survey) and international (International Fertilizer Association) organizations and manuscript(s) published in reputed academic journals. Analysis of the fertilizer use assessments (GCP/PAK/143/USA and GCP/PAK/130/USA) at farm-gate (~6,500 informants) revealed a national trend of high N (urea) fertilizer application, with limited use of micronutrients like Zn. Micronutrient use, particularly Zn, is common in rice-wheat dominated systems (*Personal communication* Country Office, FAO-UN, Pakistan). Farmers also apply organic fertilizers such as farmyard manure (FYM), but without chemical testing (Figure 5.2).

Sub-Saharan African soils are very diverse and range from the oldest to relatively young volcanic soils in the Great Rift Valley, and alluvial soils near rivers. Because of intensive cropping without nutrient replenishment and their inherent soil properties, many African soils suffer from various micronutrient deficiencies (Vanlauwe et al., 2015). Approximately 50% of agricultural soils suffer from Zn deficiency in Ethiopia, reaching over 70% in the Tigray region, posing similar risks to crop health and human nutrition (De Groote et al., 2021; Shiferaw, 2014). Furthermore, Africa demonstrates the lowest fertilizer use at global scale, with an average fertilizer consumption of ~8.3 kg ha^{-1} (Chianu et al., 2012; Sanginga & Woomer, 2009). In Sub-Saharan Africa, the deficiency of Zn is widespread in Entisols and degraded Alfisols, due to poor nutrient holding capacity and low OM of soils, particularly in sub-humid and semi-arid regions. Data in Table 5.1 shows that Alfisols in Zambia contain 0.1-9.1 mg kg^{-1} exchangeable Zn with 13–108 mg kg^{-1} total Zn, while Entisols (~Regosols) in Algeria contain 0.01-0.11 mg kg^{-1} exchangeable Zn with 0.4–4.9 mg kg^{-1} total Zn. These values demonstrate significant Zn deficiencies in the soils of Sub-Saharan Africa. Nevertheless, Vertisols found in Tanzania and Malawi have higher Zn mobility (0.2–0.6 mg kg^{-1} exchangeable Zn with 60–120 mg kg^{-1} total Zn), which significantly increases Zn availability and uptake by crop plants (Table 5.1). Recent reports suggest ~30% higher grain Zn concentrations in maize grown on Vertisols in these regions, thereby enhancing dietary Zn intake for local populations (Botoman et al., 2020, 2022). In China, nearly 50% of the agricultural soils are deficient in Zn making it a widespread micronutrient-deficiency issue in the country. North China is predominantly characterized by calcareous soils, which are particularly susceptible to Zn deficiency (Liu et al., 2016). The deficiency of Zn in the soil is reflected in severe deficiency symptoms in plants (Figure 5.3), with subsequent reduction in crop growth, and it ultimately causes adverse effects on human health, especially in children. Prior studies indicate that a considerable proportion of Chinese children, especially those under six years old, suffer from Zn deficiency due to low Zn content in their diet derived from crops on such deficient soils [Zinc Nutrient Initiative China].

FIGURE 5.3 Examples of Zn deficiency symptoms in rice, wheat and cotton crops in Pakistan.
Picture credits from left to right: rice crop (Dr. Abdul Rashid); wheat crop (Dr. Iftikhar Ahmad/Dr. Murad Ali); and cotton crop (Dr. Tariq Aziz) – used with permission.

ZN CONTENT IN CEREAL CROPS, FORAGES AND ITS BIOAVAILABILITY: IMPLICATIONS FOR LIVESTOCK AND HUMAN NUTRITION

Zn is an essential micronutrient both for animals and plants that plays a critical role in immune response, metabolic functions, and growth- and development-related processes. Zn deficiency either directly impacts human health through low Zn levels in dietary staples, or indirectly through an animal-based diet where the animals are fed on Zn-deficient forage crops. Data from several studies conducted across the globe indicate supplementing Zn through chemical fertilizers to cereal and forage crops substantially improved crop yields and nutritional quality (Tables 5.2 and 5.3). Herein, Zn exhibits variable mobility in plants affecting its bioavailability through the food chain. While the Zn variations across different forage types as well as regions indicate the complexity of maintaining adequate Zn levels in animal diets.

Intensive cropping combined with improper nutrient management and low soil OM contribute to Zn deficiency in cereals and forage crops as one of the major challenges across many regions worldwide. Data in Table 5.2 demonstrates the substantial role of Zn fertilization in increasing grain Zn content of various cereal crops grown across different geographical regions. The data highlights that the effectiveness of Zn supplementation is largely dependent on crop type, dose, and method of application (foliar vs. soil). Interestingly, soil-applied Zn (23 kg ha^{-1}) caused a notable increase of up to 190% in grain Zn content of durum wheat (Kunduru-1149) (Yilmaz et al., 1997), whereas foliar Zn supply (5.35 kg ha^{-1}) to wheat (PDW 291) grown in India showed 122% higher grain Zn content (Dhaliwal et al., 2019). Nevertheless, it is important to reiterate that, while foliar applications are more cost-effective, they face challenges such as difficulties in deployment and Zn fixation in the soil. Moreover, the variations in grain Zn content across different geographical regions and in different cereal varieties underscore the importance of using site-specific Zn-management strategies. Furthermore, extensive research work carried out in Pakistan using different application methods and management strategies highlights the significance of Zn biofortification in cereals to meet nutritional requirement of both livestock and human beings (Ahmad et al., 2021; Kanwal et al., 2010; Rashid et al., 2022) (Table 5.2).

Additionally, legume pastures contain 25–50 mg kg^{-1} Zn (on dry mass basis) contents on an average that are generally considered sufficient for grazing animals (Greene 2000; White 1993).

Nevertheless, considerable variations exist among different regions due to differences in forage species, soil Zn and moisture levels, altitude, forage growing stage, grazing intensity, microbial activity, and use of soil amendments – thereby indicating site-specific nutrient-management strategies as a sustainable approach to maintain adequate Zn levels throughout the year (Greene 2000; Khan et al., 2006a, 2006b) (Table 5.3). For instance, grass forages in Southeast Asia are reported to contain 21–35 mg kg^{-1} Zn, while Zn fertilization further improved forage quality in terms of nutritional value (Blair et al., 1986; Stür et al., 2002). Baranova et al., (2019) reported variability in Zn concentrations in Qilian mountains rangelands of northwest China, which was influenced by grazing intensity (20–100 mg kg^{-1}), forage growth stage (40–70 mg kg^{-1}), and altitude (40–100 mg kg^{-1})

While reports from South Asia also indicate substantial improvement in forage yields and nutritional quality after Zn supplementation as $ZnSO_4$ @15 kg ha^{-1} (Asif et al., 2024; Dhaliwal et al., 2023; Kumar et al., 2024). In another study, Zn application to barley, annual ryegrass, triticale, and oats in Pakistan showed similar positive results in terms of improved Zn uptake by plants (Sher et al., 2022) (Table 5.3). However, variations exist in forage crops where certain Brassica species did not respond to soil-applied Zn indicating no additional amendments may be necessary (Khan et al., 2020). In addition to forage crops, the benefits of feeding Zn-efficient cereal crops like "*Zincol-2016*" in Pakistan and "*Zinc Shakti*" in India to livestock have been extended beyond human health (Dhaliwal et al., 2023). Increase in milk and meat Zn content, through inclusion of biofortified grains in livestock feed, offers a sustainable solution to address Zn deficiency across the food chain.

TABLE 5.2
Studies Reporting the Percentage Increase in Grain Yield and/or Zn Content in Cereals after Zn Application

Region/ Country	Cereal	Variety	Zn Application Method	Dose (kg ha^{-1})	% ↑ in grain/Zn content	References
South Asia/ India	Wheat (*Triticum durum* Desf.)	PDW 291	Foliar	5.35	122	Dhaliwal et al., (2019)
-do-	Pearl Millet (*Pennisetum americanum* L.)	HHB 67	Foliar	10 (mg L^{-1})	11	–
-do-	Oat (*Avena sativa* L.)	Kent	Soil	5	45	Shivay et al., (2013)
-do-	Wheat	PBW-824	NS	NS	NS	(Mentioned for high grain yield)
South Asia/ Pakistan	Wheat	–	Soil	2	12 (grain yield)	Rafique et al., (2006)
-do-	Maize (*Zea mays* L.)	FHY-421	Soil	54	40	Kanwal et al., (2010)
-do-	Wheat	Various (Shahbaz, TJ-83, Al-Ghazi, Abdul Sattar)	Soil	NS	171	Ali et al., (2021)
-do-	Wheat	Zincol-2016	NS	NS	NS	Rashid et al., (2022)
-do-	Wheat and Rice	NS	Foliar (Zn, I, Se)	–	Biofortification established	Rashid et al., (2022)
China	Rice (*Oryza sativa* L.)	Wuyunjing 7	Foliar	0.9	20	Fang et al., (2008)
Thailand	Rice (*Oryza sativa* L.)	NS	Foliar + Soil	9.3	73	Phattarakul et al., (2012)
Sub-Saharan Africa/ Malawi	Maize	NS	Soil	30	NS	Chilimba et al., (2011)
Sub-Saharan Africa/ Zimbabwe	Cowpea	NS	Soil (ISFM)	2.5–5	26 (highest conc.)	Manzeke et al., (2017)
-do-	Maize	NS	NS	NS	27.2 mg kg^{-1} (baseline)	Manzeke et al., (2019)
Sub-Saharan Africa/ Ethiopia	Maize, Teff, Wheat	NS	Soil (granular) + Foliar		29–38% and; 32–40% reduction in Zn disease burden	Abdu et al., (2022); Tessema et al., (2019)
Europe/ Turkey	Wheat (*Triticum aestivum* L.)	Cerek-79	Soil	23	90	Yilmaz et al., (1997)

(*Continued*)

Table 5.2 (Continued)
Studies Reporting the Percentage Increase in Grain Yield and/or Zn Content in Cereals after Zn Application

Region/ Country	Cereal	Variety	Zn Application Method	Dose (kg ha^{-1})	% ↑ in grain/Zn content	References
-do-	Wheat (*Triticum aestivum* L.)	Cerek-79	Foliar	0.40 (%)	33	Yilmaz et al., (1997)
-do-	Wheat (*Triticum aestivum* L.)	Baldo	Soil	45	20	Yilmaz et al., (1997)
-do-	Wheat (*Triticum durum* Desf.)	Kunduru-1149	Soil	23	190	Yilmaz et al., (1997)

Note: NS: Not specified; ZAB: Zn agronomic biofortification under pessimistic and optimistic scenarios; ISFM: Integrated Soil Fertility Management (use of Zn, N and P along with other organic sources of nutrients).

TABLE 5.3
Zn Content in Plant Tissues (Forages) Fed to Cattle for Milk and/or Meat

Forage Types	Zn Content (mg kg^{-1} DM)	Region/ Country	Remarks	References
General pastures	25–50	Global	Worldwide average	Greene (2000); James (2023)
Forages	Improved	South Asia	Application of $ZnSO_4$ at 15 kg ha^{-1} increased forage yield, and quality.	Asif et al., (2024); Dhaliwal et al., (2023); Kumar et al., (2024)
Grass forages	~21–35	Southeast Asia	Zn supplementation improved forage quality	Blair et al., (1986)
Zn-biofortified wheat	NA	India, Pakistan	Biofortified wheat varieties (Zincol-2016 in Pakistan; 'Zinc Shakti/Chitra' in India) improved human Zn intake.	Singh & Velu (2017); Joy et al. (2017)
Grass forages (oat, barley, annual ryegrass, triticale)	NS	Pakistan	Improved with 15 kg ha^{-1} $ZnSO_4$ application.	Sher et al., (2022)
Brassica napus (winter season)	3.58	-do-	Low Zn content in forages under study,	Khan et al., (2020)
Brassica campestris	0.073	-do-	Very low Zn content in forages under study.	Khan et al., (2020)
Forage pasture	Slightly low to moderate levels 28.50–35.7	-do-	Forage Zn levels did not meet recommended dietary requirements of ruminants, and seasonal variation observed.	Khan et al., (2006a)
Forages	Winter Zn levels averaged 49.32, with values declining from 52.087 (I) to 47.67 (IV). Summer Zn averaged lower at 25.01, dropping from 26.27(I) to 23.47 (IV).	-do-	Seasonal variations observed, Zn levels sufficient in winter but marginally deficient in summer.	Khan et al., (2006b)

(*Continued*)

Table 5.3 (Continued)
Zn Content in Plant Tissues (Forages) Fed to Cattle for Milk and/or Meat

Forage Types	Zn Content (mg kg^{-1} DM)	Region/ Country	Remarks	References
Forage	32.59–42.17	South Asia/ Pakistan	Range observed in forage crops.	Chen et al., (2022)
Maize	<1%	South Africa	Widespread Zn deficiency found in sandy soils.	Alloway (2008)
Maize	~30% higher grain Zn	Malawi	Vertisols increase maize grain Zn by ~30% compared to other soils.	Chilimba et al., (2011)
Maize	~394	Tanzania	Zn deficiency widespread in grains and shoots	Wortmann et al., (2019a, 2019b); Lisuma et al., (2006)
Finger millet	665	Uganda	Zn content measured primarily in shoots.	Wortmann et al., (2019a, 2019b)
Maize	~28	Rwanda	Low Zn levels observed in shoots.	Wortmann et al., (2019a, 2019b)
Maize	6246	Nigeria	High variability in Zn content across soils.	Wortmann et al., (2019a, 2019b); African Soil Information Service (AfSIS), 2009.
Cowpea, maize, pearl millet, sorghum	1232	Nigeria	Zn content varies across crops.	Wortmann et al., (2019a, 2019b)

Note: NS: Not specified; NA: Not applicable.

Maize is grown as both grain and forage crops worldwide, however, its Zn content varies substantially across regions depending on genetic and environmental factors. For example, Alloway (2008) reported severe Zn deficiency (<1%) in maize grown on sandy soil, whereas, up to 394 mg kg^{-1} Zn content in maize grains has been reported in Tanzania (Lisuma et al., 2006; Wortmann et al., 2019a, 2019b). In contrast, grain Zn content of ~28 mg kg^{-1} has been reported in maize grown in Rwanda, whereas Zn content in maize as high as 6246 mg kg^{-1} has been reported in Nigeria (Wortmann et al., 2019a, 2019b). Interestingly, a study from Malawi reported ~30% higher grain Zn content in maize grown on Vertisols, reinforcing the role of soil type in Zn bioavailability (Chilimba et al., 2011) (Table 5.3).

The bioavailability of Zn either through plant- or animal-based food refers to the proportion of Zn from dietary sources that is absorbed and utilized by the body to perform numerous cellular functions and contributes to growth- and development-related processes in humans (Devarshi et al., 2024; Nriagu, 2019). Zn from animal-based foods (meat, fish, and dairy) is generally more bioavailable than Zn from plant-based foods (such as legumes and grains), primarily due to the presence of phytates in plants, which inhibit Zn absorption (Lönnerdal, 2000). Cooking methods can also influence Zn bioavailability. For example, soaking, fermenting, or sprouting can reduce phytate levels, thereby improving Zn absorption and its utilization (Liu et al., 2017). Certain nutrients, such as protein and amino acids, enhance Zn absorption, while high levels of competing minerals like iron (Fe) and calcium (Ca) may hinder it (Frossard et al., 2000). Additionally, factors such as gender, age, health status, and the chemical form of Zn in supplements (e.g., Zn picolinate or citrate) also affect its bioavailability (Elliott et al., 2022). In Sub-Saharan Africa as well as Pakistan and India, where such diets and cooking practices are common, deficiency of Zn is relatively high. The risks associated with Zn deficiency in humans are known to become a serious health concern, particularly when insufficient Zn intake is prevalent in more than 25% population (Gibson, 2012).

Costs and Consequences of Zn Deficiency in Humans

Micronutrients deficiency is one of the main causes of malnutrition (also referred to as hidden hunger), that affects a large proportion of developing economies. Globally, over 2 billion people are reported to suffer from malnutrition (particularly Zn and Fe deficiency), with a greater proportion of the population from those regions that are primarily dependent on cereal-based diets (Gregory et al., 2017). On a global scale, approximately 1 billion people are reported to have severe calorie deficiency (undernourishment), around 3 billion have inadequate access to micronutrients (malnourishment/hidden hunger), while >2.5 billion consume more than their daily requirements (overnutrition) (Canales & Fears, 2023). Quite often, the coexistence of these three burdens renders a major challenge to achieve food and nutrition security (Godfray and Garnett, 2014). For example, at global scale, approximately 495 million (7.3%) children suffer from low weight-for-height (also referred to as wasting/atrophy), 149 million (21.9%) are reported to be at low height-for-age (also termed stunting), and 40.1 million (5.9%) are overweight due to multifaceted malnutrition (UNICEF, 2020). Overall, undernourishment and micronutrient deficiencies (malnourishment) contribute to >6% of the global burden of morbidity and mortality (Bolan et al., 2025; Kumssa et al., 2015; Wilhelm, 2016). To this end, the severity of malnutrition issues has been primarily associated with low dietary intake of micronutrients contributed by cereal crops with low mineral density, caused by edaphic and landscape factors (Gashu et al., 2021; Ishfaq et al., 2021).

Among micronutrients, Zn is essential for human health due to its numerous structural, catalytic, and signaling functions. For instance, it acts as a catalyst for several enzymes, promotes wound healing, strengthens the immune system, and contributes to protein and DNA synthesis as well as cell division (Nriagu, 2019). The human body lacks a specialized system for storing Zn, which makes it necessary to consume adequate amounts daily (daily intake) for maintaining normal metabolic functions. The Zn requirement of the human body is relatively low, that is, approximately 11 mg day^{-1} for men, and 8 mg day^{-1} for women (Wessells & Brown, 2012). Due to the insufficient or inadequate absorption of daily Zn intake, still ~17.3% of the population at global scale suffer from serious health risks associated with insufficient Zn in their diet (Danaei et al., 2016; Supasai et al., 2017; Wessells & Brown, 2012; Wessells et al., 2012). Globally, deficiency of Zn is a serious health concern (particularly in children and women) causing multiple diseases and disorders including metabolic syndrome, childhood mortality, anorexia, hypogonadism, cognitive dysfunction, impaired immune function, stunted growth, various skin disorders, and increased prevalence of pneumonia (Kaur et al., 2014; Katimba et al., 2024; WHO e-LENA, 2023). Approximately, 4.4% of deaths in children, at global scale, are caused by Zn deficiency, with the highest burden in regions like Asia, Latin America, and Africa (Walker et al., 2009). The WHO (World Health Organization) estimates that about 0.8 million cases $year^{-1}$ are attributable to the deficiency of Zn (Caulfield & Black, 2004). Infants born prematurely or with low birth weight, as well as preschool children under the age of two, represent the groups with high vulnerability (Chaffee & King, 2012). Zinc's role in pathogenesis of diseases such as atherosclerosis, diabetes, diarrhea, age-related neurodegenerative diseases like Alzheimer's autoimmune diseases, and several types of cancer has been well established (Chasapis et al., 2012; Roohani et al., 2013; Kaur et al., 2014).

Disability-adjusted life years (DALYs) are a widely used metric to measure the burden of disease by combining the impact of premature death and disability into a single index. Developed by the World Bank in collaboration with the WHO, DALYs allow for comparisons across diseases and regions, particularly in global health and developing countries. Zn deficiency is a significant contributor to DALYs lost, with an estimated 245,000 DALYs lost annually in the Punjab and

Sindh provinces of Pakistan (Joy et al., 2017). In Africa, enriching granular fertilizers with Zn through subsidy programs, particularly in Malawi, could reduce DALYs lost due to Zn deficiency by ~10%. The cost per DALY saved ranges from US $624 to $5,893 for granular fertilizers, and $46 to $347 for foliar fertilizers (Joy et al., 2015). However, savings at global and regional scales could be far greater. These findings emphasize the importance of implementing Zn supplementation strategies to reduce the health burden caused by Zn deficiency.

Impact on Human Plasma Zn Concentration from Increased Zn Intake

Despite its essential roles, daily intake of Zn should not exceed 40 mg for adults, as excessive absorption can lead to adverse effects such as poor appetite, diarrhea, nausea, vomiting, and headaches (Nriagu, 2019). Excessive intake of Zn can also impair immune function. Consequently, high levels of Zn in edible plant tissues can pose health risks. Uptake of Zn by plants and its compartmentalization between non-edible and edible plant tissues are critical factors in determining the potential health risks of Zn intake from plants (Natasha et al., 2022). Consuming Zn-biofortified varieties of crops, such as wheat, can increase Zn concentrations in human blood. When individuals incorporate these biofortified crops into their diets, they are likely to consume higher amounts of bioavailable Zn compared to those consuming non-biofortified varieties. However, it is challenging to measure such a change since a negligible fraction (less than 1%) of whole-body Zn is in the serum, whereas most of the Zn is in soft tissues. Consequently, several investigations have been aimed at establishing a sensitive Zn biomarker in humans. Research in a resource-poor set up in Pakistan has demonstrated that biofortified crops that contribute ~21% higher Zn can improve Zn status in populations suffering from Zn deficiency. However, the studies indicate difficulty in measuring the improved status because of the negligible fraction that comes into the serum. This was observed in the BiZiFED (Biofortified Zinc Flour to Eliminate Deficiency) project studies wherein non-significant differences for plasma Zn concentration were noted in the intervention vs. control group, i.e., (643.8 ± 106.2 µg L^{-1} in intervention group vs. 641.6 ± 95.3 µg L^{-1} in control group).

SUSTAINABLE SOIL MANAGEMENT AND AGRICULTURAL PRACTICES IMPACTING ZN AVAILABILITY

Despite advancements in industry and technology, crop production in Pakistan – home to one of the largest populations in South Asia – has remained stagnant for decades. This contrasts with the need to double food availability to fulfill the food demands of an increasing world population. The Green Revolution, while improving yields in some regions, has exacerbated regional inequalities and food security challenges (Lal, 2018). According to a recent Global Hunger Index report, Pakistan currently ranks 109 out of 127 countries, has a serious level of hunger "score of 27.9", with ~20.7% of the population undernourished and 33.2% of children under five experiencing stunted growth (Global Hunger Index, 2024). The continued reliance on outdated agricultural technologies, combined with heavy consumption of imbalanced fertilizer application (primarily due to excess of N), has not led to significant improvements in productivity. The overapplication of phosphatic fertilizers, in particular, can increase Zn deficiency by reducing its solubility in soils (Bolan et al., 2025). In provinces such as Punjab and Sindh in Pakistan, the reliance on N-based fertilizers has caused N use efficiency (NUE) to decline sharply from 58% in the early 1960s to just 23% by 2013. This inefficiency not only contributes to the wastage of fertilizers but also impacts overall soil health (Raza et al., 2018).

Farmers across various provinces use a range of fertilizer combinations, but these applications often lack balance and fail to address micronutrient deficiencies adequately. For instance, while micronutrients like Zn are critical in rice-wheat systems, their application remains well below recommended levels. The addition of Zn fertilizers can significantly improve crop yields and enhance grain Zn content, which is critical for addressing human Zn deficiency (Yogi et al., 2023). In Pakistan, Punjab province exhibits a variety of cropping patterns, with farmers employing multiple fertilizer-application strategies. However, practices such as burning crop residues further exacerbate soil degradation and nutrient loss (FAO, 2017a; Ahmad et al., 2021). In Khyber Pakhtunkhwa and Balochistan, agricultural practices are shaped by local ecological conditions. Still, many farmers continue to apply fertilizers without proper soil testing or consideration of soil health. Overall, while there is a positive correlation between the application of N and P fertilizers and crop yields, the failure to optimize fertilizer use according to microclimatic conditions limits potential yield gains across Pakistan's agricultural landscape (NFDC, 2018a, 2018b). In Sindh and Punjab, 23.2 and 4.4% of farmers, respectively, reported using Zn fertilizers on wheat. This represents a total area receiving Zn fertilizer of 1.1 and 0.4 Mha in Punjab and Sindh, respectively (Joy et al., 2017). A more balanced and precise approach to fertilizer use with a focus on micronutrient management is essential to improve soil health and improve productivity.

STRATEGIES TO VALUE ZN FOR HUMAN HEALTH: SELECTED CASE STUDIES

Cereal crops inherently contain low Zn concentration in their grains (Cakmak, 2008; Rashid et al., 2022). The optimal Zn concentration in staple cereals should range between 40–60 mg kg^{-1} to meet nutritional needs for Zn (Rashid et al., 2022; Sazawal et al., 2018). However, wheat cultivars and other cereal grains in developing countries often contain much lower Zn levels. For example, wheat cultivars in Pakistan typically contain approximately 25 mg Zn kg^{-1}. To address Zn deficiency along the soil–plant–human continuum, several strategies have been identified including: (1) sustainable soil management (SSM) with concomitant practices such as the 4R nutrient stewardship; (2) use of mineral amendments, for example, to wheat flour; (3) bioengineering techniques including genetic engineering, plant breeding, and the use of plant growth-promoting rhizobacteria alongside mineral fertilization of crops including foliar sprays at specific growth stages for enrichment of grains with specific micronutrients (biofortification); and (4) nanoparticle interventions, which have become a recent area of research at global, regional, and national levels (Dimkpa et al., 2022; Irewale et al. 2024; Singh et al., 2024). Nevertheless, it is important to recognize that each of these strategies carries potential benefits and risks for crop physiology and the environment, requiring careful and informed implementation (Sharma et al., 2025).

STRATEGY-1: SOIL MANAGEMENT AND THE 4R NUTRIENT STEWARDSHIP FRAMEWORK

Soil health is fundamental to ecosystem sustainability, as soils serve as a crucial habitat for terrestrial organisms (Kopittke et al., 2019). Soils are also a major source of nutrients, providing approximately 80% of global per capita calorie consumption from food crops. Consequently, soil degradation caused by traditional and intensive farming practices is a major contributor to the declining nutritional content of crops. Thus, proper soil management can enhance Zn availability, thereby improving human nutrition through the food chain. Although the role that soil nutrients play in crop production has been well recognized, the link between SSM practices, soil health and human health has received relatively less attention (Singh & Ahmad, 2018). The 4R nutrient stewardship concept promoted by USDA together with IFA (International Fertilizer Association) aims to increase farm productivity and profitability for both subsistence and progressive farmers (FAO, 2017a, 2017b). The 4Rs are the right source (appropriate nutrient sources, both inorganic and organic, tailored to crop needs), the right rate (appropriate nutrient quantity for crop requirements), the right time (timing nutrient applications to match crop demands), and the right place (ensuring nutrients are applied in areas where crops can use them). However, in many developing countries like Pakistan, the right rate and the right source (first two Rs) are mostly followed, but the right place and right time are rarely practiced, especially by smallholders at farm-gate. This discrepancy results in poor NUE and low economic returns (FAO, 2017a, 2017b; NFDC, 2018a, 2018b).

Implementing the 4R nutrient stewardship framework can help increase Zn concentration in grains, enhance productivity, and boost economic returns. For example, the 4R Nutrient Stewardship Project (2019–2024) has been transformative for smallholder farming systems in Ethiopia, Ghana, and Senegal, improving socio-economic well-being and environmental sustainability for 60,000 beneficiaries. In Zimbabwe, a 10-year study on maize-based cropping systems showed that combining organic nutrient inputs with mineral NPK and Zn fertilizers significantly increased dietary Zn supply (Manzeke-Kangara et al., 2021). In Pakistan, Zn grain enrichment in wheat up to 40–50 mg Zn kg^{-1} has been achieved by applying Zn fertilizer to the soil and combining this with foliar Zn-sprays, particularly to Zn-efficient wheat genotypes (Zia et al., 2020). A combination of soil-applied Zn with two foliar sprays of 0.32% zinc sulphate ($ZnSO_4$) (33% Zn) before and after the heading stage significantly increased Zn concentrations in wheat grains (Rashid et al., 2022).

Strategy-2: Fortification

Fortification, the external addition of Zn to the food, typically in the form of inorganic salts like $ZnSO_4$ or zinc oxide (ZnO), is another effective strategy to combat human micronutrient deficiency (hidden hunger). This method is often used to fortify cereal grains flour. While this strategy is effective, it has numerous limitations. First, to ensure homogenous Zn distribution in flour, a well-optimized central processing is required. Second, this strategy is only effective when communities participate in government-led fortification initiatives, where designated flour mills add Zn to the flour. In regions where wheat is milled in small, local mills without fortification facilities, the benefits may not reach the intended populations (Lowe et al., 2020). Ansari et al., (2018) reported that more than 60% of the wheat harvest in Pakistan is retained by farmers for their own consumption, as well as for making in-kind payments to laborers, further complicating the widespread impact of fortification efforts.

Strategy-3: Biofortification

Biofortification, which aims to increase the nutrient density of crops through agronomic practices, or plant breeding, is a promising approach to fight micronutrient deficiencies (Ahmad et al., 2012; De Groote et al., 2021; Ofori et al., 2022; Zia et al., 2020). Studies have been carried out to evaluate changes in mineral concentration in wheat grains, over an extended period of 160 years. For example, archived soil/wheat grain samples from the Broadbalk wheat experiment (Rothamsted, UK) were used to establish trends over time in relation to harvest index, yield, and cultivar type. The results demonstrated a decrease in grain mineral concentration (including Zn), particularly in treatments receiving no fertilizer amendments. Likewise, the introduction of high-yielding cultivars during the Green Revolution era of the mid-1960s caused a dilution effect in grain mineral density, emphasizing the need for biofortification to address micronutrient deficiencies (Fan et al., 2008).

Biofortification, particularly in resource-poor settings, has gained attention as a feasible and cost-effective strategy to increase nutrient availability in staple crops (Ahmad et al., 2012; Ahmad et al., 2022; Johns and Eyzaguirre, 2007). By developing varieties with higher Zn concentrations in their grains, researchers have successfully introduced biofortified wheat varieties in countries like Pakistan, India, Bangladesh, and parts of Africa and Latin America (HarvestPlus, 2024). Under the BiZiFED project, Lowe and coworkers conducted a series of studies to evaluate the impact of using biofortified wheat flour in improving Zn status in the rural community of Pakistan (Lowe et al., 2018, 2020, 2022). The grains were produced from two wheat genotypes, the standard Galaxy-2013 variety (control) and the biofortified Zincol-2016 variety (intervention), grown under controlled conditions at a selected farm in the Punjab province. While randomized, double-blind, controlled efficacy human trials were conducted on women of reproductive age from the rural community of Khyber Pakhtunkhwa province (Lowe et al., 2022). The authors reported that regular consumption of biofortified wheat flour increased daily Zn intake by 30–60%, resulting in an initial increase in plasma Zn concentration after four weeks, though the increase was not linked to the sustained improvement in indices of Zn status. Previously, researchers had also reported a negative correlation between nutrient density and crop yield due to the "dilution effect" (Joy et al., 2017; Simmonds, 1995). Therefore, breeding programs should aim to increase both yield and nutrient density in cereal grains simultaneously.

In addition, agronomic interventions such as the application of Zn-enriched amino acids (particularly Zn tryptophan) improved the bioavailability of Zn in rice under saline-sodic conditions. This study suggested that application of Zn-enriched amino acids is one of the effective approaches to improve rice growth on marginal soils (Raza et al., 2023). In maize hybrids (DK-919 and Pioneer 30-Y-87), seed priming (2.0% Zn solution) coupled with 2.0% foliar spray of Zn solution resulted in improved grain productivity (20% increase in grain yield), increase in grain Zn density from 19.8 mg kg^{-1} to 28.5 mg kg^{-1} and overall crop growth (Mohsin et al., 2014).

Recent progress in the field of crop nutrient management such as Zn-coated urea has further optimized NUE (N and Zn) and could play a crucial role in improving both crop productivity and human health. Nitrogen in the form of urea is a favored nutrient among the farming community in Pakistan (FAO, 2017a, 2017b). However, excessive use of urea along with rapid volatilization rate causes severe economic and environmental consequences. Therefore, research has been geared towards developing slow-release urea fertilizers coated with different Zn-based coating materials including ZnO in its bulk and nano form (Beig et al., 2022, 2023, 2024; Irfan et al., 2018). Literature shows that different coating techniques such as slurry and wet coating to optimize the coating efficiency and release rate of nutrients have been employed. Further, the results revealed that urea coated with ZnO nanoparticles (NPs) (0.5%) reduced leaching and increased N and Zn uptake by wheat plants (Beig et al., 2022, 2023). Interestingly, the ZnO-NPs also improved wheat grain yield by up to 34% compared to uncoated urea, suggesting value-added fertilizers as a promising solution for improved bioavailability of Zn and N. Foliar application of Zn (0.4% Zn solution; analytical grade zinc sulfate heptahydrate) has been reported to effectively mitigate cadmium (Cd) toxicity in wheat plants grown in Cd-contaminated soils (silt loam) (Qian et al., 2023). In another study, Khan et al. (2021) reported alleviated Cd stress and overall yield of wheat with the foliar application of ZnO -NPs.

Bio-inoculants, such as Zn-solubilizing bacteria, have also demonstrated potential in improving wheat growth and Zn uptake (Ali et al., 2023). A study conducted at the NARC (National Agriculture Research Centre), Islamabad, Pakistan evaluated 69 bacterial strains for their Zn-solubilizing potential. Ten promising strains were identified, including *Pantoea* sp. and *Klebsiella* sp., which demonstrated strong plant growth-promoting traits. These strains were tested on wheat varieties Zincol-2016 and Wadaan-17, revealing that a critical Zn level of 50 mg kg^{-1} from ZnO was optimal for growth. Consortium inoculation of these strains without ZnO improved wheat growth characteristics. With ZnO, root growth and Zn content in shoots increased significantly. Thus, this study highlighted the potential of Zn-solubilizing bacteria as bio-inoculants to enhance wheat growth and combat Zn deficiency (Ali et al., 2023).

Current production systems focus more on increasing grain yields than improving nutritional quality. Simply producing more food does not solve malnutrition, as it is more about how people consume calories and nutrients rather than about just having enough food available. Further, all strategies identified above have some limitations. To address the issue of Zn deficiency effectively, and increase to optimal levels of Zn in human blood through recommended daily intake, it will be prudent to use a multipronged approach while focusing on the smallholders. It is pertinent to mention that 84% (approximately 570 million) farms are smallholdings that is, they are less than two hectares but contribute significantly to global food security. Despite their small size, these farms are estimated to produce 70–80% of the food supply (Lowder et al., 2016; Ricciardi et al., 2021; Ritchie, 2021). A shift of coordinated effort toward small-scale farming can be a sustainable strategy for the Zn management plan for soil, plant and human health improvement. In this context, the most dominant cropping system can be selected to start with, for example the rice-wheat system is quite popular in the Indo-Pak region, comprising an extensive (~16 M ha) agricultural production system spanning across Pakistan, Nepal, Bangladesh and India. Besides, the maize-potato cropping system is common in Africa. Following the success story of the BiZiFED project, biofortified cereal varieties can be grown independently by the farmers to feed the surrounding local community. Applying $ZnSO_4$ coating to commonly used urea fertilizer, a product marketed by the FFC in Pakistan, offers a practical strategy to combat widespread soil Zn deficiency, thereby increasing Zn levels in grain crops. Continuous application of this fertilizer ($ZnSO_4$ coated urea) will hopefully also rectify the negative Zn balance in calcareous soils of the country. Smallholders face diverse challenges that affect their farm output and subsequently rural livelihoods. Environmental degradation and economics are major constraints to productivity, reducing the capacity of smallholders to meet growing food demand (Touch et al., 2024). Therefore, cost-effectiveness and viability should be considered while recommending coated fertilizers to the community.

CONCLUSIONS

Soil composition, agricultural practices, dietary patterns, and socio-economic factors vary significantly between countries and even within regions. Country-scale studies are therefore essential to translate local scale Zn knowledge into effective interventions. Through a focus on specific conditions within each country, research scientists and policymakers can devise targeted strategies to improve Zn nutrition and reduce the disease burden. While soil application of Zn is important, its limited translocation to cereal grains calls for the introduction of complementary approaches in order to address the Zn deficiency in a wider population. Foliar application of Zn in cereals is two to three times more effective (than that of the soil-applied Zn) for biofortification of the grains; however, to maintain soil Zn balance and simultaneously enrich the grains both the strategies need equal attention.

The UN SDGs (Sustainable Development Goals) corresponding to zero hunger and good health and well-being for all cannot be achieved without alleviating Zn malnutrition. This chapter has highlighted the critical interdependence of Zn in soil and human well-being, emphasizing its essential role in both plant and human physiology. The widespread Zn deficiency in agricultural soils at global scale, affecting more than half of the global arable land, poses a substantial threat to agricultural productivity and nutritional security. This deficiency, compounded by inadequate dietary intake, has far-reaching consequences for human health, especially among vulnerable populations. In regions like South Asia and Sub-Saharan Africa, limited awareness of micronutrient deficiencies, insufficient access to Zn fertilizers, and reliance on imbalanced fertilization practices exacerbate this issue. Addressing this multifaceted challenge requires a comprehensive, integrated approach that strengthens the link between soil, plant, and human health through sustainable and scientifically informed management practices.

Several key areas must be prioritized to combat Zn deficiency, effectively. First, increasing awareness and providing education to farmers and the public regarding the nature, extent, and implications of Zn deficiencies is critical. Second, investment in soil-fertility research focused on Zn nutrition, alongside the enhancement of soil-testing facilities for routine Zn analysis, is essential for informed decision-making and targeted interventions. Third, improving the availability and accessibility of Zn fertilizers, while promoting their balanced and judicious use, is crucial for correcting soil Zn deficiencies and improving crop yields. Fourth, supporting biofortification efforts such as the BiZiFED project aimed at increasing Zn levels in staple crops, offers a promising pathway to enhanced dietary Zn intake, especially in regions where cereal-based diets dominate. Fifth, there is a need for the introduction of soluble Zn salt coating over popular chemical fertilizers like urea. Finally, continued research is necessary to explore the complex interactions of Zn in the soil–plant–human system, including its bio-physico-chemical roles at different concentrations, to develop sustainable interventions that balance the need for increased crop production with the protection of human health.

A collaborative effort involving researchers, policymakers, agricultural extension services, and the industry is required to achieve sustainable Zn management. Smallholders face diverse challenges because of their capacity to purchase quality inputs that can hinder their capacity to meet growing food demand. Therefore, cost-effectiveness and viability should be considered while recommending coated fertilizers to the farming community. Undoubtedly, we can enhance agricultural productivity, improve human health, and ensure a more food-secure and nutritionally balanced future by following this approach. Emphasizing biofortified crops, such as Zn-enriched wheat through initiatives like the BiZiFED project, represents a promising avenue. While biofortified wheat can improve Zn intake, the actual increase in blood Zn concentration is influenced by dietary composition, individual health status, and the quantity consumed. Further research is needed to elucidate the bio-physico-chemical roles of Zn at deficient, optimum, and toxic levels in plants, which will help develop sustainable agricultural practices that increase crop production while safeguarding human health.

RECOMMENDATIONS

R&D (Research and Development)

1. *Strengthening Research Capacity and Funding*
   a. More research funds to be dedicated for evaluation studies and projects for development of new crop varieties with increased climate resilience and sustaining Zn density and/or Zn loading. Enhancing the research capacity in relevant organizations (Research, Development, and Extension) will enable better predictions of climate impacts and the development of adaptation measures. Effectiveness of the soluble Zn-coating materials on popular fertilizers like urea and diammonium phosphate (DAP) must be evaluated so that cost-effective and viable fertilizers are made available to the farming community.
2. *Private Sector Engagement in Research and Innovation*
   a. Fertilizer companies and other stakeholders should invest at least 5% of their total revenue in research and commercialization efforts, specifically for sustainable Zn fertilizer solutions and technologies that cater to smallholders.
3. *Zn Use Scenarios Simulation*
   a. Conduct simulations of Zn use in cereal-based food production areas under varying temperature and moisture regimes. These studies should investigate the relationship between soil fertility and crop yields, offering strategies to address micronutrient deficiencies.
4. *Zn Efficient Genotypes Development*
   a. Develop Zn-efficient genotypes using advanced CRISPR-Cas9 technology to improve NUE globally. This will contribute to increased resilience and Zn density in crops.
5. *Remote Sensing for Nutrition-sensitive Mapping*
   a. Use remote sensing data to monitor soil health, crop nutrient status, and identify areas at risk of Zn deficiency. Establish Zn-sensitive hubs based on this data across various geographic scales.
6. *4Rs of Nutrient Stewardship for Zn*
   a. Develop and implement SSM practices based on the 4Rs (Right source, Right dose, Right place, Right time) for Zn application to increase crop yield and improve quality.

Policy Interventions

1. *Smallholder Support through Targeted Subsidies*
   a. Governments should provide subsidies for Zn fertilizers, biofortified seeds, or extension services to promote Zn-rich cereal production. This would help smallholders adopt Zn-enhancing practices and improve food security.

2. *Private Sector Engagement in Research and Innovation*
    a. In addition to R&D, policy measures should be implemented to stimulate the private sector, particularly fertilizer companies, to invest in research, innovation, and commercialization aimed at improving Zn availability for smallholders.
3. *Collaboration between International and Regional Agencies*
    a. International and regional agencies (e.g., FAO-UN), Global/Asian/African Soil Partnership, Scaling Up Nutrition (SUN), HarvestPlus, Global Alliance for Improved Nutrition (GAIN), Australian Centre for International Agricultural Research (ACIAR), Foreign, Commonwealth & Development Office, The Consultative Group on International Agricultural Research (CGIAR), United Nations Development Programme (UNDP), The International Food Policy Research Institute (IFPRI), International Livestock Research Institute (ILRI) among others, should collaborate to disseminate Zn-efficient varieties and other technological interventions for small, medium, and progressive growers. This collaboration should also include the design of incentive-based policy frameworks.

Awareness Raising/Outreach Campaigns

1. *Promote Farmer Education and Capacity Building*
    a. Develop educational programs to help farmers understand and implement the findings from research organizations, bridging the gap between scientific discoveries and farming practices.
2. *Civil Society Engagement in Awareness Campaigns*
    a. Involve civil society in awareness campaigns aimed at improving food security and the role of Zn in nutrition. This will help reach broader populations and promote sustainable farming practices.
3. *Scientific Knowledge Sharing and Capacity Building*
    a. Promote the global sharing of scientific knowledge and technical skills, including training on transferring laboratory-scale skills to practical applications in the field. This can help strengthen the knowledge base for improving Zn in food production.

FUNDING AND CONFLICT OF INTEREST

The authors declare that the research was conducted without any commercial or financial relationships that could be construed as a potential conflict of interest.

ACKNOWLEDGMENTS

We extend our sincere gratitude to Hassan Waqar from the Arthur Phillip High School, Parramatta, NSW, Australia for his assistance in identifying relevant abstracts and articles and providing us synthesized information. We are deeply grateful to the Food and Agriculture Organization of the United Nations (FAO-UN), Pakistan, for generously sharing open-source datasets from the "Soil Fertility Management for Sustainable Intensification in Pakistan" (GCP/PAK/130/USA) and "4R Nutrient Stewardship for Sustainable Agriculture Intensification in Pakistan" (GCP/PAK/143/USA) projects. We also acknowledge the United States Department of Agriculture (USDA) for their essential funding of these projects. Dr Muhammad Ansar Farooq (MAF) acknowledges United States Educational Foundation in Pakistan (USEFP) and Fulbright Commission, USA for Fulbright Scholar Award (2024–2025) to carry out research at The Ohio State University, USA. Financial support by the University of Aberdeen, UK under Official Development Assistance, and British Society of Soil Science is acknowledged to support research activities in MAF's lab. It is important to note that all opinions expressed herein are solely those of the authors and do not represent the views of the FAO or any other organization.

BIBLIOGRAPHY

Abdu, A. O., De Groote, H., Joy, E. J., Kumssa, D. B., Broadley, M. R., & Gashu, D. (2022). Zinc agronomic biofortification of staple crops may be a cost-effective strategy to alleviate zinc deficiency in Ethiopia. *Frontiers in Nutrition*, 9, 1037161.

African Soil Information Service (AfSIS). (2009). African Soil Information Service (AfSIS) | OER Africa. https://www.oerafrica.org/content/african-soil-information-service-afsis

Ahmad, W., Ullah, N., Xu, L., & El Sabagh, A. (2022). Global food and nutrition security under changing climates. *Frontiers in Agronomy*, 3, 799878.

Ahmad, W., Watts, M. J., Imtiaz, M., Ahmed, I., & Zia, M. H. (2012). Zinc deficiency in soils, crops and humans. A review. *Agrochimica*. 2, 65–97.

Ali, M.A., Naeem, F., Tariq, N., Ahmed, I., & Imran, A. (2021). Bioactive nutrient fortified fertilizer: A novel hybrid approach for the enrichment of wheat grains with zinc. *Frontiers in Plant Science*, 12, 743378.

Ali, M., Ahmed, I., Tariq, H., Abbas, S., Zia, M. H., Mumtaz, A., & Sharif, M. (2023). Growth improvement of wheat (*Triticum aestivum*) and zinc biofortification using potent zinc-solubilizing bacteria. *Frontiers in Plant Science*, 14, 1140454.

Alloway, B. J. (2008). *Zinc in soils and crop nutrition*, 2nd edition. Published by IZA and IFA Brussels, Belgium and Paris, France.

Ansari, N., Mehmood, R., Gazdar, H. (2018). *Going against the grain of optimism: Flour fortification in Pakistan*. The Institute of Development Studies and Partner Organizations. Report. https://hdl.handle.net/20.500.12413/13517

Arora, N. K. (2019). Impact of climate change on agriculture production and its sustainable solutions. *Environmental Sustainability*, 2, 95–96.

Asadi, K. A., & Akhlaghi A. N. (2020). Evaluation of growth rate and vegetative and physiological characteristics of Satsuma mandarin on C-35 rootstock in some calcareous soils. *Iranian Journal of Soil Research*, 34(2), 215–233.

Asif, M., Asad, M. S., Safdar, M. E., Khan, I., Akhtar, N., Hassan, M. A., ... & Ansari, M. J. (2024). Agronomic bio-fortification of zinc improves the yield and quality of fodder oat. *Journal of Ecological Engineering*, 25(6), 153–163.

Baranova, A., Oldeland, J., Wang, S.-L., & Schickhoff, U. (2019). Grazing impact on forage quality and macronutrient content of rangelands in Qilian Mountains, NW China. *Journal of Mountain Science*, 16 (1), 43–53.

Beig, B., Niazi, M. B. K., Jahan, Z., Haider, G., Zia, M., Shah, G. A., ... & Hayat, A. (2023). Development and testing of zinc sulfate and zinc oxide nanoparticle-coated urea fertilizer to improve N and Zn use efficiency. *Frontiers in Plant Science*, 13, 1058219.

Beig, B., Niazi, M. B. K., Jahan, Z., Zia, M., Shah, G. A., Iqbal, Z., & Douna, I. (2022). Facile coating of micronutrient zinc for slow-release urea and its agronomic effects on field-grown wheat (*Triticum aestivum* L.). *Science of the Total Environment*, 838, 155965.

Beig, B., Niazi, M. B. K., Ullah, B., Gondal, A. N., Jahan, Z., Zia, M., & Ahmad, N. (2024). Effect of zinc oxide and zinc oxide nanoparticles coating on urea diffusion and its release kinetics for design and development of slow-release fertilizer: An experimental and numerical investigation. *Journal of Coatings Technology and Research*, 21(1), 199–213.

Berkhout, E. D., Malan, M., & Kram, T. (2019). Better soils for healthier lives? An econometric assessment of the link between soil nutrients and malnutrition in Sub-Saharan Africa. *PloS One*, 14(1), e0210642.

Blair, G. J., Ivory, D. A., & Evans, T. R. (1986). *Forages in Southeast Asian and South Pacific agriculture. Proceedings of an international workshop held at Cisarua*, p. 202. ACIAR Proceedings Series No. 12, Indonesia, 19–23 August 1985.

Bolan, N., Srinivasarao, C., Rocco, C., Bolan, S., Mansoor, S., Wani, O. A., ... & Siddique, K. H. (2025). Zinc in soil-crop-animal-human health continuum. *Advances in Agronomy*, 189, 1–61.

Botoman, L., Chagumaira, C., Mossa, A. W., Amede, T., Ander, E. L., Bailey, E. H., ... & Nalivata, P. C. (2022). Soil and landscape factors influence geospatial variation in maize grain zinc concentration in Malawi. *Scientific Reports*, 12(1), 7986.

Botoman, L., Nalivata, P. C., Chimungu, J. G., Munthali, M. W., Bailey, E. H., Ander, E. L., ... & Broadley, M. R. (2020). Increasing zinc concentration in maize grown under contrasting soil types in Malawi through agronomic biofortification: Trial protocol for a field experiment to detect small effect sizes. *Plant Direct*, 4(10), e00277.

Cakmak I. (2008). Enrichment of cereal grains with zinc: Agronomic or genetic biofortification? *Plant and Soil*, 302(1–2), 1–7.

Canales, C., & Fears, R. (2023). The role of science, technology, and innovation for transforming food systems in Europe. In J. von Braun, K. Afsana, L. O. Fresco, M. H. A. Hassan (Eds.) *Science and innovations for food systems transformation* (pp. 763–777). Springer International Publishing, Cham.

Caulfield, L. E., & Black, R. E. (2004). Zinc deficiency. In: M. Ezzati, A. D. Lopez, A. Rodgers, C. J. L. Murray (Eds.) *Comparative quantification of health risks: Global and regional burden of disease attributable to selected major risk factors*, Volume 1 (pp. 257–279). World Health Organization, Geneva.

Chaffee, B. W., & King, J. C. (2012). Effect of zinc supplementation on pregnancy and infant outcomes: A systematic review. *Paediatric and Perinatal Epidemiology*, 26, 118–137.

Chasapis, C. T., Ntoupa, P. S. A., Spiliopoulou, C. A., & Stefanidou, M. E. (2020). Recent aspects of the effects of zinc on human health. *Archives of Toxicology*, 94, 1443–1460.

Chen, Fu, Saqlain, L., Ma, J. et al. (2022). Evaluation of potential ecological risk and prediction of zinc accumulation and its transfer in soil plants and ruminants: public health implications. *Environmental Science and Pollution Research*, 29, 3386–3393. https://doi.org/10.1007/s11356-021-15821-2

Chianu, J. N., Chianu, J. N., & Mairura, F. (2012). Mineral fertilizers in the farming systems of Sub-Saharan Africa. A review. *Agronomy for Sustainable Development*, 32, 545–566.

Chilimba, A. D. C. et al. (2011). Maize grain and soil surveys reveal suboptimal dietary selenium intake is widespread in Malawi. *Scientific Reports*, 1, 72.

Danaei, G. et al. (2016). Risk factors for childhood stunting in 137 developing countries: A comparative risk assessment analysis at global, regional, and country levels. *PLoS Medicine*, 13(11), e1002164.

De Groote, H., Tessema, M., Gameda, S., & Gunaratna, N. S. (2021). Soil zinc, serum zinc, and the potential for agronomic biofortification to reduce human zinc deficiency in Ethiopia. *Scientific Reports*, 11(1), 8770.

De Valenca, A., Bake, A., Brouwer, I. & Giller, K. (2017). Agronomic biofortification of crops to fight hidden hunger in Sub-Saharan Africa. *Global Food Security*, 12, 8–14.

Degui, N., Djili, K., Daoud, Y., & Belghemmaz, S. (2024). Speciation of Zn in arid soils with saline accumulations. Case study of the Biskra region, Algeria. *Tropical Ecology*, 42, e1877. https://doi.org/10.28940/terra.v42i0.1877

Devarshi, P. P., Mao, Q., Grant, R. W., & Hazels Mitmesser, S. (2024). Comparative absorption and bioavailability of various chemical forms of Zinc in humans: A narrative review. *Nutrients*, 16(24), 4269.

Dhaliwal, S. S., Ram, H., Shukla, A. K., & Mavi, G. S. (2019). Zinc biofortification of bread wheat, triticale, and durum wheat cultivars by foliar zinc fertilization. *Journal of Plant Nutrition*, 42(8), 813–822.

Dhaliwal, S. S., Sharma, V., Kumar Shukla, A., Singh Shivay, Y., Hossain, A., Verma, V., . . . & Singh, P. (2023). Agronomic biofortification of forage crops with zinc and copper for enhancing nutritive potential: A systematic review. *Journal of the Science of Food and Agriculture*, 103(4), 1631–1643.

Dimkpa, C. O., Campos, M. G., Fugice, J., Glass, K., Ozcan, A., Huang, Z., . . . & Santra, S. (2022). Synthesis and characterization of novel dual-capped Zn–urea nanofertilizers and application in nutrient delivery in wheat. *Environmental Science: Advances*, 1(1), 47–58.

Elliott, H., Woods, P., Green, B. D., & Nugent, A. P. (2022). Can sprouting reduce phytate and improve the nutritional composition and nutrient bioaccessibility in cereals and legumes?. *Nutrition Bulletin*, 47(2), 138–156.

Fan, M. S., Zhao, F. J., Fairweather-Tait, S. J., Poulton, P. R., Dunham, S. J., & McGrath, S. P. (2008). Evidence of decreasing mineral density in wheat grain over the last 160 years. *Journal of Trace Elements in Medical Biology*, 22(4), 315–324.

Fang, Y., Wang, L., Xin, Z., Zhao, L., An, X., & Hu, Q. (2008). Effect of foliar application of zinc, selenium, and iron fertilizers on nutrients concentration and yield of rice grain in China. *Journal of Agriculture and Food Chemistry*, 56, 2079–2084.

Food and Agriculture Organization of the United Nations. (2017a). *Soil fertility atlas of Pakistan: The Punjab province*. In W. Ahmad, Y. Niino, M. H. Zia, K. Mahmood, A. Ashraf, N. Ahmad, M. Salim, & M. A. Shakir (Eds.). Islamabad, Pakistan. https://ffc.com.pk/marketing-group/kashtkar-desk/soil-fertility-atlas-of-pakistan-punjab/

Food and Agriculture Organization of the United Nations. (2017b). *Soil fertility atlas of Pakistan: The Sindh Province*. In W. Ahmad, Y. Niino, M. H. Zia, K. Mahmood, A. Ashraf, K. S. Memon, Z. H. Shah, M. Aslam, N. Ahmad, M. Salim, & M. A. Shakir (Eds.). Islamabad, Pakistan. https://ffc.com.pk/marketing-group/kashtkar-desk/soil-fertility-atlas-of-pakistan-sindh/

Frossard, E., Bucher, M., Mächler, F., Mozafar, A., & Hurrell, R. (2000). Potential for increasing the content and bioavailability of Fe, Zn and Ca in plants for human nutrition. *Journal of the Science of Food and Agriculture*, *80*(7), 861–879.

Gashu, D., Nalivata P. C., Amede, T., Ander, E. L., Bailey, E. H. et al. (2021). The nutritional quality of cereals varies geospatially in Ethiopia and Malawi. *Nature*, 594(7861),71–76.

Gibson, R. S. (2012). A historical review of progress in the assessment of dietary zinc intake as an indicator of population zinc status. *Advances in Nutrition*, 3(6), 772–782.

Global Hunger Index. (2024). How gender justice can advance climate resilience and zero hunger. *Concern Worldwide, Welthungerhilfe 2024*. https://www.globalhungerindex.org

Godfray, H. C. J., & Garnett, T. (2014). Food security and sustainable intensification. *Philosophical Transactions of the Royal Society of London*, 369(1639), 20120273.

Greene, L. W. (2000). Designing mineral supplementation of forage programs for beef cattle. *Journal of Animal Science*, 77(1), 10–2527.

Gregory, P. J., Wahbi, A., Adu-Gyamfi, J., Heiling, M., Gruber, R., Joy, E. J. et al. (2017). Approaches to reduce zinc and iron deficits in food systems. *Global Food Security*, 15, 1–10. doi:10.1016/j.gfs.2017.03.003. https://crops.zinc.org/china/?form=MG0AV3

Gupta, S., Brazier, A. K. M., & Lowe, N. M. (2020). Zinc deficiency in low-and middle-income countries: Prevalence and approaches for mitigation. *Journal of Human Nutrition and Dietetics*, 33(5), 624–643.

HarvestPlus. (2024). 20-years of impact for a food systems transformation. 2023 Annual Report. Accessible https://www.HarvestPlus.org. (Accessed on February 23, 2025).

Irewale, A. T., Dimkpa, C. O., Agunbiade, F. O., Oyetunde, O. A., Elemike, E. E., & Oguzie, E. E. (2024). Unlocking sustainable agricultural development in Africa via Bio-Nanofertilizer application-challenges, opportunities and prospects. *Scientific African*, 25, e02276.

Irfan, M., Khan Niazi, M. B., Hussain, A., Farooq, W., & Zia, M. H. (2018). Synthesis and characterization of zinc-coated urea fertilizer. *Journal of Plant Nutrition*, 41(13), 1625–1635.

Ishfaq, M., Wakeel, A., Shahzad, M. N. et al. (2021). Severity of zinc and iron malnutrition linked to low intake through a staple crop: A case study in east-central Pakistan. *Environmental Geochemistry and Health* 43, 4219–4233.

James, A. S. (2023). Effects of supplementing sources of zinc on digestibility parameters of beef steers (Masters Thesis, Mississippi State University). Mississippi State University Digital Commons. https://scholarsjunction.msstate.edu/td/6039

Johns, T., & Eyzaguirre, P. B. (2007). Biofortification, biodiversity and diet: A search for complementary applications against poverty and malnutrition. *Food Policy*, 32(1), 1–24.

Joy, E. J., Stein, A. J., Young, S. D., Ander, E. L., Watts, M. J., & Broadley, M. R. (2015). Zinc-enriched fertilizers as a potential public health intervention in Africa. *Plant and Soil*, 389, 1–24.

Joy, E. J. M., Ahmad, W., Zia, M. H., Kumssa, D. B., Young, S. D., Ander, E. L., Watts, M. J., Stein, A. J., and Broadley, M. R. (2017). Valuing increased zinc (Zn) fertilizer use in Pakistan. *Plant Soil*, 411(1), 139–150.

Kabata-Pendias, A. (2004). Soil–plant transfer of trace elements—an environmental issue. *Geoderma*, 122, 143–149.

Kabata-Pendias, A., & Pendias, H. (2001). *Trace elements in soils and plants*, 3rd edition. CRC Press, Boca Raton, FL.

Kanwal, S., Rahmatullah, A. R., & Ahmad, R. (2010). Zinc partitioning in maize grain after soil fertilization with zinc sulfate. *International Journal of Agriculture & Biology*, 12(2), 299–302.

Katimba, H. A., Wang, R., Cheng, C., Zhang, Y., Lu, W., & Ma, Y. (2024). Zinc absorption & homeostasis in the human body: A general overview. *Food Reviews International*, 40(2), 715–739.

Katyal, J.C. & Sharma, B.D. (1991). DTPA-extractable and total Zn, Cu, Mn, and Fe in Indian soils and their association with some soil properties. *Geoderma*, 49:165–179.

Kaur, K., Gupta, R., Saraf, S. A., & Saraf, S. K. (2014). Zinc: The metal of life. *Comprehensive Reviews in Food Science & Food Safety*, 13, 358–376

Khan, A., Waqas, M., Ali, B., Adnan, M., Zahir, Z. A., & Rizvi, R. (2021). Foliar application of zinc oxide nanoparticles alleviates cadmium stress and improves growth and yield of wheat. *Plant Physiology & Biochemistry*, 166, 109–119.

Khan, S. T., Malik, A., Alwarthan, A., & Shaik, M. R. (2022). The enormity of the zinc deficiency problem and available solutions; an overview. *Arabian Journal of Chemistry*, 15(3), 103668.

Khan, Z., Ahmad, K., Safdar, H., Uğulu, İ., Wajid, K., Nadeem, M., . . . & Doğan, Y. (2020). Monitoring of zinc profile of forages irrigated with city effluent. *Pakistan Journal of Analytical & Environmental Chemistry*, 21(2), 303–313.

Khan, Z. I., Ashraf, M., & Valeem, E. E. (2006a). Forage mineral status evaluation: The influence of pastures. *Pakistan Journal of Botany*, 38(4), 1043.

Khan, Z. I., Hussain, A., Ashraf, M., & McDowell, L. R. (2006b). Mineral status of soils and forages in Southwestern Punjab-Pakistan: Micro-minerals. *Asian-Australasian Journal of Animal Sciences*, 19(8), 1139–1147.

Khokhar, J. S., Broadley, M. R., & Ander, E. L. (2024). Soil zinc surveillance frameworks can inform human nutrition studies: Opportunities in India. *Frontiers in Soil Science*, 4, 1421652.

Kihara, J., Bolo, P., Kinyua, M., Rurinda, J., & Piikki, K. (2020). Micronutrient deficiencies in African soils and the human nutritional nexus: Opportunities with staple crops. *Environmental Geochemistry & Health*, 42(9), 3015–3033.

Kopittke, P. M., Menzies, N. W., Wang, P., McKenna, B. A., & Lombi, E. (2019). Soil and the intensification of agriculture for global food security. *Environment International*, 132, 105078.

Kumar, B., Ram, H., & Schoenau, J. (2024). Agronomic biofortification of fodder maize (Zea mays L.) with Zn for improving herbage productivity and its quality. *Agronomy*, 14(5), 912.

Kumssa, D. B., Joy, E. J., Ander, E. L., Watts, M. J., Young, S. D., Walker, S., & Broadley, M. R. (2015). Dietary calcium and zinc deficiency risks are decreasing but remain prevalent. *Scientific Reports*, 5(1), 10974.

Lal, R. (2018). Managing agricultural soils of Pakistan for food and climate. *Soil & Environment*, 37(1),1–10. doi:10.25252/SE/18/61527

Lin, P. H., Sermersheim, M., Li, H., Lee, P. H., Steinberg, S. M., & Ma, J. (2017). Zinc in wound healing modulation. *Nutrients*, 10(1), 16.

Lisuma, J. B., Semoka, J. M. R., & Semu, E. (2006). Maize yield response and nutrient uptake after micronutrient application on a volcanic soil. *Agronomy Journal*, 98(2), 402–406.

Liu, H., Gan, W., Rengel, Z., & Zhao, P. (2016). Effects of zinc fertilizer rate and application method on photosynthetic characteristics and grain yield of summer maize. *Journal of Soil Science & Plant Nutrition*, 16(2), 550–562.

Liu, T., Hou, G. G., Cardin, M., Marquart, L. & Dubat, A. (2017). Quality attributes of whole-wheat flour tortillas with sprouted whole-wheat flour substitution. *LWT-Food Science & Technology*, 77, 1–7.

Lönnerdal, B. (2000). Dietary factors influencing zinc absorption. *The Journal of Nutrition*, 130(5S), 1378S–1383S.

Lowder, S. K., Sánchez, M. V. & Bertini, R. (2019). *Farms, family farms, farmland distribution and farm labour: What do we know?* FAO Agricultural Development Economics Working Paper 19-08. Rome, FAO.

Lowder, S. K., Skoet, J., & Raney, T. (2016). The number, size, and distribution of farms, smallholder farms, and family farms worldwide. *World Development*, 87, 16–29.

Lowe, N. M., Khan, M. J., Broadley, M. R. et al (2018). Examining the effectiveness of consuming flour made from agronomically biofortified wheat (Zincol-2016/NR-421) for improving Zn status in women in a low-resource setting in Pakistan: Study protocol for a randomised, double-blind, controlled crossover trial (BiZiFED). *British Medical Journal Open*, e021364. doi:10.1136/bmjopen-2017-021364

Lowe, N. M., Zaman, M., Khan, M. J., Brazier, A. K., Shahzad, B., Ullah, U., . . . & Tishkovskaya, S. (2022). Biofortified wheat increases dietary zinc intake: A randomised controlled efficacy study of zincol-2016 in rural Pakistan. *Frontiers in Nutrition*, 8, 809783.

Lowe, N. M., Zaman, M., Moran, V. H., Ohly, H., Sinclair, J., Fatima, S., . . . & Khan, M. J. (2020). Biofortification of wheat with zinc for eliminating deficiency in Pakistan: Study protocol for a cluster-randomised, double-blind, controlled effectiveness study (BIZIFED2). *British Medical Journal Open*, 10(11), e039231.

Manzeke, G. M., Mtambanengwe, F., Watts, M. J., Hamilton, E. M., Lark, R. M., Broadley, M. R., & Mapfumo, P. (2019). Fertilizer management and soil type influence grain zinc and iron concentration under contrasting smallholder cropping systems in Zimbabwe. *Scientific Reports*, 9, 6445. doi:10.1038/s41598-019-42828-0

Manzeke, M. G., Mtambanengwe, F., Nezomba, H., Watts, M. J., Broadley, M. R., & Mapfumo, P. (2017). Zinc fertilization increases productivity and grain nutritional quality of cowpea (*Vigna unguiculata* L. Walp.) under integrated soil fertility management. *Field Crops Research*, 213, 231–244.

Manzeke-Kangara, M. G., Joy, E. J. M., Mtambanengwe, F. et al. (2021). Good soil management can reduce dietary zinc deficiency in Zimbabwe. *CABI Agriculture & Bioscience*, 2(1), 1–12.

Martínez-Ríos, O., Bravo-Vinaja, Á., San-Martín-Hernández, C., Hidalgo-Moreno, C. I., Sánchez-de-Jesús, M. A., Llampallas-Díaz, J. D., . . . & García-Preciado, J. C. (2024). Zinc deficiency in calcareous soils: A bibliometric analysis from 1989 to 2024. *Agriculture*, 14(12), 2285.

Mohsin, A. U., Ahmad, A. U. H., Farooq, M., and Ullah, S. (2014). Influence of zinc application through seed treatment and foliar spray on growth, productivity and grain quality of hybrid maize. *Journal of Animal & Plant Sciences*, 24, 1494–1503.

Natasha, N., Shahid, M., Bibi, I., Iqbal, J., Khalid, S., Murtaza, B., . . . & Arshad, M. (2022). Zinc in soil-plant-human system: A data-analysis review. *Science of the Total Environment*, 808, 152024.

NFDC (National Fertilizer Development Centre). (2018a). In W. Ahmad, M. J. Khan, Z. Mustafa, K. Mahmood, B. Singh, A. Ali, M. H. Zia, & A. J. Marwat (Eds.) *Soil fertility and inputs use atlas of Pakistan: The Balochistan Province*, (p. 120). ISBN 978-969-7888-01-6. Islamabad, Pakistan.

NFDC (National Fertilizer Development Centre). (2018b). In W. Ahmad, M. J. Khan, Z. Mustafa, K. Mahmood, B. Singh, M. H. Zia, A. Ali, & A. Jalil (Eds.) *Soil fertility and inputs use atlas of Pakistan: The Khyber Pakhtunkhwa Province* (p. 115). ISBN 978-969-7888-01-6. Islamabad, Pakistan. https://ffc.com.pk/marketing-group/kashtkar-desk/soil-fertility-atlas-of-pakistan-kpk/

Nriagu, J. O. (2019). *Encyclopedia of environmental health*. Elsevier.

Ofori, K. F., Antoniello, S., English, M. M., & Aryee, A. N. (2022). Improving nutrition through biofortification–A systematic review. *Frontiers in Nutrition*, 9, 1043655.

Phattarakul, N., Rerkasem B, Li LJ, Wu LH, Zou CQ, Ram H, Kalayci M, & Yazici, A. (2012). Biofortification of rice grain with zinc through zinc fertilization in different countries. *Plant & Soil*, 361(1–2), 131–141.

Podewils, LJ., Mintz ED, Nataro JP, & Parashar, UD. (2004). Acute, infectious diarrhea among children in developing countries. *Seminars in Pediatric Infectious Diseases*., 15(3), 155–168.

Qian, L., Dawar, K., Ullah, I., Irfan, M., Zhang, Z., Mian, I. A., . . . & Alarfaj, A. A. (2023). Zinc foliar application mitigates cadmium-induced growth inhibition and enhances wheat growth, chlorophyll contents, and yield. *ACS Omega*, 8(36), 32372–32381.

Rafique, E., Rashid, A., Ryan, J., and Bhatti, A. U. (2006). Zinc deficiency in rainfed wheat in Pakistan: Magnitude, spatial variability, management, and plant analysis diagnostic norms. *Communications in Soil Science & Plant Analysis*, 37, 181–197.

Rashid, A., Zia, M., & Ahmad, W. (2022). *Micronutrient fertilizer use in Pakistan: Historical perspective and 4R nutrient stewardship*, 1st edition (p. 340). CRC Press (Taylor & Francis Group), Boca Raton, Florida. eBook ISBN-9781003314226.

Ray, D. K., Gerber, J. S., Macdonald, G. K., & West, P. C. (2015). Climate variation explains a third of global crop yield variability. *Nature Communication*, 6(1), 1–9.

Raza, S., Zhou, J., Aziz, T., Afzal, M. R., Ahmed, M., Javaid, S., & Chen, Z. (2018). Piling up reactive nitrogen and declining nitrogen use efficiency in Pakistan: A challenge not challenged (1961–2013). *Environmental Research Letters*, 13(3), 034012.

Raza, S., Zia-ur-Rehman, M., Alghamdi, S. A., Alghanem, S. M. S., Usman, M., Ahmed, R., . . . & Al-Haithloul, H. A. S. (2023). Effects of zinc-enriched amino acids on rice plants (Oryza sativa L.) for adaptation in saline-sodic soil conditions: Growth, nutrient uptake and biofortification of zinc. *South African Journal of Botany*, 162, 370–380.

Ricciardi, V., Mehrabi, Z., Wittman, H., James, D., & Ramankutty, N. (2021). Higher yields and more biodiversity on smaller farms. *Nature Sustainability*, 4(7), 651–657.

Ritchie, H. (2021). Smallholders produce one-third of the world's food, less than half of what many headlines claim. Published online at https://OurWorldInData.org. Accessed on February 26, 2025.

Roohani N, Hurrell R, Kelishadi R, & Schulin, R. (2013). Zinc and its importance for human health: An integrative review. *Journal of Research in Medical Sciences*, 18(2),144–157

Rudani, K., Vishal, P., & Kalavati, P. (2018). The importance of zinc in plant growth-A review. *International Research Journal of Natural & Applied Science*, 5(2), 38–48.

Sanginga, N., & Woomer, PL. (2009). *Integrated soil fertility management in Africa: Principles, practices and developmental process* (p. 263). Tropical Soil Biology and Fertility Institute of the International Centre for Tropical Agriculture, Nairobi.

Sazawal, S., Dhingra, U., Dhingra, P., Dutta, A., Deb, S., Kumar, J., Devi, P., & Prakash, A. (2018). Efficacy of high zinc biofortified wheat in improvement of micronutrient status, and prevention of morbidity among preschool children and women — a double masked, randomized, controlled trial. *Nutrition Journal*, 17, Article 86. https://doi.org/10.1186/s12937-018-0391-5

Sharma, V., Maurya, A., Kandhol, N., Singh, V. P., Sharma, S., Peralta-Videa, J., & Tripathi, D. K. (2025). Zinc deficiency in plants: An insight into fortification strategies. In *Zinc in plants* (pp. 249–264). Academic Press.

Sher, A., Ul-Allah, S., Sattar, A., Ijaz, M., Qayyum, A., Manaf, A., & Suleman, M. (2022). Zinc sulfate application to grass forages (oat, barley, annual ryegrass and triticale) for increasing their yield, quality and profitability. *Crop & Pasture Science*, 73(5), 473–483.

Shiferaw, H. (2014). *Tigray region's soil fertility status and new fertilizer recommendation.* Paper presented at a seminar at ATA, October 24, 2014. The Ethiopia Strategy Support Program and the Ethiopian Development Research Institute, Addis Ababa.

Shivay, Y. S., Prasad, R., & Pal, M. (2013). Zinc fortification of oat grains through zinc fertilization. *Agricultural Research*, *2*(4), 375–381.

Simmonds, N. W. (1995). The relation between yield and protein in cereal grain. *Journal of the Science of Food & Agriculture*, 67(3), 309–315.

Singh, B. & Ahmad, W. (2018). Micronutrients in soils and human health. *3rd Global Soil Security Conference*, Darlington, NSW Australia, 4–6 December, 2018.

Singh, M., Goswami, S. P., Sachan, P., Sahu, D. K., Beese, S., & Pandey, S. K. (2024). Nanotech for fertilizers and nutrients-improving nutrient use efficiency with nano-enabled fertilizers. *Journal of Experimental Agriculture International*, 46(5), 220–247.

Singh, R., & G. Velu. (2017). *Zinc-biofortified wheat: Harnessing genetic diversity for improved nutritional quality.* Science Brief: Biofortification No. 1 (May 2017). CIMMYT, HarvestPlus, and the Global Crop Diversity Trust. Bonn, Germany.

Sommer, A. L. & Lipman, C. B. (1926). Evidence on the indispensable nature of zinc and boron for higher green plants. *Plant Physiology*, 1, 231–249.

Stür, W. W., Horne, P. M., Gabunada Jr, F. A., Phengsavanh, P., & Kerridge, P. C. (2002). Forage options for smallholder crop–animal systems in Southeast Asia: Working with farmers to find solutions. *Agricultural Systems*, 71(1–2), 75–98.

Supasai, S., Aimo, L., Adamo, A. M., Mackenzie, G. G. & Oteiza, P. I. (2017). Zinc deficiency affects the STAT1/3 signaling pathways in part through redox-mediated mechanisms. *Redox Biology*, 11, 469–481. doi:10.1016/j.redox.2016.12.027

Takkar (1996). Micronutrient research and sustainable productivity in India. *Journal of Indian Society of Soil Science*, 44, 562–581.

Tandon, H. L. S. (2013). Micronutrient Handbook –from research to application. *Fertilizer development and consultation organization* (p. 234), New Delhi, India.

Tessema, M., De Groote, H., Brouwer, I. D., Feskens, E. J. M., Belachew, T., Zerfu, D., Belay, A., Demelash, Y., & Gunaratna, N. S. (2019). Soil zinc is associated with serum zinc but not with linear growth of children in Ethiopia. *Nutrients*, 11, 221.

Touch, V., Tan, D. K., Cook, B. R., Li Liu, D., Cross, R., Tran, T. A., . . . & Cowie, A. (2024). Smallholder farmers' challenges and opportunities: Implications for agricultural production, environment and food security. *Journal of Environmental Management*, 370, 122536.

UNICEF. (2020). Global nutrition report. Retrieved October 12, 2021, from https://globalnutritionreport.org/reports/2020-global-nutrition-report/

Vanlauwe, B., Descheemaeker, K., Giller, KE., Huising J, Merckx R, Nziguheba G et al. (2015). Integrated soil fertility management in Sub-Saharan Africa: Unravelling local adaptation. *Soil*, 1, 491–508.

Walker, C. L. F., Ezzati, M. & Black, R. E. (2009). Global and regional child mortality and burden of disease attributable to zinc deficiency. *European Journal of Clinical Nutrition*, 63, 591–597.

Wessells, K. R., & Brown, K. H. (2012). Estimating the global prevalence of zinc deficiency: Results based on zinc availability in national food supplies and the prevalence of stunting. *PloS One*, *7*(11), e50568.

Wessells, K. R., Singh, G. M., & Brown, K. H. (2012). Estimating the global prevalence of inadequate zinc intake from national Food Balance Sheets: Effects of methodological assumptions. *PLoS One* 7: e50565. doi:10.1371/journal.pone.0050565

White, C.L. (1993). The zinc requirements of grazing ruminants. In A. D. Robson (Eds.) *Zinc in soils and plants. Developments in plant and soil sciences*, Volume 55. Springer, Dordrecht. https://doi.org/10.1007/978-94-011-0878-2_14

WHO e-LENA (Electronic Library on Evidence for Nutrition Actions). (2023). *Zinc supplementation and growth in children.* World Health Organization. https://www.who.int/tools/elena/interventions/zinc-stunting (Accessed on December 23, 2025).

Wilhelm, N. (2016). Detecting and managing trace element deficiencies in crops – Grains Research and Development Corporation. *Grains Research and Development Corporation.* https://grdc.com.au/resources-and-publications/grdc-update-papers/tab-content/grdc-update-papers/2016/02/detecting-and-managing-trace-element-deficiencies-in-crops?form=MG0AV3

Wortmann, C. S., Kaizzi, K. C., Maman, N., Cyamweshi, A., Dicko, M., Garba, M., Milner, M., Senkoro, C., Tarfa, B., Tettah, F., & Kibunja, C. (2019a). Diagnosis of crop secondary and micronutrient deficiencies in Sub-Saharan Africa. *Nutrient Cycling in Agroecosystems*, 113(2), 127–140. doi:10.1007/s10705-018-09968-7

Wortmann, C. S., Kaizzi, K. C., Maman, N., Cyamweshi, A., Dicko, M., Garba, M., Milner, M., Senkoro, C., Tarfa, B., Tettah, F., Kibunja, C., Munthali, M., Nalivata, P., Nkonde, D., Nabahungu, L., Ouattara, K., Serme, I., & Wortmann, C. (2019b). Tropical Africa crop nutrient diagnosis data set. doi:10.5061/dryad.5t065bb/1

Yerokun, O. A., & Chirwa, M. (2014). Soil and foliar application of zinc to maize and wheat grown on a Zambian Alfisol. *African Journal of Agricultural Research*, 9(11), 963–970. https://doi.org/10.5897/AJAR2013.7623

Yilmaz, A., Ekiz, H., Torun, B., Gultekin, I., Karanlik, S., Bagci, S. A., & Cakmak, I. (1997). Effect of different zinc application methods on grain yield and zinc concentration in wheat cultivars grown on zinc-deficient calcareous soils. *Journal of Plant Nutrition*, 20(4–5), 461–471.

Yogi, A. K., Ram Swaroop, B., Bamboriya, S. D., Choudhary, R. L., Laing, A. M., Singh, D., Godara, S., Babu, S., & Chaudhary, A. (2023). Foliar zinc fertilization improves yield, biofortification and nutrient-use efficiency of upland rice. *Nutrient Cycling in Agroecosystems*, 125(3), 453–469. doi:10.1007/s10705-023-10270-4

Yoshida, S., & Tanaka, A. (1969). Zinc deficiency of the rice plant in calcareous soils. *Soil Science Plant Nutrition*, 15, 75–80. doi:10.1080/00380768.1969.10432783

Zia, M. H., Ahmed, I., Bailey, E. H., Lark, R. M., Young, S. D., Lowe, N. M., ... & Broadley, M. R. (2020). Site-specific factors influence the field performance of a Zn-biofortified wheat variety. *Frontiers in Sustainable Food Systems*, 4, 135.

Zia, M. H., Lodhi, A. H., Aziz, M. Z., Naseem, A., Ahmad, W., Lark, R. M., & Ander, E. L. (2021). Soil fertility mapping and agronomic advice at the regional scale using private sector data. *AgriRxiv*, 20210511543.

Zinc in Health – Zinc.org India [Internet]. (2020, March 23). Available from: http://zinc.org.in/programs-and-activities/zinc-in-health/

6 The Role of the Microbiome in Modulating Human Health

Dennis J. Grencewicz, Shiva Jahanbakhshi, Jessica Fulton, Caroline Wheeler, Julia Chalif, Laura Chambers, and Daniel Spakowicz

The microbiome is the collection of bacteria, fungi, archaea, and viruses that live on and in the human body. There are roughly equal numbers of microbes as human cells, and collectively, two orders of magnitude more gene functions than humans. When microbes are mentioned in relation to humans, the most common association is with the gut microbiome, which colloquially refers to all microorganisms living within the gastrointestinal tract of humans. The gut microbiome is most abundant by number, but microbes reside on every surface and membrane exposed to the environment.

Most research associating microbes with health and disease is completed either on a case-by-case basis of a particular microbe or by comparing the whole microbiome of healthy individuals to that of individuals with a specific illness. The latter case is computationally and experimentally challenging due to the network effects of groups of microbes and disease development. However, recent studies in preclinical models and humans have developed causal links to show that states of perturbation to the microbiome, commonly called dysbiosis, can predispose and cause individuals to develop an illness.

Early studies relating microbes to health and disease relied on microscopy and the cultivation of specific taxa associated with a disease state. As DNA-reading technologies advanced, the field shifted to molecular identification, most commonly reading part of the only gene known to be conserved across all bacteria: the ribosome. Further advances in DNA sequencing enabled entire communities to be counted at once. This led to an explosion of observations of the gut microbiomes in patients with a particular illness, which were different from patients of similar demographics without that illness. In addition, it became clear that many more microbes were present in the gut than could be readily cultivated. Advances in culturing techniques have led the field back to microbe cultivation, and thereby, the study of whether manipulating single or small numbers of organisms in a complex community causally contributes to the development or progression of a disease.

Within human populations, such investigations of changes to the microbiome in disease states have unveiled noteworthy variations in microbiota composition throughout the lifespan. Notably, common microbiota are seeded at birth and known to change at different stages of life and development, including disparities between pre-weaned infants, adults, and the frail elderly.[1,2] The microbes are affected by human interactions and behaviors such as diet and physical activity. For instance, the infant gut microbiome is seeded during its passage through the birth canal and directed by human milk oligosaccharides in breast milk to increase the abundance of the genus *Bifidobacterium*.[3] Conversely, limited dietary fiber intake may explain the alterations observed in the gut microbiota of frail elderly individuals.[1]

These findings support the hypothesis that the gut microbiome changes as individuals age, significantly affecting the immune system, particularly in regulating inflammatory responses and contributing to developing age-related diseases.[4,5] With age, the diversity of the gut microbiome tends to decrease, including reduced numbers of specific bacterial groups like *Bifidobacteria*,

DOI: 10.1201/9781003469742-6

Firmicutes, *Faecalibacterium prausnitzii*, *Clostridium* cluster XIV, and *Blautia coccoides-Eubacterium rectale*,[6] and increased *Bacteroides*, *Alistipes*, and *Parabacteroides*.[7] Others have used the ratio of pro- and anti-inflammatory taxa to describe age-related shifts, including the *Escherichia* to *Methanobrevibacter*, *Faecalibacterium* to *Akkermansia*, *Bacteroidetes* to *Firmicutes*, and *Bacteroides* to *Prevotella* ratios.[8] Similarly, exceptional longevity has been associated with high biodiversity, a significant representation of short-chain fatty acid (SCFA) producers, and a low representation of pathogens.[9,10] In particular, *Akkermansia* and *Bifidobacterium* are recognized as health-related genera that support immune regulation, prevent inflammation, and maintain healthy metabolic balance.[11,12] These age-related alterations in the gut microbiome are associated with changes in skeletal muscle and the brain, though studies to define causality for these complex relationships are ongoing.[13,14]

Outside of the tractable changes throughout life and the range of normal microbiota associated with any individual, there are specific states, diseases, and behaviors associated with changes to the microbiome. This chapter reviews how the human microbiome interacts with its host in health and disease, focusing on the dominant mechanisms of metabolism, the immune system, the nervous system, and cancer. Next, we describe how the microbiome can be modulated to affect human health, including by drugs, diet, exercise, and direct microbiome design. Finally, prospects for leveraging the microbiome to predict health and disease and modulating the microbiome to improve health are discussed. Ultimately, this chapter aims to clarify how microbes and humans interact and establish the important relationship that microbes play in our overall well-being.

THE ROLE OF THE MICROBIOME IN NORMAL PHYSIOLOGY AND MEDIATING DISEASE

Overview

Microbes, their metabolites, and their influence on the immune system have been associated with many human disease states. Often, these relationships are difficult to parse, given the multifactorial features of health and disease. The following section explains the relationship between the microbiome, normal physiological processes, and distinct disease states outside of infectious diseases (Figure 6.1). It is important to recognize the complex interplay between human physiology and the role of colonizing microbes, as many disease states involve microbes but are not solely driven by the presence or activity of microbes. However, like many multifaceted disease processes, the role of the microbiome in facilitating illness can function as a key foothold to initiate disease or play a role in disease progression.

Physiologic and Pathologic States Associated with the Microbiome

The following examples do not represent every known relationship between the microbiome and the human body in health and systemic disease. Still, they are provided as clear examples of how changes to the microbiome can elicit or exacerbate disease states.

Metabolism, Metabolic Disease, and the Microbiome

Microbes play an integral role in mammalian metabolism, given that the primary home for microbes in mammals is the gut. The small and large bowels are organs specifically tasked with the digestion of food and nutrient absorption, and microbes that live in the lumens of these organs help facilitate these processes. The environment within the gut is constantly fluctuating, with changes in function, water content, pH, and oxygen levels throughout, indicating that unique microbes can inhabit these niches and play particular roles in the breakdown of food to support themselves and the metabolism of their host. As described above, research has established a core microbiome in the human gut, yet significant intra- and inter-human variability exists between gut-colonizing microbes. Given this

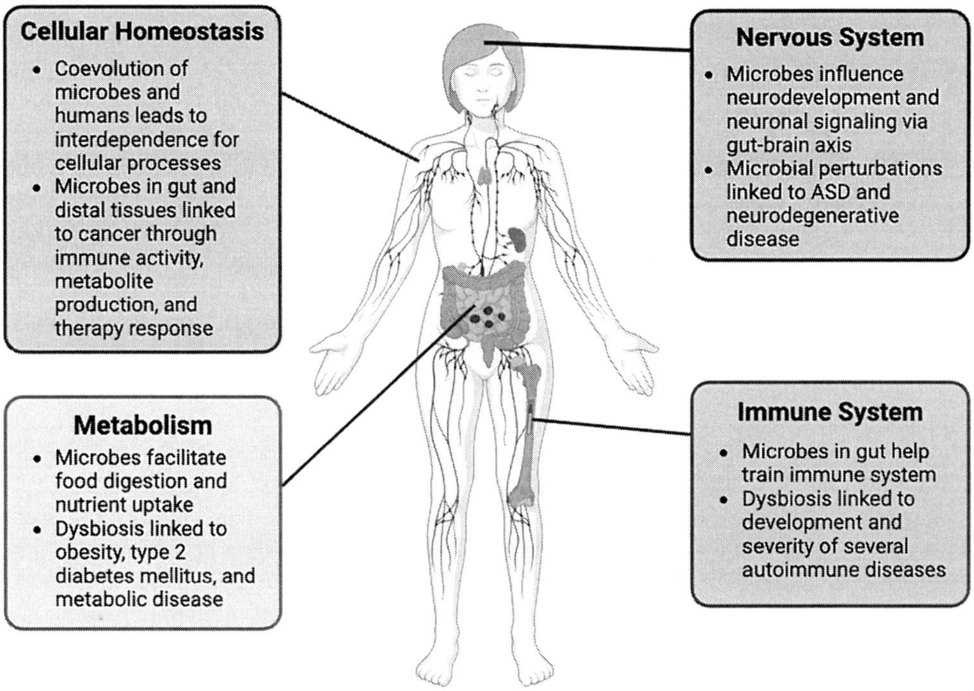

FIGURE 6.1 Overview of the mechanisms by which the microbiome interacts with humans, described through physiological systems, including metabolism, the immune system, the nervous system, and cellular homeostasis. Each interaction is associated with disease states.

variability in the microbiome, gut functions, including nutrient extraction, metabolic stability, and energy storage, are all subject to influence from gut microbes.

More recently, distinct connections have been established between the gut microbiome and metabolic disease. Determining the causal role of microbes in metabolic diseases and metabolic syndrome has been primarily achieved through germ-free (GF) preclinical models. A landmark study that provided causal evidence that microbes could impact energy storage in mice occurred in 2004. It showed that introducing gut microbes from healthy mice to GF mice led to increased body fat content and insulin resistance despite decreased eating patterns.[15] This work showed that microbes could directly influence liver function and increase triglyceride storage, increasing body fat content.

Subsequently, several types of metabolic diseases, including obesity and metabolic syndrome, have associations with the gut microbiome.[16] In a landmark study, GF mice received fecal microbiota transplantation (FMT) from four human twin pairs discordant for obesity.[17] After transplantation, GF mice receiving the microbiome from an obese twin grossly showed greater body mass and adiposity increases. These phenotypic changes were associated with decreased capacity to ferment short chain fatty acids (SCFAs), increased metabolism of branched-chain amino acids, and decreased bile acid metabolism. Upon cohousing of mice receiving FMT from obese or lean twins, the obese phenotype was ameliorated, showing that the lean microbiome was transferred within cages and preferential. Further studies have expanded on this work to identify obesity-associated and obesity-depleted microbial species and circulating metabolites.[18] The prevalence of a particular microbe, *Bacteroides thetaiotaomicron*, appears depleted in obese individuals, and gavage of *B. theta* into obese mice can reduce diet-associated adiposity increases.

Beyond obesity, metabolic syndrome features, including the disease type 2 diabetes mellitus (T2DM), have been tied to the microbiome. Metabolic syndrome is a grouping of risk factors that predispose individuals to increased risk of cardiovascular disease and all-cause mortality. The processes involved in metabolic syndrome are directly linked to the development of insulin resistance and, thus, the clinical diagnosis of T2DM. Patients with T2DM have also been shown to have abnormal microbiomes, with increased abundances of pathobionts and decreased butyrate-producing microbes identified in preclinical and clinical studies.[19] Further, it appears that microbes within T2DM patients produce fewer SCFAs, which have been shown to suppress fat accumulation and promote energy metabolism throughout the body.[20] Regarding individual microbes known to impact metabolic syndrome and T2DM, *Akkermansia muciniphila* has been shown to improve insulin sensitivity, reduce insulinemia, and affect plasma total cholesterol.[21]

Given the proximity of digestion to the home of most microbes in the human body, it is no surprise that microbes can affect nutrient availability, metabolism, and metabolic disease. The above examples of metabolic pathologies highlight that microbes are subject to selection in metabolic disease and, therefore, change as the illness progresses. They are also capable of driving illness or health based on host-microbe interactions and host signaling.

The Immune System, Autoimmune Disease, and the Microbiome

The human immune system has evolved incredibly complex methods of identifying and dispatching materials in the body that are found to be "non-self." Leveraging a more primitive innate immune system, pathogen-associated molecular patterns (PAMPs) from infiltrative microbes and damage-associated molecular patterns (DAMPs) from damaged host cells tip off innate immune cells to initiate cellular killing via the complement system. Simultaneously, the adaptive immune system builds bespoke responses to particular "non-self" markers called antigens. These include cellular and antibody-based responses from T and B cells, respectively. Autoimmune diseases arise when the immune system inaccurately assesses a normal, healthy body part as foreign. Notably, microbes throughout the body are all capable of causing an immune response due to the inherent properties of the molecular targets immune cells respond to. The human immune system is consistently sampling and killing microbes found in barrier organs, damaged tissues, and the bloodstream. This process allows for the prevention of widespread infection of the body and the training of the immune system to attack all potential invaders.

Subsequently, aberrant immune activation toward this tissue leads to a loss of function and the development of symptoms of hypofunction. Although the human immune system undergoes exquisite clonal deletion of cells that respond to "self-antigens," certain circumstances can lead to mistargeting of immune responses that affect healthy host tissues and cells.

Interestingly, a perturbed microbiome or gut barrier that is abnormally permeable has been linked to the development of autoimmunity. All microbes contain molecules recognized by immune cells as foreign and thus develop immune responses, a crucial defense mechanism against infections. Under certain circumstances, particular microbes can trigger the immune system to attack the invading microbe and nearby or distant host tissue. Several autoimmune diseases have been linked to the gut microbiome, including systemic lupus erythematosus (SLE). SLE typically affects women and can present with various symptoms in several organs, including the gastrointestinal tract, joints, lungs, and heart. Notably, a link between an altered gut microbiome in SLE patients was made in 2014, with a reported decrease in microbial diversity and enrichment of *Proteobacteria*, *Bacteroidetes*, and *Actinobacteria* in SLE patients.[22] Subsequently, mouse models of SLE have identified causal relationships between increases in pro-inflammatory microbes and increased gut permeability with worsened SLE symptoms.[23,24] Given the difficulty of managing the complex and lifelong symptoms associated with SLE, modulation of the microbiome as a treatment method would be advantageous compared to immunosuppressive medications, the current standard of care. Data are currently sparse on clinical interventions that target the microbiome to treat SLE, but early studies suggest that changes in diet may impact the presence of particular pro-inflammatory microbes, such as *Lactobacillus reuteri*, leading to improvements in symptoms.[25]

Beyond SLE, a perturbed microbiome has been linked to several other autoimmune diseases. These include but are not limited to rheumatoid arthritis,[26] type 1 diabetes,[27] and multiple sclerosis.[28] More research is needed to determine whether therapeutic strategies may be modified to target or manipulate the microbiome to treat these difficult illnesses. However, patterns of lower microbial diversity in the gut, increased mouth and gut permeability, and increased pathobiont abundance are consistently observed regardless of the specific autoimmune disease. Continuing to study the relationship between immune sampling and activity toward the microbiome may offer novel methods to assist the medical treatment of these chronic and debilitating diseases.

The Nervous System, Neurologic Diseases, and the Microbiome

The nervous system in mammals consists of a central nervous system (CNS), the brain and spinal cord, and a peripheral nervous system (PNS), including nerves for the rest of the body, such as spinal, cranial, and visceral nerves. In recent decades, research has established a strong relationship between the brain and PNS nerves in the gut, called the enteric nervous system. This relationship, known as the gut-brain axis, is driven by information disseminated through the enteric nervous system by interactions with the immune system, hormones, and the gut microbiome. Further, certain microbes can produce neurotransmitters and SCFAs, thus driving neuronal signaling.[29]

Bidirectional communication between the PNS and the gut microbiome appears crucial to normal physiological and neurologic development in mammals.[30] A meta-analysis of births by vaginal delivery versus Cesarean section identified a 23% increase in relative risk of a child developing an autism spectrum disorder (ASD).[31] Further, babies born by Cesarean section had altered gut microbiomes through early infancy.[32] Notably, the gut microbiome has also been linked to several specific neural processes, including the hypothalamic-pituitary-adrenal (HPA) axis[33,34] and the development of vagal tone.[35,36]

With the established relationship between neurodevelopment and the gut microbiome, perturbations to the gut microbiome have been linked to several disease states, including several diseases. Changes to the gut microbiome that affect the HPA axis have been linked to increased risk of anxiety and depression. A unique gut microbiome has also been established in Parkinson's disease patients. An increase in pathobionts, which are microbes associated with disease states in certain circumstances, as well as a relative decrease in SCFA-producing microbes, are observed in these patients. These results support the idea that changes in the gut-brain axis may be at play in the disease process or may signal that Parkinson's is progressing.

Given the difficulty of studying neurologic disease in humans, many studies thus far rely on preclinical models to establish causality in the functional relationships between microbes and the nervous system. This is an inherent weakness in the field, as neurologic diseases in mouse models remain difficult to correlate to human presentations, especially in the context of autism spectrum disorders and other behavioral and personality disorders.

Cellular Homeostasis, Cancer, and the Microbiome

Cancer is a disease of unfettered cell division and expansion, leading to the development of self-sufficient cells that maintain unique characteristics from their initial parent tissue of origin. To be considered cancerous, cells must accomplish several tasks, including hijacking cell cycle signaling to replicate independently and avoid cell death, adapting their microenvironment to create a pro-inflammatory and angiogenic niche, and developing capabilities to avoid the immune system.[37]

Communities of microbes and individual taxa are known to affect these processes, with the link between microbes and cancer therapy dating back to the late 19th century. In 1891, Dr. William Coley began treating aggressive soft-tissue sarcomas with injections of live *Streptococci* as an innovative type of immunotherapy, aptly named Coley's toxins.[38] Although this treatment was not universally successful, some patients with this otherwise fatal, inoperable disease were cured of their widespread malignancy. Touted as the first immunotherapy, microbes are necessary to elicit the immune response observed in these patients.

Although using injections of live pathogens and microbes as cancer therapy fell out of favor in the mid-1950s, more recent research has reestablished the link between microbes and cancer. Since the early 2000s, case-control studies have associated several classes of antimicrobial drugs, most commonly antibiotics, with both increasing or decreasing cancer risks in a disease-dependent manner.[39] For instance, in a case-control study that was matched for age, sex, body mass index (BMI), smoking and alcohol use, metabolic disease status, and nonsteroidal anti-inflammatory drug (NSAID) and aspirin use, antibiotic use was associated with an increased risk of primary colon cancer but a reduced risk of primary rectal cancer.[40]

Beyond associative studies, several microbes have been causally linked to cancer development. In epithelial cell-derived cancers, particularly colorectal cancer, *Fusobacterium nucleatum*, *Escherichia coli PKS+*, and *Bacteroides fragilis* can directly stimulate tumorigenesis through defined biochemical mechanisms.[41] microbes have recently been identified within tumors by culturing, sequencing, and staining methods, establishing an intratumoral microbiome.[42,43] Preclinical mouse models show that introducing different microbes can both causally increase immunotherapy response in certain conditions[44] and stimulate tumor progression and therapy resistance in others.[45] Microbial-derived metabolites have been directly linked to immune checkpoint inhibitor (ICI) response in mouse models of colorectal cancer, with mice colonized by inosine-producing microbes showing stronger responses to ICI drugs.[46] Ongoing research looks to establish whether preclinical findings of microbes in specific cancer types may predict therapy response or future disease progression in human patients.

Importantly, steps remain to determine the strength of the microbiome's influence on tumorigenesis and cancer progression. Further, it remains unclear how microbes colonize tumors and how to leverage these observations clinically.

Summary

As seen from the multisystem interactions between the microbiome and the body, microbes can be integral contributors to health and disease. Based on the example systems above, microbes commonly interact with the body in the following ways: (1) production of metabolites that influence host cellular functions; (2) direct and indirect interactions with the host immune system, such as immune activation, chronic inflammation, and antigen mimicry; and (3) generation of unique microenvironments that foster changes in host cellular microenvironments and host metabolism. At every tissue barrier, microbes are checked by the local immune system and the barrier itself. However, changes to these systems can act positively or negatively for the host, and these changes are consistently context- and microbe-specific. For instance, certain microbes can increase immune response in their host, which may be positive in the context of increasing response to immunotherapies for patients with cancer but can be negative in the progression of autoimmune disease.

MODULATING THE MICROBIOME

Changes to the microbiome have been effectively associated with several human disease states. Given the complexity of treatment for some of these illnesses, including medication side effects, research has begun to study how modulating the microbiome may affect disease development, progression, and treatment. Further, many current health interventions and daily activities have been linked to changes in the microbiome. This section includes methods to modulate the microbiome, including medications, diet, exercise, pre/pro/post/syn-biotics, and fecal microbiota transplants (Figure 6.2).

Medications and the Human Microbiome

Overview

The composition, population dynamics, and functions of microbial communities within the human body are constantly in flux. Perturbations in the microbiome, such as those induced by medications,

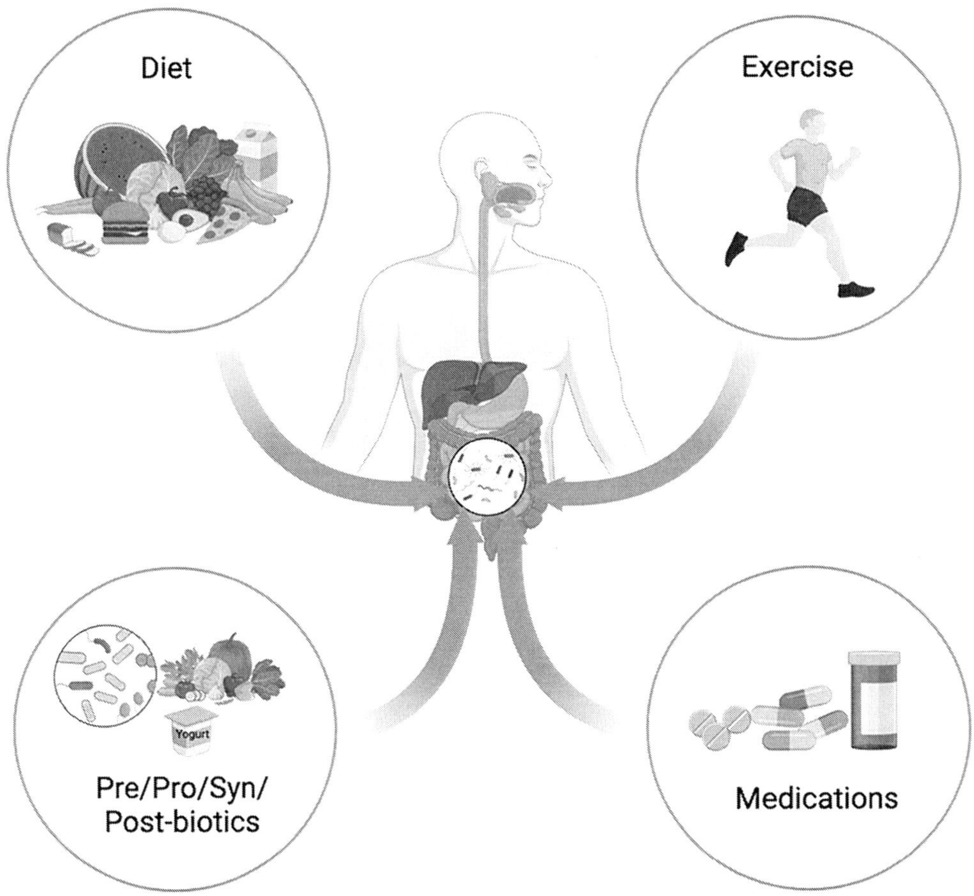

FIGURE 6.2 Methods to modify the microbiome.

topical treatments, and diverse therapies, form a fundamental aspect of modern medicine. Notably, in microbial pathogenesis, scientists have developed antimicrobial medications, including antibiotics and antifungals, to combat infections that directly threaten human health. However, these pharmaceuticals and a myriad of natural and synthetic chemicals can influence the presence, density, and diversity of specific domains within the human microbiome.

This section delves into microbiome modulation via medications, offering specific examples of how broad-spectrum and targeted therapies can selectively deplete particular microbial taxa. Furthermore, we explore the implications of antimicrobial medication usage, including how microbes repopulate common domains and reestablish communities. Additionally, we address the pressing issue of antibiotic resistance and its profound implications for the efficacy of existing antimicrobial medications.

Medications to Prevent and Cure Microbial Infections

Living organisms are susceptible to infection from microbes, which can subsequently lead to the death of the host organism if not addressed. Humans are not unique to this process, even though our immune system consistently works to eliminate all non-self organisms within the body. However, our immune systems are not perfect, and until the advent of antimicrobials in the 1900s, humans were susceptible to deadly infections from pathogenic microbes. The Black Death, a pandemic occurring from 1346 to 1353, was responsible for the death of about one-third of Europeans, killing anywhere between 75 and 200 million people. This pandemic was caused by the microbe *Yersinia pestis*, which is non-infectious in small mammals and rodent fleas. A member of the family *Enterobacteriaceae*, *Y. pestis* is a gram-negative, nonmotile, non-spore-forming, facultative anaerobic coccobacillus with several virulence factors that allow the microbe to invade eukaryotic cells, evade immune responses, and survive and grow in the bloodstream, causing the human disease called bubonic plague. The plague can cause fever, headache, chills, and tender lymph nodes in the first days of infection, followed by nausea, vomiting, painful skin lesions, bacteremia, septic shock, and death if untreated. However, due to the breakthrough development of medications against bacterial pathogens like *Y. pestis*, humans can recover and survive infections using antibiotics such as streptomycin and tetracycline. Although these medications are incredibly effective, overuse can lead to adverse outcomes such as endotoxic shock in humans or the development of microbial strains that evolve resistance to the medications. Therefore, proper stewardship of antimicrobial drug prescriptions is crucial to delivering effective healthcare that maintains our ability to fight pathogens in the future.

Antibiotics

First developed as modern medicines in 1910, antibiotics are considered one of the greatest scientific breakthroughs in human history. Antibiotics are usually defined as being bacteria-specific. They target bacteria to quell infections in the body, with differing levels of microbial specificity based on the microbial molecular targets the drugs antagonize. For instance, some classes of antibiotics can act against many types of bacteria, deemed broad-spectrum, versus drugs that target specific taxa, deemed narrow-spectrum. The first recorded use of microbes to treat disease dates back to ancient Egypt. The Ebers papyrus, which dates to around 1550BCE, was noted using special soil called ochre clay to treat intestinal and eye illnesses.

The utilization of antimicrobials in healthcare settings displays significant variability in terms of urgency for how quickly an infection needs to be resolved and accuracy regarding the identification of the causative microorganism in an infection. In cases where patients develop bacterial sepsis, a condition marked by a severe bloodstream infection and multiple organ failure, a common approach involves the administration of multiple broad-spectrum antibiotics. This administration is completed intravenously with the utmost urgency in an attempt to swiftly halt the progression of the infection. However, clinicians hope to avoid these scenarios by delivering antibiotics guided by first establishing cultures of the pathogen causing the infection. For patients with infections similar to bacterial endocarditis, which require weeks to months to clear, establishing which microbe is causing the infection before administering any antimicrobial medications is crucial. In the case of bacterial endocarditis, a potentially lethal infection of the heart valves, identifying the infective pathogen is accomplished by culturing a patient's blood for up to 72 hours and selecting an appropriate medication based on the identified microbes.[47] The development and implementation of antibiotics have also facilitated many other medical breakthroughs outside of infection prevention, including safer and more invasive procedures. Routine surgeries can now be completed with an understanding of sterile technique and without major risk of infection due to the implementation of post-operative antimicrobials.

To treat the diverse range of diseases that pathogenic bacteria can cause, many antibiotic medications have been developed. Given the variability in pathogen characteristics, including whether the microbes are gram-positive or negative, whether they contain certain enzymes, or if they use specific methods to infect cells, a diverse array of antibiotic classes are currently in use. Additionally, several drug delivery techniques have been developed for antibiotics given the indication, including pills, creams, and intravenous injections.

Antifungals

Beyond having an established bacterial microbiome, the human body harbors many fungi on all epidermal and mucosal surfaces, including skin and throughout our gastrointestinal tract. Although, similar to bacteria, most fungi are commensal to humans, several fungal strains can become pathogenic. One of the most well-known fungal opportunistic pathogens, *Candida albicans*, has been linked to several disease states, including oropharyngeal candidiasis,[48] exacerbations of inflammatory bowel diseases,[49] and several pulmonological illnesses, including cystic fibrosis.[50] Beyond *C. albicans*, many other opportunistic pathogenic fungi can lead to infection, including species within the *Malassezia*, *Cryptococcus*, *Blastomyces*, *Aspergillus*, and *Histoplasma* genera.

Similar to antimicrobial medications, antifungals have been developed to deliver broad- and narrow-spectrum coverage of potential fungal infections.

Microbiome–Drug Interactions beyond Antimicrobials

The extensive array of medications employed in treating human diseases has the potential to influence the microbiome indirectly. Additionally, specific therapies exhibit varying effectiveness depending on an individual's gut microbiome composition.[51] While the examples provided are not exhaustive, they serve as a starting point for contemplating whether disturbances in the gut microbiome can be regarded as potential side effects of medications. This raises significant questions about whether, in the future, clinical decision-making will factor in microbiome effects when devising treatment strategies. This prompts consideration of a future where understanding and addressing the impacts of the microbiome become integral components of medical decision-making processes.

Proton Pump Inhibitors

The use of proton pump inhibitors (PPIs) for acid-related gastrointestinal disorders has become widespread, with millions of people using drugs like pantoprazole and omeprazole, both by prescription and over the counter. While these drugs are highly effective and generally safe, their chronic use raises concerns, most recently implicating changes in the gut microbiome.[51] PPIs significantly alter the gut microbiome, decreasing the abundance of beneficial intestinal bacteria and increasing the presence of oral bacteria. These changes are linked to reduced resistance to harmful pathogens, potentially leading to infections from pathogens, including *Clostridium difficile*.[52] However, not all indications of PPI use are deleterious to the gut microbiome, as PPIs are currently used to treat *Helicobacter pylori* infections.[53] *H. pylori* is a known gastric pathogen responsible for the exacerbation and recurrence of gastric ulcers and for creating a pro-tumorigenic environment in the stomach.[54] The observation that PPIs can lead to general gut dysbiosis and extinguish particular infections highlights the necessity for more judicious use of these medications.

Metformin

Metformin, a widely used oral medication for type 2 diabetes, was recently shown to impact the gut microbiota outside of its well-established therapeutic benefits and side effects. Unlike drugs with a variety of indications, such as PPIs, metformin's primary use in diabetes makes it challenging to distinguish its effects on the microbiome from disease-related changes. However, a significant study revealed that alterations previously attributed to diabetes were caused by metformin itself.[55] In interventions involving healthy volunteers, metformin shifted over 80 species, elevating *Escherichia coli* and reducing *Intestinibacter* abundance. In subsequent murine studies, fecal samples from metformin-treated human donors were transplanted into mice and lowered their blood glucose, indicating a direct microbiome influence.[56] Metformin's impact involves promoting SCFA-producing bacteria, like *Akkermansia muciniphilia*. Metformin users often experience gastrointestinal side effects linked to changes such as increased virulence factors and gas metabolism genes, likely from elevated *E. coli* species. This interconnection illustrates how a widely used medication alters the gut microbiome, explaining its therapeutic benefits and specific side effects. It underscores the importance of meticulous control for factors like medication use in microbiome studies examining specific diseases or conditions.

Broader Impact and Consequences of Medications that Modulate the Microbiome

Although antibiotics are quintessential to the management of local and systemic infections and have radically changed healthcare, they can have off-target effects and impacts on human health if misused. Clinicians prescribe antimicrobial medications quite commonly to manage acute microbial infections, but systemic antimicrobials directly impact all internal microbes and can completely deplete the gut microbiome for a period of time.

Antimicrobial Resistance

In the realm of infectious disease medicine, remarkable progress has been made in managing and treating infections with the help of antimicrobial drugs. However, a significant and pressing concern remains the rise of antimicrobial resistance against our current therapeutic options. Various factors have contributed to the emergence of microbes that persist despite treatment with existing antimicrobials. These factors include the misuse and overuse of antimicrobial agents, inadequate infection control measures in healthcare settings, the scarcity of novel antimicrobial discoveries, and the widespread use of antimicrobials in agriculture.[57–59]

Antimicrobial-resistant (AMR) superbugs, primarily bacteria, evolve due to selective pressures favoring variants resistant to commonly used medications. Considering the immense global consumption of antibiotics since their inception, microbes have undergone numerous selection events. These events have driven surviving strains to retain AMR genes, rendering standard treatments ineffective.[57–59] The continued evolution of these resistant strains underscores the critical importance of understanding and addressing the complex interplay between antimicrobials and microbial adaptation.

What may be ironic is that a majority of multidrug-resistant (MDR) microbial strains are now present in hospital settings. One of the most well-known hospital-acquired infections that exhibit MDR is from methicillin-resistant *Staphylococcus aureus* (MRSA). Since 2005, however, hospital infection rates have dropped due to the implementation of stricter antibiotic stewardship and sterility practices.[60] Unfortunately, MRSA has since moved out into the community, with community-acquired MRSA (CA-MRSA) cases rising since 2012.

Effects of Antimicrobial Medications on the Environment

Although this section specifically focuses on the human microbiome, it is important to discuss how antimicrobial medications developed for humans have impacted the microbial communities in the environment. The widespread use of antibiotics in healthcare, agriculture, and animal husbandry has led to the release of these compounds into natural ecosystems through various pathways, such as wastewater discharge and agricultural runoff. This environmental exposure to antibiotics and their residues has profound consequences on environmental microbial communities. Antimicrobials in the environment can alter the composition and function of microbial communities in soil and water, affecting nutrient cycling and ecosystem stability.[61] Further, studies have also demonstrated that antibiotics can promote the horizontal transfer of resistance genes among environmental bacteria, exacerbating the problem of AMR.[62]

Adopting sustainable practices in human medicine, agriculture, and wastewater management is crucial to mitigate the environmental impact of antimicrobial medications. Additionally, ongoing research is essential to understand the long-term effects of antimicrobial medications on environmental microbiota and ecosystems, enabling the development of effective strategies to preserve environmental health.

Summary

In this section, we discussed how medications can affect microbiomes in humans. Antimicrobial medications have revolutionized healthcare and extended the average human lifespan by decades. Drug classes have been developed to target numerous pathogens; as our understanding of pharmacomicrobiomics improves, researchers and clinicians may temper medication choices based on how certain drugs impact the microbiome more systemically. However, we face new challenges with antibiotic overuse and increasing cases of AMR infections. Additionally, our impact on the global microbiome through antimicrobials continues to grow as antibiotic runoff shifts microbial communities in our water sources and soil.

Outside of antimicrobial medications, research is blossoming on how common medications used for other illnesses may affect the gut microbiome. First-line drugs for T2DM and gastric reflux have been causally linked to perturbations in the gut microbiome. As this field develops, the relationships between other common gastrointestinal- and metabolism-related medications will likely be described.

Diet

Overview

The interconnection between diet, the microbiome, and overall health has been recognized for over a century.[63] This section examines the relationship between food and the microbiome, highlighting how dietary patterns and macronutrients influence gut microbial diversity and health. It underscores the potential of precision nutrition to harness these relationships for health benefits despite existing knowledge gaps on dietary impacts across time and cultures. Analyzing overarching dietary patterns and the metabolic influence of specific macronutrients is critical to optimize precision nutrition, which may help lead to personalized diet strategies.

Macro- and Micronutrients and the Microbiome

The acceptable macronutrient distribution ratios (AMDR), expressed as percentages of total energy for protein (ranging from 10% to 35%), carbohydrates (ranging from 45% to 65%), and fat (ranging from 20% to 35%), were set by the former Institute of Medicine of the National Academies in North America. These guidelines offer valuable insight into the recommended ranges of macronutrient intake that ensure adequate essential nutrient levels while reducing the risk of chronic disease.[64] Notably, when these recommendations were originally set, one aspect that was not considered was the potential for macronutrient intake to influence the gut microbiota composition.[65] Longitudinal studies, which involve continuous observation of healthy individuals' dietary patterns over an extended period, are instrumental in understanding the nuanced relationships between daily food choices and human gut microbiome composition. In a recent study, researchers integrated 24-hour food records with microbiome data from 34 healthy human subjects collected daily over 17 days. This research found that microbiome composition depended on multiple days of dietary history and was more strongly associated with food choices than with conventional nutrient profiles, highlighting the importance of monitoring daily dietary patterns in microbiome studies.[66]

These data support the concept that distinct dietary habits of individuals contribute to the establishment of unique microbiota diversity. Reciprocally, changes to the gut microbiome then affect potential health advantages or disadvantages derived from dietary interventions involving specific foods.[66] Research has subsequently been completed to determine the outcomes of particular dietary interventions on health and disease, described below.

Carbohydrates

Carbohydrates are a diverse group of nutrients essential for human health and are categorized by how they are digested. Sugars, or digestible carbohydrates, are absorbed and utilized for energy, playing a significant role in insulin signaling and glucose homeostasis.[67,68] Indigestible carbohydrates, known as dietary fibers, microbiota-accessible carbohydrates (MACs), or resistant starches (RS), pass through the small intestine to the colon, where they are the dominant carbon source of the microbiome. The downstream products of this fermentation, notably butyrate, have been associated with intestinal health and function.[69,70]

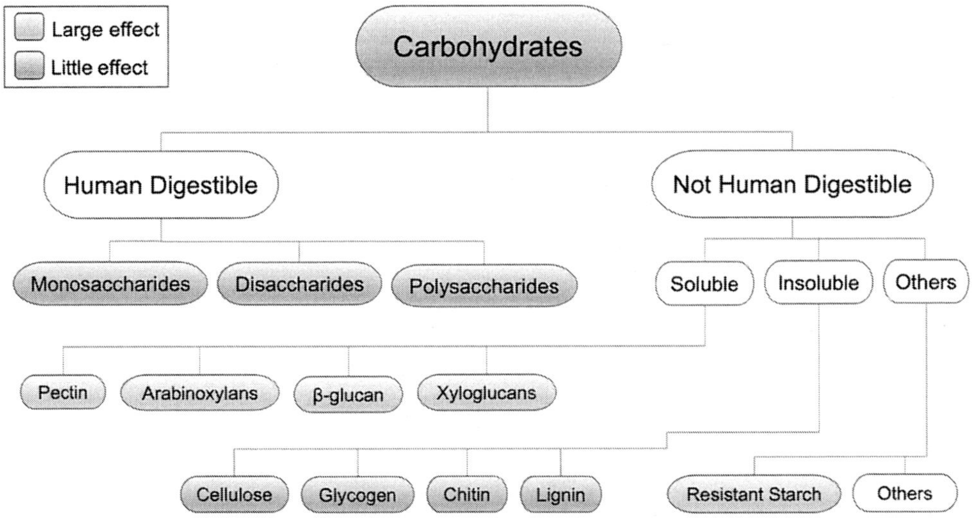

FIGURE 6.3 Different carbohydrate types and their effects on the human microbiome.

Different classification systems have been used to describe the diversity of MACs. For example, the RS-system was established by Englyst et al. in 1982, and its most commonly described five categories.[71] The RS1 is physically inaccessible (legumes, pasta, seeds), RS2 is ungelatinized (un-ripe bananas), RS3 is amylose that has been chemically modified (cooked, then cooled pasta), RS4 is chemically modified by industrial processes, and RS5 is amylose-lipid complexes.[72] The RS1–3 are most commonly associated with effects on the microbiome. For example, RS3 has been associated with increased *Actinobacteria* and *Bacteroidetes*, while RS2 has been associated with increased *Ruminococcus bromii* and *Eubacterium rectale*. These changes underscore the importance of RS in promoting a healthy gut environment. Moreover, studies have highlighted the disparity between the expected and observed effects of certain fibers like inulin on the production of SCFAs, revealing the intricate nature of the gut ecosystem and the need for further research to unravel these complex interactions (Figure 6.3).[73–75]

Fat

Dietary fat, which dietary guidelines typically recommend should constitute around 30% of the daily diet,[76] encompasses various subgroups, including saturated, polyunsaturated fatty acids (PUFA), and monounsaturated (MUFA) fatty acids. In murine studies, high saturated fat diets have been associated with weight gain, insulin resistance, and the accumulation of visceral adiposity.[76] While it remains uncertain whether these effects directly translate to humans, such diets seem to impact microbiome diversity and compromise gut barrier integrity.[68,77] Saturated fats have been observed, in various models, to decrease the presence of beneficial bacteria such as *Bifidobacterium* and *Lactobacillus*, as well as certain Bacteroidetes like Bacteroides and Prevotella, while increasing Firmicutes such as *Faecalibacterium prausnitzii* and Proteobacteria like Bilophila, as reported by various sources.[76] On the other hand, observational studies suggest that diets rich in PUFA and MUFA promote a lower Firmicutes/Bacteroidetes ratio, which has been associated with normal weight individuals.[76]

MUFAs, known for their cardioprotective benefits and prominently featured in the Mediterranean diet, have been associated with a decrease in overall microbial abundance. However, unlike saturated fats, they do not impact microbiome diversity negatively.[77] PUFAs can be further classified into omega-3 and omega-6 fatty acids. Omega-3 fatty acids have been linked to increased levels of beneficial bacteria such as *Bifidobacterium*, *Roseburia*, and *Lactobacillus*, as well as the production of SCFAs.[78] Conversely, a higher intake of omega-6 relative to omega-3 PUFAs has been associated with increased gut barrier permeability and metabolic endotoxemia, primarily driven by the gut microbiota.[68]

Overall, dietary fats and the microbiome have a complex relationship. Specific dietary fats, including PUFAs and MUFAs, have been associated with positive health outcomes in certain cardiovascular conditions, but their relationship to shifts in the microbiome still requires further research.

Protein

Current dietary recommendations suggest that protein intake should constitute 10–20% of the daily diet.[76] When examining protein sources, it becomes evident that where protein is sourced from affects the microbiota and gut fermentation.[68] In murine models, plant-based soy protein intake has been associated with increased production of SCFAs, improved microbial diversity, and enhanced metabolic health compared to animal-based casein protein.[79] Conversely, a human diet rich in beef protein has been found to lead to a lower abundance of *Bifidobacterium* than a meat-free diet.[76] Additionally, animal-based proteins have been shown to negatively impact microbial composition, contributing to the proatherogenic microbe community[80] and increased production of trimethylamine N-oxide.[81] It is worth noting that the total quantity of protein consumed also plays a significant role. In mice, high total protein intake has been linked to increased pathogenic microorganisms such as *Streptococcus*, *E. coli*, *Shigella*, and *Enterococcus*.[70] Therefore, aside from the source of protein, excessive protein intake may potentially lead to an increase in pathogenic microorganisms and associated disease states.[70,82] Dietary amino acids in mammals are utilized by both enterocytes and gut microbiota; they act as precursors for various metabolic products that influence host immunity and the microbiota through multiple signaling pathways. Microbial metabolites associated with dietary amino acids can cross the epithelial barrier into the host's circulatory system, where they interact with immune cells via diverse receptors, impacting the gut-microbiome-immune axis and contributing to intestinal homeostasis. The signaling roles of specific amino acids, such as tryptophan, glutamine, and branched-chain amino acids, are highlighted for their importance in regulating gut immunity and potentially improving gut health and systemic well-being.[83]

Food Antioxidants and the Microbiome

Food antioxidants, such as carotenoids, polyphenols, vitamins C and E, zinc, and selenium, can also influence gut microbiota interactions. While some studies suggest antioxidants contribute positively by aiding in gut microbiota restoration after antibiotic treatment and improving disease states, including liver damage, hypertension, diabetes, obesity, and inflammatory bowel disease (IBD), it is important to acknowledge that the effects may vary among individuals and require further research for a comprehensive understanding.[84] Common sources of polyphenols include blueberries, raspberries, oranges, onions, soybeans, and green tea leaves.[85,86] Research has indicated that a diet rich in polyphenols offers protection against various chronic diseases, including cancer, cardiovascular diseases, and inflammatory conditions.[87]

Carotenoids are a diverse group of pigments found in nature and are significant to humans as precursors to vitamin A.[88] Key dietary sources of carotenoids include carrots, spinach, corn, tomatoes, and bell peppers.[89] These compounds encompass a variety of forms, including β-carotene, α-carotene, β-cryptoxanthin, lutein, zeaxanthin, neoxanthin, capsanthin, bixin, and lycopene.[90] Their antioxidant properties stem from the presence of conjugated double bonds, making them susceptible targets for free radicals.[91] In the context of interactions with the gut microbiome, carotenoids have emerged as key players in modulating human health. Several studies have delved into how different types of carotenoids engage with the gut microbiome. One such carotenoid is astaxanthin, commonly found in seafood and known for its robust antioxidant properties.[92] A recent investigation using a mouse model lacking β-carotene oxygenase (C57BL/6 mice) revealed that astaxanthin treatment led to a significant increase in the populations of *Mucispirillum schaedleri* and *Akkermansia muciniphila* when compared to wild-type mice. It is worth noting that the gut microbiota structures exhibited variations between male and female mice.[93]

Essential Nutrients and Their Impact on the Gut Microbiome

Vitamin C (ascorbic acid) and vitamin E, found in various fruits and nuts, are crucial for their antioxidant properties and roles in immune function. Vitamin C, particularly known for aiding collagen synthesis and combating oxidative stress-related diseases,[94,95] and vitamin E, essential for cardiovascular and neurodegenerative health,[96] interact with the gut microbiome. These interactions influence diverse bacterial populations, potentially aiding conditions like IBD[97] and regulating blood pressure.[98]

Zinc and selenium, essential for reducing oxidative stress and vital in immune and thyroid function,[99] also significantly interact with gut microbes.[100] Zinc impacts bacteria such as *Lactobacillus reuteri*,[101] whereas selenium's impact is broader, enhancing the diversity of beneficial bacterial species.[102] However, these interactions are complex and can vary depending on several factors, indicating the need for further research to fully understand these relationships. In summary, these nutrients are vital for human health and play a significant role in modulating the gut microbiome, underscoring the intricate connection between diet, nutrient intake, and overall health.

Dietary Patterns

Diet and other factors like geographic location, genetic relatedness, ethnicity, and lifestyle play a significant role in shaping gut microbial communities. This influence has been extensively studied and is well-documented in research.[103] Reduced susceptibility to chronic diseases is linked to wholesome dietary patterns, such as the Mediterranean diet[104] and plant-based diets.[105] Given the intricacies of dietary consumption and that people consume entire foods rather than isolated nutrients, exploring how dietary patterns affect the microbiome may offer valuable insights into the underlying mechanisms.

The Western-style dietary pattern is typically characterized by elevated consumption of red meat, saturated fats, processed grains, added sugars, and a fiber deficiency. Notably, the Western diet lacks fruits, vegetables, legumes, nuts, seeds, and whole grains, which American Dietary Guidelines recommend and promote globally.[65] The Mediterranean diet is a traditional eating pattern historically consumed in countries bordering the Mediterranean Sea, including Spain, Greece, and Italy. While each country has unique food traditions, this diet typically emphasizes plant-sourced foods such as fruits, vegetables, whole grains, and legumes. Dairy consumption is limited, and olive oil is the primary fat source. Moreover, moderate amounts of fish, poultry, and wine are commonly included.[106] Vegetarian and vegan diets exhibit variations in their food composition, with vegan diets excluding all animal-origin foods, with certain health-enhancing attributes commonly associated.[107] Vegetarian diets, predominantly plant-based, provide elevated levels of dietary fiber and tend to be lower in protein content than omnivore diets.[108] One study examined the associations between vegan or vegetarian diets and microbiota composition. Their findings revealed no significant differences in the microbiome between consumers of vegetarian diets, vegan diets, and omnivores. They concluded that the substantial inter-individual variability made it challenging to describe a characteristic microbiome profile for vegetarians or vegans.[109]

In summary, a distinct microbiome profile does not appear to be specifically attributed to consuming a vegetarian diet. This may be due, in part, to the diversity of dietary practices within the vegetarian category. Hence, current data suggest that gut microbial profile alone is not the primary mechanism through which this dietary pattern imparts health benefits.[110]

EXERCISE

Overview of the Gut Microbiome and Physical Activity

The symbiotic relationship between exercise and the gut's microbial inhabitants has recently captured the attention of researchers. This section explores the reciprocal influence of exercise and the gut microbiome, highlighting their dual significance in fostering overall well-being, including cognitive health. This part aims to shed light on the dynamic interplay between exercise and the gut microbiome, focusing on how physical activity can modulate the composition and function of these microorganisms. By dissecting how exercise might influence these microorganisms, we strive to offer valuable insights into the role of lifestyle choices in shaping our microbiome and, consequently, our overall well-being.

Recent research has highlighted the connection between the gut microbiome and physical activity.[111] Distinct differences in gut microbiome composition are observed between individuals who exercise regularly and those who do not. Notably, individuals engaged in exercise, particularly elite athletes or those with a moderate exercise regimen, tend to harbor a gut microbiome that leans toward health promotion and showcases heightened species diversity.[112] Among exercisers, there is a higher prevalence of gut microbes adept at producing advantageous metabolites like SCFAs and secondary bile acids. These microbial components are pivotal in thwarting chronic inflammation and fortifying the immune system.[113]

Importantly, the relationship between the gut microbiome and exercise is bidirectional. A healthy gut microbiome enhances exercise capacity by regulating energy, muscle maintenance, metabolism, and nutrient pathways.[114] As a result, it becomes evident that maintaining a healthy gut microbiome and engaging in regular exercise are of paramount significance for preserving overall human health.[115]

Acute Effects

The link between exercise, cognitive function, and the microbiome has predominantly centered on aerobic exercise.[116] Exercise increases key antioxidant enzymes, anti-inflammatory cytokines, and antiapoptotic proteins, reducing inflammation.[117,118] Aerobic exercise has been shown to enhance microbial diversity and reinforce the integrity of the intestinal barrier in humans.[111,119,120] Exercise elevates brain-derived neurotrophic factor (BDNF), which is known to regulate gastrointestinal tight junction proteins crucial for maintaining epithelial integrity, thereby reducing the translocation of pro-inflammatory endotoxins like lipopolysaccharide (LPS) into the bloodstream.[121] Greater exercise and cardiorespiratory fitness levels are positively linked with bacteria that produce SCFAs and increased fecal SCFA concentrations, independently of diet.[122,123]

However, the impact of aerobic exercise on gut health appears to be closely tied to exercise intensity. Moderate-intensity aerobic exercise has been associated with maintaining healthy intestinal blood flow, which, in turn, positively influences gastrointestinal motility while also reducing inflammation.[124]Several studies have shown that moderate-intensity exercise can also induce significant changes in the gut microbiome. For example, exercise can lead to increased representation of species such as *Akkermansia* and *Prevotella*, which increase biodiversity and promote metabolic activity.[125,126]

In contrast, engaging in prolonged, high-intensity aerobic exercise (exceeding 60%–70% of VO_2 max, the maximum rate of oxygen the body can use during exercise) triggers an acute stress response within the body. This response leads to increased levels of stress hormones like cortisol and epinephrine.[127] High levels of gastrointestinal disorders and a compromise in performance, which may impact overall health, are prevalent among endurance athletes such as ultramarathoners and triathletes, accompanied by a rise in pro-inflammatory cytokines and proteins.[128]

Aerobic exercise interventions have been shown to have a positive impact on gut microbiota and brain health in rodents, yet there is a dearth of research examining this relationship in humans.[129]

An example in rodents involved a 16-week running wheel exercise regimen, which resulted in notable alterations in the microbiota composition. This included an increase in the relative abundance of *Firmicutes* and a decrease in the relative abundance of *Bacteroidetes* and *Tenericutes*. Concurrently, the exercise group exhibited enhanced contextual memory compared to control groups in adult male C57BL/6J mice. The memory performance level was correlated with specific bacterial abundances, particularly those from the Ruminococcaceae and Lachnospiraceae families.[130]

One proposed mechanism in which the microbiome interacts with increased health via exercise is through a relationship between the increasing prevalence of microbes producing short-chained fatty acids. SCFA-producing bacteria can lead to increased acetate, butyrate, and propionate concentrations.[125,126] SCFAs such as butyrate can enter the systemic circulation and increase insulin sensitivity through interaction with free fatty acid receptors 2 and 3 (FFAR-2 and FFAR-3) and activate regulators of mitochondrial biosynthesis.[131] Butyrate affects various regulatory pathways, increasing ATP production and myofiber metabolic efficiency.[132]

Probiotics, including *Faecalibacterium prausnitzii*, have shown promise in reducing systemic inflammation and promoting muscle health.[133] This underscores the potential therapeutic role of SCFA-producing bacteria and SCFAs in maintaining muscle health and combating age-related fragility and opens the door for future research.

Habituated Effects

Engagement in regular physical activity is increasingly recognized as a therapeutic intervention that helps reduce inflammation.[134] Current studies suggest that such physical activities may influence the composition of gut microbiota in healthy individuals and those with inflammatory bowel disease.[134,135] Exercise programs that span a minimum of 12 weeks have been shown to induce changes in the composition of the gut microbiota, triggering immuno-metabolic pathways that bring about anti-inflammatory benefits.[136] A recent systematic review has indicated a correlation between the abundance of *Prevotella* in the gut and the exercise duration.[115] Consistent with these findings, it has also been observed that *Methanobrevibacter smithii* is more prevalent in professional cyclists' guts than in their amateur counterparts.[137] Similarly, elite female endurance runners were found to have a greater quantity of *Faecalibacterium*, which has been linked to succinate accumulation.[138]

In conclusion, while introductory evidence supports a positive relationship between exercise, the microbiome, and positive health outcomes, the precise exercise duration that best supports a healthful shift in gut bacteria composition remains to be established.[139] Further investigation is required to illuminate the relationship between exercise duration and its overall impact on gut health.

Relationship between Exercise and the Gut-Brain Axis

The concept of frailty in older individuals is evolving beyond physical frailty, with a growing interest in cognitive frailty, characterized by the combination of physical frailty and cognitive dysfunction.[140] Recent research has unveiled a significant connection between an imbalanced gut microbiome and cognitive dysfunction, particularly in the context of conditions like dementia.[113] For instance, patients with Alzheimer's disease (AD) exhibited higher percentages of microbial byproducts such as mannitol, succinic acid, and 3,4-dihydroxy benzeneacetic acid, which differ from those in healthy controls.[141] The presence of *Clostridium difficile* was also significantly higher in patients with dementia compared to those without.[128]

The gut-brain axis represents a bidirectional communication pathway influenced by hormones, immunity, and nerve signals. Exercise-induced changes in the gut environment can potentially influence cognitive function, although the precise mechanisms are not yet fully understood. SCFAs, such as butyrate, have increased the expression of satiety-inducing hormones and BDNF.[142] Regular aerobic exercise can lead to prolonged parasympathetic nerve stimulation, potentially mimicking the effects of vagal nerve stimulation (VNS) and enhancing the cholinergic anti-inflammatory pathway, which may hold promise in preventing brain diseases.[13]

Summary

An intricate relationship exists between exercise, the gut microbiome, cognitive health, and aging. Exercise and the gut microbiome share a mutually beneficial connection, with exercise influencing the microbiome's composition and function and the microbiome impacting various aspects of health. Aging changes the gut microbiome, affecting immunity and contributing to age-related diseases. Additionally, the gut-muscle axis reveals the connection between the microbiome and physical frailty, suggesting that exercise may help counteract this condition. The gut–brain axis is relevant to cognitive health, and exercise potentially promotes a healthy gut–brain connection. Future research in this area promises valuable insights into leveraging lifestyle choices for better overall well-being.

Probiotics, Prebiotics, Synbiotics, and Postbiotics

Probiotics

Probiotics, defined by the International Scientific Association for Probiotics and Prebiotics (ISAPP) in 2014, are live microorganisms that offer health benefits when administered in adequate amounts.[143] Commonly, they are ingested in fermented foods containing complex microbial communities or as supplements in the form of specific microbial colonies (e.g., *Bifidobacterium* spp., *Lactobacillus* spp.).[144,145] The mechanisms in which they influence health are incredibly diverse, ranging from inhibiting the growth of pathogens to synthesizing essential vitamins.[145] Several studies underscore the benefit of probiotics in gut health, including their ability to reduce gut inflammation, specifically in ulcerative colitis.[144,146,147] Regarding neoplastic processes, a case-control study found that increased yogurt consumption was associated with decreased odds of developing colorectal hyperplastic polyps.[148] In another study, a prospective cohort demonstrated that one or more servings of yogurt per week were associated with a lower incidence of colon cancer in individuals with similar fiber intake.[149] Several studies have also shown a benefit from probiotics in preventing cavities by limiting oral adhesion of pathogens and reducing periodontal inflammation.[150,151]

While probiotics have been shown to have several benefits, they do not always improve every disease state.[152] In one case, probiotics may directly impact the efficacy of immune checkpoint inhibitors (ICIs), a class of cancer drugs.[152] Using a murine melanoma model, researchers compared treatment responses between mice receiving probiotics and those receiving sterile water.[152] The mice that received probiotics exhibited an impaired drug response, leading to the development of larger tumors compared to the control group.[152] However, when assessing cancer recurrence and survival in human patients with late-stage melanoma, there was no significant difference between patients who reported taking probiotics within one month of starting ICI treatment.[152]

Regulating these substances is essential with such a vast array of probiotic options and wide variation of effects. In the United States (US), if live microbes are included in food, such as yogurt, industries may label their product as containing live and active cultures if they have at least 10^7 colony-forming units of lactic acid-producing bacteria per gram (CFU/g).[153,154] Surprisingly, probiotics can also be sold without US Food and Drug Administration (FDA) approval if labeled as dietary supplements instead of food or drugs.[155] Probiotics have multifaceted applications, but more research is needed to harness their potential, limit their disadvantages, and regulate their use.

Prebiotics

Prebiotics are defined by the ISAPP as substrates selectively utilized by microorganisms to confer health benefits by modifying the composition and function of the gut microbiota.[156] This classification includes substances that forego digestion in the stomach and are then fermented by beneficial bacteria in the intestines.[157] Excellent dietary sources of prebiotics include carbohydrates, wheat, beans, onions, and tomatoes; additional information can be found in the macronutrients section of this chapter.[157] Products of intestinal fermentation then modulate gut acidity and reduce inflammatory responses.[157] For instance, in lowering the pH of the gut, there was a decrease in the growth of *Bacteroides* spp., a bacteria associated with inflammatory bowel disease.[158,159] A particular prebiotic, germinated barley foodstuff (GBF), has also demonstrated decreased symptoms of ulcerative colitis, such as severe diarrhea and bloody stools, in individuals suffering from inflammatory bowel disease. Further, when comparing the effects of a prebiotic-based supplement in children with acute malnutrition, research supported that participants receiving lower-calorie prebiotic-based supplements had better outcomes than those receiving a high-calorie supplement without prebiotics.[160] Specifically, the children who received microbiota-directed prebiotics showed improvement in growth markers, such as changes in weight-for-length and weight-for-age.[160] These examples highlight that through modulation of the gut microbiota, prebiotics present a spectrum of promising health benefits, an area of study that continues to expand.

Synbiotics

Synbiotics represent a unique combination of probiotic microorganisms and prebiotic substrates, working in synergy to enhance the survival of probiotic microorganisms in the gastrointestinal tract.[161,162] In other words, the microorganisms are administered along with the particular prebiotics they ferment. Administration of a synbiotic capsule containing friendly *Bifidobacterium*, *Lactobacillus*, *Streptococcus* spp., and fructo-oligosaccharides (FOS) has been shown to improve fasting blood sugar, decrease triglyceride and cholesterol levels, and reduce insulin resistance.[163] These data were the first to show synbiotics' efficacy in improving markers of metabolic syndrome in conjunction with diet-based treatment.[163] Further experiments have demonstrated that synbiotic supplementation with microorganisms of *Bifidobacterium* and *Lactobacillus* spp. in conjunction with the prebiotic trans-galactooligosaccharides was associated with an increase in the beneficial bacteria in fecal samples.[164] Lastly, the effects of synbiotic supplementation in obese individuals has found a significant reduction of BMI, body fat, and visceral fat compared to the placebo group.[165] These findings point to the overall benefit of the gut microbiome when supplementing with synbiotics. Synbiotics, which combine probiotic microorganisms and prebiotic substrates, have been linked to improved metabolic health, gut microbiome, and markers of obesity.

Postbiotics

Postbiotics are defined by the ISAPP as inanimate microorganisms or their components that confer a health benefit to the host.[166,167] In other words, postbiotics include bacteria that have been killed or inactivated, as well as their cell components and fermentation products. A commonly referenced postbiotic is SCFA, a fermentation product that includes butyric, acetic, and propionic acid.[168] Research shows that SCFAs have numerous health benefits, especially in autoimmune, metabolic, neurologic, and carcinogenic processes.[168–170] A group studying the effects of diet in the development of allergic airway disease (AAD) fed mice different diets and then induced AAD through intranasal house-dust mites. They found that the mice given acetate in their drinking water never developed signs of AAD, such as lung inflammation and elevated IgE levels.[171] In a mouse colitis model, the mice that received oral butyrate had reduced inflammatory markers, weight loss, and colitis scores.[164]

As for administering the inanimate organisms, a randomized controlled trial showed that a strain of heat-inactivated *Bifidobacterium* spp. administered to individuals with IBS helped alleviate symptoms the same as the active microorganisms.[172] It has been proposed that the addition of postbiotics to baby formula has the potential to replicate the immune-strengthening effects seen in breastfeeding.[173,174] In a mouse model study, the administration of skim milk containing heat-inactivated *Lactobacillus paracasei* and its fermentation products indicated a protective effect over *Salmonella*-induced colitis by decreasing the release of inflammatory mediators and reducing weight loss.[173,174] This finding suggests that incorporating postbiotics into baby formula has the potential to help build an infant's host defense without exposure to live pathogens. Postbiotics show promising health benefits, with studies demonstrating their effectiveness in conditions like IBS and their potential to enhance immunity.

Summary

Understanding the differences and functions of probiotics, prebiotics, synbiotics, and postbiotics is crucial to fully comprehend their importance in microbiome regulation and human health. Clinical trials in this area of research have grown tremendously since the 1990s and continue to explore the benefit of -biotics in various disease states (see Figure 6.4). Collectively, these elements contribute to maintaining a balanced and healthy gut ecosystem, influencing multiple aspects of human well-being, from digestive health to immune function. An overview of the diverse array of clinical trials being conducted in the US during the creation of this chapter is presented in Table 6.1.

Fecal Microbiota Transplant to Modulate Human Disease States

Overview

Fecal microbiota transplant (FMT), also colloquially called fecal transplant, has been an established treatment option for individuals with certain gastrointestinal diseases for millennia. With the first documented in China in the 400s CE and remaining through the centuries in other societies, FMT increases microbial diversity in the gut, thus generating increased competition for pathogens that may exist within the gut. Notably, coprophagia, the act of purposely eating stool, is a common practice amongst mammals and is associated with better overall health in several species. Therefore, the intentional transfer of microbes between individuals, particularly through stool, is an established part of the animal kingdom.

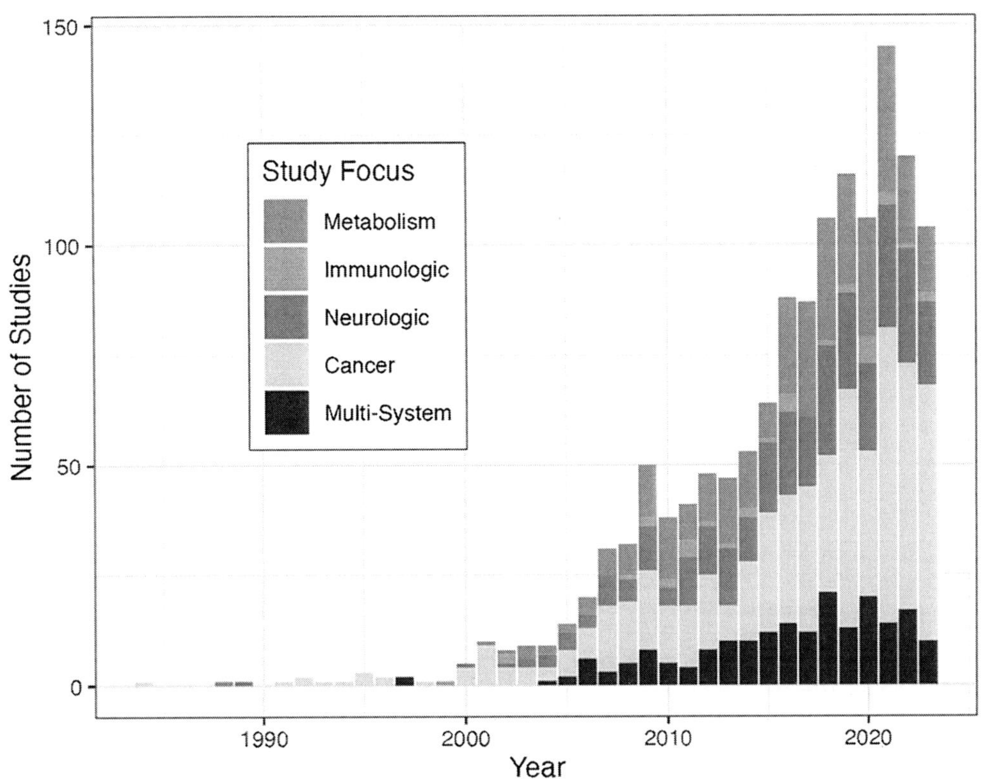

FIGURE 6.4 Number of new US-registered studies in clinicaltrials.gov that seek to modify the microbiome.

TABLE 6.1
Representative Clinical Trials that Seek to Modify the Microbiome to Alter Metabolic, Immunologic, or Neurologic Diseases, or Cancer

	Title	Phase	Intervention
Autoimmune + Multi System			
NCT03550885	Diet Modulation of Bacterial Sulfur and Bile Acid Metabolism and Colon Cancer Risk	NA	OTHER: High taurine and saturated fat diet I OTHER: Low in taurine and saturated fat diet
NCT03926286	Fecal Microbial Transplant (FMT) for Sjogrens Syndrome	PHASE1	DRUG: FMP-30
NCT03423121	A Trial of Bile Acid Supplementation in Patients With Multiple Sclerosis	PHASE1 I PHASE2	DRUG: Tauroursodeoxycholic Acid I DRUG: Placebo oral capsule
NCT03751007	A Study to Assess the Safety and Tolerability of Different Doses of AG019 Administered Alone or in Combination With Teplizumab in Participants With Recent-onset Diagnosed Type 1 Diabetes (T1D)	PHASE1 I PHASE2	BIOLOGICAL: AG019 – Low Dose I DRUG: Teplizumab I DRUG: Placebo-AG019 I DRUG: Placebo-Teplizumab I BIOLOGICAL: AG019 – High Dose I BIOLOGICAL: AG019 - High Dose
NCT04096443	Oral FMT (Fecal Microbial Transplant) in Subjects With Multiple Sclerosis	EARLY_PHASE1	BIOLOGICAL: Fecal microbial transplant (FMT)
NCT03594487	Fecal Microbiota Transplantation (FMT) of FMP30 in Relapsing-Remitting Multiple Sclerosis	PHASE1	DRUG: FMP30 Donor Stool I PROCEDURE: Fecal Microbiota Transplantation (FMT) of FMP30 Donor Stool I OTHER: Observational Control
NCT03539094	Intermittent Fasting in Multiple Sclerosis	NA	OTHER: Intermittent fasting
NCT03919435	TEMPO Study: Trimethoprim-Sulfamethoxazole in Granulomatosis With Polyangiitis	PHASE1 I PHASE2	DRUG: Trimethoprim Sulfamethoxazole
NCT03420482	Fecal Microbiota Transplant as Treatment of Hepatic Encephalopathy	PHASE2	DRUG: Fecal Microbiota Transplant (FMT) oral capsules I DRUG: Placebo oral capsule
NCT03324399	A Study of Protein Metabolism, Microbiome and Investigational Probiotic Use in Patients With ALS	NA	DIETARY_SUPPLEMENT: probiotic
Cancer			
NCT04530812	Fermented Milk Supplementation on Symptoms of Disease and Treatment in Patients With Multiple Myeloma	NA	OTHER: Best Practice I DIETARY_SUPPLEMENT: Kefir I OTHER: Quality-of-Life Assessment I OTHER: Questionnaire Administration
NCT02843425	The Beans to Enrich the Gut Microbiome vs. Obesity's Negative Effects (BE GONE) Trial	NA	OTHER: Regular Diet I OTHER: Navy Beans (Canned)
NCT04985565	Neoadjuvant Dietary Intervention in Intermediate Risk Prostate Cancer	NA	OTHER: Dietary Intervention I PROCEDURE: Radical Prostatectomy
NCT04857697	Effects of Probiotics on the Gut Microbiome and Immune System in Operable Stage I–III Breast or Lung Cancer	EARLY_PHASE1	PROCEDURE: Biopsy I PROCEDURE: Biospecimen Collection I DRUG: Probiotic I PROCEDURE: Therapeutic Conventional Surgery

(*Continued*)

Table 6.1 (Continued)
Representative Clinical Trials that Seek to Modify the Microbiome to Alter Metabolic, Immunologic, or Neurologic Diseases, or Cancer

	Title	Phase	Intervention
NCT05083416	Effect of Prolonged Nightly Fasting on Immunotherapy Outcomes in HNSCC - Role of Gut Microbiome	NA	BEHAVIORAL: Prolonged Nightly Fasting I BEHAVIORAL: Regular Eating Pattern
NCT04056026	A Single Dose FMT Infusion as an Adjunct to Keytruda for Metastatic Mesothelioma	EARLY_PHASE1	BIOLOGICAL: Fecal Microbiota Transplant
NCT03358511	Engineering Gut Microbiome to Target Breast Cancer	NA	DIETARY_SUPPLEMENT: Probiotic
NCT04267874	Black Raspberry Nectar for the Prevention of Lung Cancer, BE WELL Study	EARLY_PHASE1	PROCEDURE: Biospecimen Collection I DIETARY_SUPPLEMENT: Nutritional Supplementation I DRUG: Placebo Administration I OTHER: Questionnaire Administration
NCT04875728	The Impact of an Antibiotic (Cefazolin) before Surgery on the Microbiome in Patients with Stage I–II Melanoma	PHASE1	DRUG: Cefazolin I PROCEDURE: Resection
NCT02780284	Microbiome, Exercise Tracking Study: Among Individuals at High Risk for Colorectal Cancer	NA	BEHAVIORAL: Physical activity intervention
Metabolic			
NCT02800707	Sucralose, Stevia, Gut Microbiome and Glucose Metabolism	NA	OTHER: Sucralose I OTHER: Stevia
NCT03989882	Wheat Germ Supplementation Will Improve Markers of Gut Health, Inflammation, and Insulin Resistance in Overweight Adults	NA	OTHER: Wheatgerm Energy Ball I OTHER: Control
NCT02762708	The Effect of Bariatric Surgery on Metabolism, the Metabolome and Microbiome in Patients With Type 2 Diabetes	PHASE2	PROCEDURE: Roux-en-Y Gastric Bypass (RYGB) I BEHAVIORAL: Caloric Restriction I DRUG: Exendin-9,39 I DRUG: Normal Saline
NCT03201068	Probiotic Supplement and Microbiome, Immune System and Metabolic Syndrome	NA	DIETARY_SUPPLEMENT: Probiotic supplement I DIETARY_SUPPLEMENT: Placebo
NCT03582826	Microbial Basis of Systemic Malodor and "People Allergic To Me" Conditions	NA	BEHAVIORAL: Nutritional counselling I BEHAVIORAL: Stress-reduction counseling
NCT03973996	Gut-level Antiinflammatory Activities of Green Tea in Metabolic Syndrome	NA	DIETARY_SUPPLEMENT: Green Tea Extract I DIETARY_SUPPLEMENT: Placebo
NCT03283969	Influence of Freeze Dried Strawberry Consumption on Gut Microbiota and Cardiovascular Health.	NA	OTHER: Freeze Dried Strawberry Powder I OTHER: Isocaloric powder
NCT04431960	Blackcurrants Modify Gut Microbiota and Reduce Osteoporosis and CVD Risk	PHASE1	DRUG: blackcurrant (BC) extract
NCT05762380	Effect of Iron Supplements on the Growth of Enteric Pathogens	NA	DIETARY_SUPPLEMENT: Ferrous sulfate I DIETARY_SUPPLEMENT: Ferrous sulfate-enriched Aspergillus oryzae (Ao iron)
NCT03207841	Low Carbohydrate/High Protein Diet to Improve Metabolic Health	NA	OTHER: LC/HP

(Continued)

Table 6.1 (Continued)
Representative Clinical Trials that Seek to Modify the Microbiome to Alter Metabolic, Immunologic, or Neurologic Diseases, or Cancer

	Title	Phase	Intervention
Neurologic			
NCT03667404	Resistant Maltodextrin for Gut Microbiome in Parkinson's Disease: Safety and Tolerability Study	PHASE2	DIETARY_SUPPLEMENT: Resistant maltodextrin I DIETARY_SUPPLEMENT: maltodextrin
NCT02748317	Impact of Probiotics on Urinary Symptoms in Spinal Cord Injury SCI and SB	PHASE2	DRUG: Lactobacillus rhamnosus GG
NCT03856359	Trial of Rifaximin in Probable Alzheimer's Disease	PHASE2	DRUG: Rifaximin 550 milligrams (MG)
NCT03671785	Study of the Fecal Microbiome in Patients With Parkinson's Disease	PHASE1	DRUG: PRIM-DJ2727 I DRUG: Placebo oral capsule
NCT03153397	Effect of Prebiotic Fiber- Enriched (scFOS) Enteral Feeding on the Microbiome in Neurological Injury Trauma Patients (PreFEED Microbiome Trial)	NA	OTHER: Nutraflora scFOS I OTHER: Osmolite
NCT04315714	Impact of a Yoga Intervention on Pain and Multiomics in Participants With IBS	NA	OTHER: Yoga Program
NCT03554265	Brain and Gut Plasticity in Mild TBI or Post-acute COVID Syndrome Following Growth Hormone Therapy	PHASE3	DRUG: Somatropin
NCT04683900	Mediterranean Diet Intervention to Improve Gastrointestinal Function in Parkinson's Disease: A Randomized, Controlled, Clinical Trial	NA	OTHER: Standard of care + Mediterranean diet (intervention) I OTHER: Standard of care (control)
NCT04512599	Prebiotics in the Parkinson's Disease Microbiome	NA	OTHER: Prebiotic Bar
NCT02723344	Biological Signatures, Probiotic among Those With mTBI and PTSD	PHASE1	DIETARY_SUPPLEMENT: of Lactobacillus reuteri (L. reuteri; DSM 17938) I OTHER: Sunflower and medium chain triglyceride oils

More recent scientific studies have shown relationships between heavily controlled FMT interventions and improved response to therapy in certain diseases. The FDA classified FMT as an investigative drug, thus making it the subject of clinical approval by the FDA for use in medical settings based on clinical indication. Therefore, implementation of FMT must pass clinical trial testing throughout the US, controlling the conditions and speed at which FMT can be implemented in disease. This section will discuss the current implementation of controlled FMT in humans, and disease states that respond to FMT treatment will be reviewed.

Clinical Indications for Fecal Microbiota Transplants

Gastrointestinal Disease

Fecal microbiota transplant has been most associated with restoring the gut microbiome after infection or inflammatory disease. Chinese physicians have used FMT to treat diarrheal disease since the 4th century, with physician Ge Hong providing patients with therapeutic "yellow soup" consisting of a fecal slurry.[175] Current theory relates the use of FMT to restoring a complex community of microbes within the gut that is relatively impervious to opportunistic pathogens, which may expand when infection or antibiotic use opens environmental niches in the gut.[176]

The first clinical trial using FMT for gastrointestinal disease focused on treating *Clostridium difficile* infection (CDI) and was published in 2013.[177] *C. difficile* lives well on surfaces and is an opportunistic pathogen, meaning that the microbe is a common commensal that has the capacity to generate infections in certain dysbiotic circumstances. Unfortunately, CDI is a major source of hospital-related deaths in the US, causing severe diarrhea and fluid and nutrient loss. Particularly, CDI commonly appears in patients on long-term antibiotics and those with known gut microbial perturbations. The clinical trial using FMT to treat CDI was created in response to the difficulty in treating CDI, as many strains of *C. difficile* have developed drug resistance. In this trial, patients with recurrent *C. difficile* infections were randomly assigned to receive standard antibiotics, initial antibiotics followed by FMT, or standard antibiotics and bowel lavage. Notably, the trial had to be stopped early, as the group receiving FMT treatment had such positive results that it was unethical to continue to treat other patients with the previous standard of care.[177] Upon analysis of the microbiota of patients before and after treatment, patients receiving donor FMTs significantly increased their gut microbial diversity, which was identified to be notably low at initial CDI diagnosis. Many studies have followed this initial clinical trial, and the FDA approved the first licensed FMT therapy in 2022. This therapy, named Rebyota, is now prescribed to adult patients with recurrent *C. difficile* infections.

Outside of treating CDI, FMT has shown efficacy as additive therapy for patients with IBD, including Crohn's disease and ulcerative colitis.[178–180] FMT has been shown to induce ulcerative colitis remission without increased risk of side effects compared to placebo, which is a critical piece of longitudinal care for ulcerative colitis patients. Further trials in ulcerative colitis patients looked to assess what gut microbial changes occur after FMT, finding that microbial diversity significantly increased in donor samples and feces from patients who responded to the FMT.[181] Additionally, in several trials, an abundance of opportunistic pathogens was associated with a lack of remission after FMT for ulcerative colitis patients.[181,182] Use of FMT in Crohn's disease has also shown positive results in inducing and maintaining remission. Active larger clinical trials are still ongoing for this disease, but initial studies suggest that effective engraftment of diverse microbes leads to potential disease remission for Crohn's patients.[183]

An overarching theme in the efficacy of FMT for gastrointestinal-related diseases is the re-establishment of a diverse and complex gut microbial community. In gastrointestinal infections and inflammatory bowel disease, pathology is associated with a constrained community of microbes in the gut related to antimicrobial drug use or health states that lead to a lesser representation of microbial taxa.

Cancer Treatment

The relationship between cancer and the microbiome has continued to expand with modern scientific experimental methods and treatments, as established earlier in this chapter. Notably, the gut and tumor microbiomes have been shown to affect tumorigenesis, tumor progression, and therapy response and can be bidirectionally affected by cancer treatment. Given this complex relationship, research has looked to determine whether supplementation or transplantation of particular microbes can impact disease development or treatment outcomes in cancer.

Within the past two decades, immunotherapy has revolutionized cancer treatment in many cancers. Particularly, ICIs, including anti-PD-1, anti-PD-L1, and anti-CTLA4 antibody therapies, have radically improved cancer outcomes. These drugs were initially shown to be effective in patients with melanoma, a malignant and deadly skin cancer, when the disease metastasized to organs including the lungs, brain, and liver. Although these medications drastically improved outcomes for patients with advanced disease, about 60% of patients with advanced melanoma unfortunately do not respond to immunotherapy. Recently, the efficacy of anti-PD-1 therapy was correlated to the gut microbiome in preclinical models and human patients.[184] Notably, higher oral and gut microbial diversity was observed in melanoma patients who responded to PD-1 therapy, and FMT of human microbes in germ-free mice conferred response to PD-1 therapy in mouse models of melanoma.

Clinical trials in this space providing FMT to melanoma patients who responded to anti-PD-1 therapy are now ongoing.[185,186] A recent phase I trial assessing the efficacy of FMT plus anti-PD-1 immunotherapy in melanoma patients showed that adding FMT to anti-PD-1 treatment did not lead to increased adverse events. Further, longitudinal microbial engraftment was observed in treatment responders, with 13 of 20 patients showing therapeutic response. Additionally, more favorable immune cell profiles were observed in patients who received FMT plus anti-PD-1 therapy, indicating that the FMT may relate to inducing an immune environment that supports immunotherapy use. This result recapitulated findings from a smaller phase I study in 2021 that showed similar clinical outcomes and immune cell changes.

The field of microbial intervention in modulating cancer therapy remains quite young, but early results, as evidenced by preclinical models and clinical trials, suggest that a relationship between the microbiome and immunotherapy response exists in certain circumstances. Upcoming research looks to establish more direct relationships between particularly impactful microbes or microbial communities that can most reliably increase treatment response in specific cancer types.

Summary

The use of FMT as a treatment for several human disease states has recently been revived through modern clinical trial strategies. This section reviewed the utility of FMT in *C. difficile* infection and inflammatory bowel diseases and as a potential adjunct to cancer treatments. As noted by the recency of clinical trials in this space, the field is active and rapidly expanding, with the exact utility of FMT in many conditions still yet to be determined. For instance, trials of FMT plus common therapies will need to occur based on individual cancer types to determine which conditions would benefit from adding FMT. Additionally, there may be particular aspects of IBD conditions in which FMT is more efficacious than others, which will only be determined through further studies in human patients. Therefore, this field remains a source of untapped potential, especially in a society where microbial gut diversity is decreasing,[187] individuals are living longer, and environmental influences are becoming more associated with disease states. Moving forward, FMT as therapy or supplementation may be considered in various health and disease states.

CONCLUSIONS

Research has identified a set of normal physiologic conditions affected by the microbiome that, when disrupted, lead to disease. The broad categories of these conditions include metabolism, the immune system, the nervous system, and cancer, which touch on all aspects of health. While experiments have demonstrated that, for several diseases, the arrow of causality points from microbes to the host, all diseases include feedback mechanisms by which negative effects exacerbate over time. However, we also described research into how the microbiome can be modified, giving a potential lever by which we could disrupt or modify these feedback loops.

Methods emphasized here include medications, diet, exercise, microbiome-targeted supplements (pre-, pro-, syn-, and postbiotics), and FMT. As the field uncovers additional mechanisms by which microbes interact with the host, these methods will increasingly be used to repair the microbiome to alleviate disease. Each method is active on different timescales, which may require combining multiple methods. For example, a cancer patient with a new diagnosis of metastatic melanoma may find their microbiome is not optimal for treatment response. To start immunotherapy as soon as possible, this patient may need medications (antibiotics) to optimize FMT engraftment, followed by a probiotic to seed more of the most beneficial organisms, and then a diet and exercise regimen optimized to maintain a pro-response community. On the other hand, disease prevention may take a less intense form, more like tending a microbial garden, with consumption of diverse fruits and vegetables, regular exercise, regular monitoring of the gut microbiome, and the use of autologous FMT following disruptions. While most of the effort in this space is associated with increasing the abundance of specific taxa, the selective depletion of others without disrupting the entire community is also desirable. Microbe-specific antibiotics hold promise in this regard and may be more limited by financial considerations than science. These and other tools will be needed to use the microbiome to rationally modulate human health and disease.

ACKNOWLEDGEMENTS

This work was partially supported by an Alpha Omega Alpha Carolyn L. Kuckein Student Research Fellowship (D. Grencewicz); a Samuel J. Roessler Memorial Scholarship (D. Grencewicz), the National Institute of Allergy and Infectious Diseases (T32AI106704, J. Chalif), the National Institute on Aging (K01AG070310, D. Spakowicz), the American Lung Association (D. Spakowicz), the American Cancer Society (D. Spakowicz), and the Lung Cancer Foundation of America (D. Spakowicz).

REFERENCES

1. Claesson, M. J., Jeffery, I. B., Conde, S., Power, S. E., O'Connor, E. M., Cusack, S., Harris, H. M. B., Coakley, M., Lakshminarayanan, B., O'Sullivan, O., Fitzgerald, G. F., Deane, J., O'Connor, M., Harnedy, N., O'Connor, K., O'Mahony, D., Van Sinderen, D., Wallace, M., Brennan, L., Stanton, C., Marchesi, J. R., Fitzgerald, A. P., Shanahan, F., Hill, C., Ross, R. P. & O'Toole, P. W. Gut microbiota composition correlates with diet and health in the elderly. *Nature* **488**, 178–184 (2012).
2. Yatsunenko, T., Rey, F. E., Manary, M. J., Trehan, I., Dominguez-Bello, M. G., Contreras, M., Magris, M., Hidalgo, G., Baldassano, R. N., Anokhin, A. P., Heath, A. C., Warner, B., Reeder, J., Kuczynski, J., Caporaso, J. G., Lozupone, C. A., Lauber, C., Clemente, J. C., Knights, D., Knight, R. & Gordon, J. I. Human gut microbiome viewed across age and geography. *Nature* **486**, 222–227 (2012).
3. Zivkovic, A. M., German, J. B., Lebrilla, C. B. & Mills, D. A. Human milk glycobiome and its impact on the infant gastrointestinal microbiota. *Proc. Natl. Acad. Sci.* **108**, 4653–4658 (2011).
4. O'Toole, P. W. & Jeffery, I. B. Gut microbiota and aging. *Science* **350**, 1214–1215 (2015).
5. Ticinesi, A., Tana, C. & Nouvenne, A. The intestinal microbiome and its relevance for functionality in older persons. *Curr. Opin. Clin. Nutr. Metab. Care* **22**, 4–12 (2019).
6. Rondanelli, M., Giacosa, A., Faliva, M. A., Perna, S., Allieri, F. & Castellazzi, A. M. Review on microbiota and effectiveness of probiotics use in older. *World J. Clin. Cases WJCC* **3**, 156 (2015).
7. Claesson, M. J., Cusack, S., O'Sullivan, O., Greene-Diniz, R., de Weerd, H., Flannery, E., Marchesi, J. R., Falush, D., Dinan, T. & Fitzgerald, G. Composition, variability, and temporal stability of the intestinal microbiota of the elderly. *Proc. Natl. Acad. Sci.* **108**, 4586–4591 (2011).
8. Wang, F., Yu, T., Huang, G., Cai, D., Liang, X., Su, H., Zhu, Z., Li, D., Yang, Y. & Shen, P. Gut microbiota community and its assembly associated with age and diet in Chinese centenarians. *J. Microbiol. Biotechnol.* **25**, 1195–1204 (2015).
9. Biagi, E., Rampelli, S., Turroni, S., Quercia, S., Candela, M. & Brigidi, P. The gut microbiota of centenarians: signatures of longevity in the gut microbiota profile. *Mech. Ageing Dev.* **165**, 180–184 (2017).
10. Odamaki, T., Kato, K., Sugahara, H., Hashikura, N., Takahashi, S., Xiao, J., Abe, F. & Osawa, R. Age-related changes in gut microbiota composition from newborn to centenarian: A cross-sectional study. *BMC Microbiol.* **16**, 1–12 (2016).
11. Biagi, E., Franceschi, C., Rampelli, S., Severgnini, M., Ostan, R., Turroni, S., Consolandi, C., Quercia, S., Scurti, M. & Monti, D. Gut microbiota and extreme longevity. *Curr. Biol.* **26**, 1480–1485 (2016).
12. Goodrich, J. K., Waters, J. L., Poole, A. C., Sutter, J. L., Koren, O., Blekhman, R., Beaumont, M., Van Treuren, W., Knight, R. & Bell, J. T. Human genetics shape the gut microbiome. *Cell* **159**, 789–799 (2014).
13. Shin, H. E., Kwak, S. E., Lee, J.-H., Zhang, D., Bae, J. H. & Song, W. Exercise, the gut microbiome, and frailty. *Ann. Geriatr. Med. Res.* **23**, 105 (2019).
14. van Tongeren, S. P., Slaets, J. P., Harmsen, H. J. M. & Welling, G. W. Fecal microbiota composition and frailty. *Appl. Environ. Microbiol.* **71**, 6438–6442 (2005).
15. Bäckhed, F., Ding, H., Wang, T., Hooper, L. V., Koh, G. Y., Nagy, A., Semenkovich, C. F. & Gordon, J. I. The gut microbiota as an environmental factor that regulates fat storage. *Proc. Natl. Acad. Sci.* **101**, 15718–15723 (2004).
16. Fan, Y. & Pedersen, O. Gut microbiota in human metabolic health and disease. *Nat. Rev. Microbiol.* **19**, 55–71 (2021).
17. Ridaura, V. K., Faith, J. J., Rey, F. E., Cheng, J., Duncan, A. E., Kau, A. L., Griffin, N. W., Lombard, V., Henrissat, B., Bain, J. R., Muehlbauer, M. J., Ilkayeva, O., Semenkovich, C. F., Funai, K., Hayashi, D. K., Lyle, B. J., Martini, M. C., Ursell, L. K., Clemente, J. C., Van Treuren, W., Walters, W. A., Knight, R., Newgard, C. B., Heath, A. C. & Gordon, J. I. Gut microbiota from twins discordant for obesity modulate metabolism in mice. *Science* **341**, 1241214 (2013).

18. Liu, R., Hong, J., Xu, X., Feng, Q., Zhang, D., Gu, Y., Shi, J., Zhao, S., Liu, W., Wang, X., Xia, H., Liu, Z., Cui, B., Liang, P., Xi, L., Jin, J., Ying, X., Wang, X., Zhao, X., Li, W., Jia, H., Lan, Z., Li, F., Wang, R., Sun, Y., Yang, M., Shen, Y., Jie, Z., Li, J., Chen, X., Zhong, H., Xie, H., Zhang, Y., Gu, W., Deng, X., Shen, B., Xu, X., Yang, H., Xu, G., Bi, Y., Lai, S., Wang, J., Qi, L., Madsen, L., Wang, J., Ning, G., Kristiansen, K. & Wang, W. Gut microbiome and serum metabolome alterations in obesity and after weight-loss intervention. *Nat. Med.* **23**, 859–868 (2017).
19. Zhou, Z., Sun, B., Yu, D. & Zhu, C. Gut microbiota: An important player in type 2 diabetes mellitus. *Front. Cell. Infect. Microbiol.* **12**, 834485 (2022).
20. Kimura, H. Histone modifications for human epigenome analysis. *J. Hum. Genet.* **58**, 439–445 (2013).
21. Depommier, C., Everard, A., Druart, C., Plovier, H., Van Hul, M., Vieira-Silva, S., Falony, G., Raes, J., Maiter, D., Delzenne, N. M., De Barsy, M., Loumaye, A., Hermans, M. P., Thissen, J.-P., De Vos, W. M. & Cani, P. D. Supplementation with Akkermansia muciniphila in overweight and obese human volunteers: A proof-of-concept exploratory study. *Nat. Med.* **25**, 1096–1103 (2019).
22. Hevia, A., Milani, C., López, P., Cuervo, A., Arboleya, S., Duranti, S., Turroni, F., González, S., Suárez, A., Gueimonde, M., Ventura, M., Sánchez, B. & Margolles, A. Intestinal dysbiosis associated with systemic lupus erythematosus. *mBio* **5**, e01548–14 (2014).
23. Luo, X. M., Edwards, M. R., Mu, Q., Yu, Y., Vieson, M. D., Reilly, C. M., Ahmed, S. A. & Bankole, A. A. Gut microbiota in human systemic lupus erythematosus and a mouse model of lupus. *Appl. Environ. Microbiol.* **84**, e02288–17 (2018).
24. Ma, Y., Xu, X., Li, M., Cai, J., Wei, Q. & Niu, H. Gut microbiota promote the inflammatory response in the pathogenesis of systemic lupus erythematosus. *Mol. Med.* **25**, 35 (2019).
25. Zegarra-Ruiz, D. F., El Beidaq, A., Iñiguez, A. J., Lubrano Di Ricco, M., Manfredo Vieira, S., Ruff, W. E., Mubiru, D., Fine, R. L., Sterpka, J., Greiling, T. M., Dehner, C. & Kriegel, M. A. A diet-sensitive commensal lactobacillus strain mediates TLR7-dependent systemic autoimmunity. *Cell Host Microbe* **25**, 113–127. (2019).
26. Zhao, T., Wei, Y., Zhu, Y., Xie, Z., Hai, Q., Li, Z. & Qin, D. Gut microbiota and rheumatoid arthritis: From pathogenesis to novel therapeutic opportunities. *Front. Immunol.* **13**, 1007165 (2022).
27. Zheng, P., Li, Z. & Zhou, Z. Gut microbiome in type 1 diabetes: A comprehensive review. *Diabetes Metab. Res. Rev.* **34**, e3043 (2018).
28. Ochoa-Repáraz, J., Kirby, T. O. & Kasper, L. H. The gut microbiome and multiple sclerosis. *Cold Spring Harb. Perspect. Med.* **8**, a029017 (2018).
29. Chen, Y., Xu, J. & Chen, Y. Regulation of neurotransmitters by the gut microbiota and effects on cognition in neurological disorders. *Nutrients* **13**, 2099 (2021).
30. Heijtz, R. D., Wang, S., Anuar, F., Qian, Y., Björkholm, B., Samuelsson, A., Hibberd, M. L., Forssberg, H. & Pettersson, S. Normal gut microbiota modulates brain development and behavior. *Proc. Natl. Acad. Sci.* **108**, 3047–3052 (2011).
31. Curran, E. A., O'Neill, S. M., Cryan, J. F., Kenny, L. C., Dinan, T. G., Khashan, A. S. & Kearney, P. M. Research review: Birth by caesarean section and development of autism spectrum disorder and attention-deficit/hyperactivity disorder: A systematic review and meta-analysis. *J. Child Psychol. Psychiatry* **56**, 500–508 (2015).
32. Rosenfeld, C. S. Microbiome disturbances and autism spectrum disorders. *Drug Metab. Dispos.* **43**, 1557–1571 (2015).
33. Sudo, N., Chida, Y., Aiba, Y., Sonoda, J., Oyama, N., Yu, X., Kubo, C. & Koga, Y. Postnatal microbial colonization programs the hypothalamic–pituitary–adrenal system for stress response in mice. *J. Physiol.* **558**, 263–275 (2004).
34. Vuong, H. E., Yano, J. M., Fung, T. C. & Hsiao, E. Y. The microbiome and host behavior. *Annu. Rev. Neurosci.* **40**, 21–49 (2017).
35. Bravo, J. A., Forsythe, P., Chew, M. V., Escaravage, E., Savignac, H. M., Dinan, T. G., Bienenstock, J. & Cryan, J. F. Ingestion of *Lactobacillus* strain regulates emotional behavior and central GABA receptor expression in a mouse via the vagus nerve. *Proc. Natl. Acad. Sci.* **108**, 16050–16055 (2011).
36. Foster, J. A., Rinaman, L. & Cryan, J. F. Stress & the gut-brain axis: Regulation by the microbiome. *Neurobiol. Stress* **7**, 124–136 (2017).
37. Hanahan, D. Hallmarks of cancer: New dimensions. *Cancer Discov.* **12**, 31–46 (2022).
38. McCarthy, E. F. The toxins of William B. Coley and the treatment of bone and soft-tissue sarcomas. *Iowa Orthop. J.* **26**, 154–158 (2006).

39. Petrelli, F., Ghidini, M., Ghidini, A., Perego, G., Cabiddu, M., Khakoo, S., Oggionni, E., Abeni, C., Hahne, J. C., Tomasello, G. & Zaniboni, A. Use of antibiotics and risk of cancer: A systematic review and meta-analysis of observational studies. *Cancers* **11**, 1174 (2019).
40. Zhang, J., Haines, C., Watson, A. J. M., Hart, A. R., Platt, M. J., Pardoll, D. M., Cosgrove, S. E., Gebo, K. A. & Sears, C. L. Oral antibiotic use and risk of colorectal cancer in the United Kingdom, 1989–2012: A matched case–control study. *Gut* **68**, 1971–1978 (2019).
41. Cullin, N., Azevedo Antunes, C., Straussman, R., Stein-Thoeringer, C. K. & Elinav, E. Microbiome and cancer. *Cancer Cell* **39**, 1317–1341 (2021).
42. Nejman, D., Livyatan, I., Fuks, G., Gavert, N., Zwang, Y., Geller, L. T., Rotter-Maskowitz, A., Weiser, R., Mallel, G., Gigi, E., Meltser, A., Douglas, G. M., Kamer, I., Gopalakrishnan, V., Dadosh, T., Levin-Zaidman, S., Avnet, S., Atlan, T., Cooper, Z. A., Arora, R., Cogdill, A. P., Khan, M. A. W., Ologun, G., Bussi, Y., Weinberger, A., Lotan-Pompan, M., Golani, O., Perry, G., Rokah, M., Bahar-Shany, K., Rozeman, E. A., Blank, C. U., Ronai, A., Shaoul, R., Amit, A., Dorfman, T., Kremer, R., Cohen, Z. R., Harnof, S., Siegal, T., Yehuda-Shnaidman, E., Gal-Yam, E. N., Shapira, H., Baldini, N., Langille, M. G. I., Ben-Nun, A., Kaufman, B., Nissan, A., Golan, T., Dadiani, M., Levanon, K., Bar, J., Yust-Katz, S., Barshack, I., Peeper, D. S., Raz, D. J., Segal, E., Wargo, J. A., Sandbank, J., Shental, N. & Straussman, R. The human tumor microbiome is composed of tumor type–specific intracellular bacteria. *Science* **368**, 973–980 (2020).
43. Dohlman, A. B., Klug, J., Mesko, M., Gao, I. H., Lipkin, S. M., Shen, X. & Iliev, I. D. A pan-cancer mycobiome analysis reveals fungal involvement in gastrointestinal and lung tumors. *Cell* **185**, 3807–3822 (2022).
44. Bender, M. J., McPherson, A. C., Phelps, C. M., Pandey, S. P., Laughlin, C. R., Shapira, J. H., Medina Sanchez, L., Rana, M., Richie, T. G., Mims, T. S., Gocher-Demske, A. M., Cervantes-Barragan, L., Mullett, S. J., Gelhaus, S. L., Bruno, T. C., Cannon, N., McCulloch, J. A., Vignali, D. A. A., Hinterleitner, R., Joglekar, A. V., Pierre, J. F., Lee, S. T. M., Davar, D., Zarour, H. M. & Meisel, M. Dietary tryptophan metabolite released by intratumoral Lactobacillus reuteri facilitates immune checkpoint inhibitor treatment. *Cell* **186**, 1846–1862. (2023).
45. Aykut, B., Pushalkar, S., Chen, R., Li, Q., Abengozar, R., Kim, J. I., Shadaloey, S. A., Wu, D., Preiss, P., Verma, N., Guo, Y., Saxena, A., Vardhan, M., Diskin, B., Wang, W., Leinwand, J., Kurz, E., Kochen Rossi, J. A., Hundeyin, M., Zambrinis, C., Li, X., Saxena, D. & Miller, G. The fungal mycobiome promotes pancreatic oncogenesis via activation of MBL. *Nature* **574**, 264–267 (2019).
46. Mager, L. F., Burkhard, R., Pett, N., Cooke, N. C. A., Brown, K., Ramay, H., Paik, S., Stagg, J., Groves, R. A., Gallo, M., Lewis, I. A., Geuking, M. B. & McCoy, K. D. Microbiome-derived inosine modulates response to checkpoint inhibitor immunotherapy. *Science* **369**, 1481–1489 (2020).
47. Vilcant, V. & Hai, O. In StatPearls (StatPearls Publishing) (2023). At http://www.ncbi.nlm.nih.gov/books/NBK470547/
48. Klein, R. S., Harris, C. A., Small, C. B., Moll, B., Lesser, M. & Friedland, G. H. Oral candidiasis in high-risk patients as the initial manifestation of the acquired immunodeficiency syndrome. *N. Engl. J. Med.* **311**, 354–358 (1984).
49. Hoarau, G., Mukherjee, P. K., Gower-Rousseau, C., Hager, C., Chandra, J., Retuerto, M. A., Neut, C., Vermeire, S., Clemente, J., Colombel, J. F., Fujioka, H., Poulain, D., Sendid, B. & Ghannoum, M. A. Bacteriome and mycobiome interactions underscore microbial dysbiosis in familial Crohn's disease. *mBio* **7**, e01250–16 (2016).
50. Magee, L. C., Louis, M., Khan, V., Micalo, L. & Chaudary, N. Managing fungal infections in cystic fibrosis patients: Challenges in clinical practice. *Infect. Drug Resist.* **14**, 1141–1153 (2021).
51. Weersma, R. K., Zhernakova, A. & Fu, J. Interaction between drugs and the gut microbiome. *Gut* **69**, 1510–1519 (2020).
52. Buffie, C. G., Bucci, V., Stein, R. R., McKenney, P. T., Ling, L., Gobourne, A., No, D., Liu, H., Kinnebrew, M., Viale, A., Littmann, E., Van Den Brink, M. R. M., Jenq, R. R., Taur, Y., Sander, C., Cross, J. R., Toussaint, N. C., Xavier, J. B. & Pamer, E. G. Precision microbiome reconstitution restores bile acid mediated resistance to Clostridium difficile. *Nature* **517**, 205–208 (2015).
53. Chey, W. D., Leontiadis, G. I., Howden, C. W. & Moss, S. F. ACG clinical guideline: Treatment of helicobacter pylori infection. *Am. J. Gastroenterol.* **112**, 212–239 (2017).
54. Wroblewski, L. E., Peek, R. M. & Wilson, K. T. *Helicobacter pylori* and Gastric Cancer: Factors That Modulate Disease Risk. *Clin. Microbiol. Rev.* **23**, 713–739 (2010).
55. Falony, G., Joossens, M., Vieira-Silva, S., Wang, J., Darzi, Y., Faust, K., Kurilshikov, A., Bonder, M. J., Valles-Colomer, M., Vandeputte, D., Tito, R. Y., Chaffron, S., Rymenans, L., Verspecht, C., De Sutter,

L., Lima-Mendez, G., Jonckheere, K., Homola, D., Garcia, R., Tigchelaar, E. F., Eeckhaudt, L., Fu, J., Henckaerts, L., Zhernakova, A., Wijmenga, C. & Raes, J. Population-level analysis of gut microbiome variation. *Science* **352**, 560–564 (2016).
56. Wu, H., Esteve, E., Tremaroli, V., Khan, M. T., Caesar, R., Mannerås-Holm, L., Ståhlman, M., Olsson, L. M., Serino, M., Planas-Fèlix, M., Xifra, G., Mercader, J. M., Torrents, D., Burcelin, R., Ricart, W., Perkins, R., Fernàndez-Real, J. M. & Bäckhed, F. Metformin alters the gut microbiome of individuals with treatment-naive type 2 diabetes, contributing to the therapeutic effects of the drug. *Nat. Med.* **23**, 850–858 (2017).
57. Laxminarayan, R., Duse, A., Wattal, C., Zaidi, A. K. M., Wertheim, H. F. L., Sumpradit, N., Vlieghe, E., Hara, G. L., Gould, I. M., Goossens, H., Greko, C., So, A. D., Bigdeli, M., Tomson, G., Woodhouse, W., Ombaka, E., Peralta, A. Q., Qamar, F. N., Mir, F., Kariuki, S., Bhutta, Z. A., Coates, A., Bergstrom, R., Wright, G. D., Brown, E. D. & Cars, O. Antibiotic resistance—the need for global solutions. *Lancet Infect. Dis.* **13**, 1057–1098 (2013).
58. Van Boeckel, T. P., Brower, C., Gilbert, M., Grenfell, B. T., Levin, S. A., Robinson, T. P., Teillant, A. & Laxminarayan, R. Global trends in antimicrobial use in food animals. *Proc. Natl. Acad. Sci.* **112**, 5649–5654 (2015).
59. Davies, J. & Davies, D. Origins and evolution of antibiotic resistance. *Microbiol. Mol. Biol. Rev.* **74**, 417–433 (2010).
60. Kourtis, A. P., Hatfield, K., Baggs, J., Mu, Y., See, I., Epson, E., Nadle, J., Kainer, M. A., Dumyati, G., Petit, S., Ray, S. M., Emerging Infections Program MRSA author group, Ham, D., Capers, C., Ewing, H., Coffin, N., McDonald, L. C., Jernigan, J. & Cardo, D. *Vital Signs*: Epidemiology and recent trends in methicillin-resistant and in methicillin-susceptible *Staphylococcus aureus* Bloodstream infections – United States. *MMWR Morb. Mortal. Wkly. Rep.* **68**, 214–219 (2019).
61. Zhu, Y.-G., Zhao, Y., Zhu, D., Gillings, M., Penuelas, J., Ok, Y. S., Capon, A. & Banwart, S. Soil biota, antimicrobial resistance and planetary health. *Environ. Int.* **131**, 105059 (2019).
62. Bengtsson-Palme, J., Kristiansson, E. & Larsson, D. G. J. Environmental factors influencing the development and spread of antibiotic resistance. *FEMS Microbiol. Rev.* **42**, (2018).
63. Penfold, W. J. Variability in the gas-forming power of intestinal bacteria. *Proc. R. Soc. Med.* **4**, 97–109 (1911).
64. Trumbo, P., Schlicker, S., Yates, A. A., Poos, M. & Food and Nutrition Board of the Institute of Medicine, The National Academies. Dietary reference intakes for energy, carbohydrate, fiber, fat, fatty acids, cholesterol, protein and amino acids. *J. Am. Diet. Assoc.* **102**, 1621–1630 (2002).
65. Dahl, W. J., Rivero Mendoza, D. & Lambert, J. M. Diet, nutrients and the microbiome. *Prog. Mol. Biol. Transl. Sci.* **171**, 237–263 (2020).
66. Johnson, A. J., Vangay, P., Al-Ghalith, G. A., Hillmann, B. M., Ward, T. L., Shields-Cutler, R. R., Kim, A. D., Shmagel, A. K., Syed, A. N., Walter, J., Menon, R., Koecher, K. & Knights, D. Daily sampling reveals personalized diet-microbiome associations in humans. *Cell Host Microbe* **25**, 789–802. (2019).
67. Ludwig, D. S. Carbohydrate-insulin model: does the conventional view of obesity reverse cause and effect? *Philos. Trans. R. Soc. Lond. B. Biol. Sci.* **378**, 20220211 (2023).
68. Rinninella, E., Cintoni, M., Raoul, P., Lopetuso, L. R., Scaldaferri, F., Pulcini, G., Miggiano, G. A. D., Gasbarrini, A. & Mele, M. C. Food components and dietary habits: Keys for a healthy gut microbiota composition. *Nutrients* 11, 2393 (2019).
69. Macfarlane, G. T. & Macfarlane, S. Bacteria, colonic fermentation, and gastrointestinal health. *J. AOAC Int.* **95**, 50–60 (2012).
70. Mills, S., Lane, J. A., Smith, G. J., Grimaldi, K. A., Ross, R. P. & Stanton, C. Precision nutrition and the microbiome Part II: Potential opportunities and pathways to commercialisation. *Nutrients* **11**, 1468 (2019).
71. Englyst, H., Wiggins, H. S. & Cummings, J. H. Determination of the non-starch polysaccharides in plant foods by gas-liquid chromatography of constituent sugars as alditol acetates. *The Analyst* **107**, 307–318 (1982).
72. Sullivan, W. R. in Chem. Prop. Starch (IntechOpen, 2020). doi:10.5772/intechopen.90159
73. Martínez, I., Kim, J., Duffy, P. R., Schlegel, V. L. & Walter, J. Resistant starches types 2 and 4 have differential effects on the composition of the fecal microbiota in human subjects. *PloS One* **5**, e15046 (2010).
74. Yang, X., Darko, K. O., Huang, Y., He, C., Yang, H., He, S., Li, J., Li, J., Hocher, B. & Yin, Y. Resistant starch regulates gut microbiota: Structure, biochemistry and cell signalling. *Cell. Physiol. Biochem. Int. J. Exp. Cell. Physiol. Biochem. Pharmacol.* **42**, 306–318 (2017).

75. Le Bastard, Q., Chapelet, G., Javaudin, F., Lepelletier, D., Batard, E. & Montassier, E. The effects of inulin on gut microbial composition: A systematic review of evidence from human studies. *Eur. J. Clin. Microbiol. Infect. Dis. Off. Publ. Eur. Soc. Clin. Microbiol.* **39**, 403–413 (2020).
76. Moszak, M., Szulińska, M. & Bogdański, P. You are what you eat-the relationship between diet, microbiota, and metabolic disorders-a review. *Nutrients* **12**, 1096 (2020).
77. Wolters, M., Ahrens, J., Romaní-Pérez, M., Watkins, C., Sanz, Y., Benítez-Páez, A., Stanton, C. & Günther, K. Dietary fat, the gut microbiota, and metabolic health - A systematic review conducted within the MyNewGut project. *Clin. Nutr. Edinb. Scotl.* **38**, 2504–2520 (2019).
78. Watson, H., Mitra, S., Croden, F. C., Taylor, M., Wood, H. M., Perry, S. L., Spencer, J. A., Quirke, P., Toogood, G. J., Lawton, C. L., Dye, L., Loadman, P. M. & Hull, M. A. A randomised trial of the effect of omega-3 polyunsaturated fatty acid supplements on the human intestinal microbiota. *Gut* **67**, 1974–1983 (2018).
79. Tamura, K., Sasaki, H., Shiga, K., Miyakawa, H. & Shibata, S. The timing effects of soy protein intake on mice gut microbiota. *Nutrients* **12**, 87 (2019).
80. Nasser, A., Dallal, M. M. S., Jahanbakhshi, S., Azimi, T. & Nikouei, L. Staphylococcus aureus: Biofilm formation and strategies against it. *Curr. Pharm. Biotechnol.* **23**, 664–678 (2022).
81. Barrea, L., Annunziata, G., Muscogiuri, G., Laudisio, D., Di Somma, C., Maisto, M., Tenore, G. C., Colao, A. & Savastano, S. Trimethylamine N-oxide, mediterranean diet, and nutrition in healthy, normal-weight adults: Also a matter of sex? *Nutr. Burbank Los Angel. Cty. Calif* **62**, 7–17 (2019).
82. Zhao, J., Zhang, X., Liu, H., Brown, M. A. & Qiao, S. Dietary protein and gut microbiota composition and function. *Curr. Protein Pept. Sci.* **20**, 145–154 (2019).
83. Ma, N. & Ma, X. Dietary amino acids and the gut-microbiome-immune axis: Physiological metabolism and therapeutic prospects. *Compr. Rev. Food Sci. Food Saf.* **18**, 221–242 (2019).
84. Naliyadhara, N., Kumar, A., Kumar Gangwar, S., Nair Devanarayanan, T., Hegde, M., Alqahtani, M. S., Abbas, M., Sethi, G. & Kunnumakkara, A. Interplay of dietary antioxidants and gut microbiome in human health: What has been learnt thus far? *J. Funct. Foods* **100**, 105365 (2023).
85. Wan, M. L. Y., Co, V. A. & El-Nezami, H. Dietary polyphenol impact on gut health and microbiota. *Crit. Rev. Food Sci. Nutr.* **61**, 690–711 (2021).
86. Zamora-Ros, R., Knaze, V., Rothwell, J. A., Hémon, B., Moskal, A., Overvad, K., Tjønneland, A., Kyrø, C., Fagherazzi, G., Boutron-Ruault, M.-C., Touillaud, M., Katzke, V., Kühn, T., Boeing, H., Förster, J., Trichopoulou, A., Valanou, E., Peppa, E., Palli, D., Agnoli, C., Ricceri, F., Tumino, R., De Magistris, M. S., Peeters, P. H. M., Bueno-de-Mesquita, H. B., Engeset, D., Skeie, G., Hjartåker, A., Menéndez, V., Agudo, A., Molina-Montes, E., Huerta, J. M., Barricarte, A., Amiano, P., Sonestedt, E., Nilsson, L. M., Landberg, R., Key, T. J., Khaw, K.-T., Wareham, N. J., Lu, Y., Slimani, N., Romieu, I., Riboli, E. & Scalbert, A. Dietary polyphenol intake in Europe: the European Prospective Investigation into Cancer and Nutrition (EPIC) study. *Eur. J. Nutr.* **55**, 1359–1375 (2016).
87. Scalbert, A., Johnson, I. T. & Saltmarsh, M. Polyphenols: Antioxidants and beyond. *Am. J. Clin. Nutr.* **81**, 215S–217S (2005).
88. Nishino, A., Maoka, T. & Yasui, H. Preventive effects of β-cryptoxanthin, a potent antioxidant and provitamin a carotenoid, on lifestyle-related diseases—A central focus on its effects on Non-Alcoholic Fatty Liver Disease (NAFLD). *Antioxidants* **11**, 43 (2021).
89. Abdel-Aal, E.-S., Akhtar, H., Zaheer, K. & Ali, R. Dietary sources of lutein and zeaxanthin carotenoids and their role in eye health. *Nutrients* **5**, 1169–1185 (2013).
90. Saini, R. K., Prasad, P., Lokesh, V., Shang, X., Shin, J., Keum, Y.-S. & Lee, J.-H. Carotenoids: Dietary sources, extraction, encapsulation, bioavailability, and health benefits—A review of recent advancements. *Antioxidants* **11**, 795 (2022).
91. Cesarino, I. Going red but not mad: Efficient astaxanthin production in tobacco without yield penalty. *Plant Physiol.* **188**, 35–37 (2022).
92. Zuluaga, M., Gueguen, V., Letourneur, D. & Pavon-Djavid, G. Astaxanthin-antioxidant impact on excessive reactive oxygen species generation induced by ischemia and reperfusion injury. *Chem. Biol. Interact.* **279**, 145–158 (2018).
93. Wu, L., Lyu, Y., Srinivasagan, R., Wu, J., Ojo, B., Tang, M., El-Rassi, G. D., Metzinger, K., Smith, B. J., Lucas, E. A., Clarke, S. L., Chowanadisai, W., Shen, X., He, H., Conway, T., Von Lintig, J. & Lin, D. Astaxanthin-shifted gut microbiota is associated with inflammation and metabolic homeostasis in mice. *J. Nutr.* **150**, 2687–2698 (2020).
94. Pehlivan, F. E. in *Vitam. C* (ed. Hamza, A. H.) (InTech, 2017). doi:10.5772/intechopen.69660

95. Ngo, B., Van Riper, J. M., Cantley, L. C. & Yun, J. Targeting cancer vulnerabilities with high-dose vitamin C. *Nat. Rev. Cancer* **19**, 271–282 (2019).
96. Aggarwal, V., Tuli, H., Varol, A., Thakral, F., Yerer, M., Sak, K., Varol, M., Jain, A., Khan, M. D. & Sethi, G. Role of reactive oxygen species in cancer progression: Molecular mechanisms and recent advancements. *Biomolecules* **9**, 735 (2019).
97. Huang, K., Dong, W., Liu, W., Yan, Y., Wan, P., Peng, Y., Xu, Y., Zeng, X. & Cao, Y. 2-*O*-β-d-Glucopyranosyl- l -ascorbic acid, an ascorbic acid derivative isolated from the fruits of *Lycium Barbarum* L., modulates gut microbiota and palliates colitis in dextran sodium sulfate-induced colitis in mice. *J. Agric. Food Chem.* **67**, 11408–11419 (2019).
98. Li, Y., Zafar, S., Salih Ibrahim, R. M., Chi, H.-L., Xiao, T., Xia, W., Li, H.-B. & Kang, Y.-M. Exercise and food supplement of vitamin C ameliorate hypertension through improvement of gut microflora in the spontaneously hypertensive rats. *Life Sci.* **269**, 119097 (2021).
99. Zhang, Y., Roh, Y. J., Han, S.-J., Park, I., Lee, H. M., Ok, Y. S., Lee, B. C. & Lee, S.-R. Role of selenoproteins in redox regulation of signaling and the antioxidant system: A review. *Antioxidants* **9**, 383 (2020).
100. Nasser, A., Jahanbakhshi, S., Soltan Dallal, M. M., Banar, M., Sattari-Maraji, A. & Azimi, T. Staphylococcus aureus dormancy: Waiting for insurgency. *Curr. Pharm. Biotechnol.* **24**, 1898–1915 (2023).
101. Chen, J., Zhang, S., Chen, C., Jiang, X., Qiu, J., Qiu, Y., Zhang, Y., Wang, T., Qin, X., Zou, Z. & Chen, C. Crosstalk of gut microbiota and serum/hippocampus metabolites in neurobehavioral impairments induced by zinc oxide nanoparticles. *Nanoscale* **12**, 21429–21439 (2020).
102. Takahashi, K., Suzuki, N. & Ogra, Y. Effect of gut microflora on nutritional availability of selenium. *Food Chem.* **319**, 126537 (2020).
103. Parizadeh, M. & Arrieta, M.-C. The global human gut microbiome: genes, lifestyles, and diet. *Trends Mol. Med.* **29**, 789–801 (2023).
104. Tosti, V., Bertozzi, B. & Fontana, L. Health benefits of the mediterranean diet: Metabolic and molecular mechanisms. *J. Gerontol. Ser. A* **73**, 318–326 (2018).
105. Kahleova, H., Levin, S. & Barnard, N. Cardio-metabolic benefits of plant-based diets. *Nutrients* **9**, 848 (2017).
106. Widmer, R. J., Flammer, A. J., Lerman, L. O. & Lerman, A. The mediterranean diet, its components, and cardiovascular disease. *Am. J. Med.* **128**, 229–238 (2015).
107. Parker, H. W. & Vadiveloo, M. K. Diet quality of vegetarian diets compared with nonvegetarian diets: A systematic review. *Nutr. Rev.* **77**, 144–160 (2019).
108. Patel, K. P., Luo, F. J.-G., Plummer, N. S., Hostetter, T. H. & Meyer, T. W. The production of p-cresol sulfate and indoxyl sulfate in vegetarians versus omnivores. *Clin. J. Am. Soc. Nephrol.* **7**, 982–988 (2012).
109. Trefflich, I., Jabakhanji, A., Menzel, J., Blaut, M., Michalsen, A., Lampen, A., Abraham, K. & Weikert, C. Is a vegan or a vegetarian diet associated with the microbiota composition in the gut? Results of a new cross-sectional study and systematic review. *Crit. Rev. Food Sci. Nutr.* **60**, 2990–3004 (2020).
110. De Angelis, M., Ferrocino, I., Calabrese, F. M., De Filippis, F., Cavallo, N., Siragusa, S., Rampelli, S., Di Cagno, R., Rantsiou, K., Vannini, L., Pellegrini, N., Lazzi, C., Turroni, S., Lorusso, N., Ventura, M., Chieppa, M., Neviani, E., Brigidi, P., O'Toole, P. W., Ercolini, D., Gobbetti, M. & Cocolin, L. Diet influences the functions of the human intestinal microbiome. *Sci. Rep.* **10**, 4247 (2020).
111. Allen, J. M., Mailing, L. J., Niemiro, G. M., Moore, R., Cook, M. D., White, B. A., Holscher, H. D. & Woods, J. A. Exercise alters gut microbiota composition and function in lean and obese humans. *Med. Sci. Sports Exerc.* **50**, 747–757 (2018).
112. Liu, Z., Liu, H.-Y., Zhou, H., Zhan, Q., Lai, W., Zeng, Q., Ren, H. & Xu, D. Moderate-intensity exercise affects gut microbiome composition and influences cardiac function in myocardial infarction mice. *Front. Microbiol.* **8**, 1687 (2017).
113. Ticinesi, A., Tana, C., Nouvenne, A., Prati, B., Lauretani, F. & Meschi, T. Gut microbiota, cognitive frailty and dementia in older individuals: A systematic review. *Clin. Interv. Aging* **13**, 1497–1511 (2018).
114. Wang, J., Zhang, Q., Xia, J. & Sun, H. Moderate treadmill exercise modulates gut microbiota and improves intestinal barrier in high-fat-diet-induced obese mice via the AMPK/CDX2 signaling pathway. *Diabetes Metab. Syndr. Obes. Targets Ther.* **15**, 209–223 (2022).
115. Cataldi, S., Bonavolontà, V., Poli, L., Clemente, F. M., De Candia, M., Carvutto, R., Silva, A. F., Badicu, G., Greco, G. & Fischetti, F. The relationship between physical activity, physical exercise, and human gut microbiota in healthy and unhealthy subjects: A systematic review. *Biology* **11**, 479 (2022).

116. Blondell, S. J., Hammersley-Mather, R. & Veerman, J. L. Does physical activity prevent cognitive decline and dementia?: A systematic review and meta-analysis of longitudinal studies. *BMC Public Health* **14**, 510 (2014).
117. Hoffman-Goetz, L., Pervaiz, N., Packer, N. & Guan, J. Freewheel training decreases pro- and increases anti-inflammatory cytokine expression in mouse intestinal lymphocytes. *Brain. Behav. Immun.* **24**, 1105–1115 (2010).
118. Packer, N. & Hoffman-Goetz, L. Exercise training reduces inflammatory mediators in the intestinal tract of healthy older adult mice. *Can. J. Aging Rev. Can. Vieil.* **31**, 161–171 (2012).
119. Morita, Yokoyama, Imai, Takeda, Ota, Kawai, Hisada, Emoto, Suzuki, & Okazaki. Aerobic exercise training with brisk walking increases intestinal bacteroides in healthy elderly women. *Nutrients* **11**, 868 (2019).
120. Taniguchi, H., Tanisawa, K., Sun, X., Kubo, T., Hoshino, Y., Hosokawa, M., Takeyama, H. & Higuchi, M. Effects of short-term endurance exercise on gut microbiota in elderly men. *Physiol. Rep.* **6**, e13935 (2018).
121. Li, C., Cai, Y.-Y. & Yan, Z.-X. Brain-derived neurotrophic factor preserves intestinal mucosal barrier function and alters gut microbiota in mice. *Kaohsiung J. Med. Sci.* **34**, 134–141 (2018).
122. Estaki, M., Pither, J., Baumeister, P., Little, J. P., Gill, S. K., Ghosh, S., Ahmadi-Vand, Z., Marsden, K. R. & Gibson, D. L. Cardiorespiratory fitness as a predictor of intestinal microbial diversity and distinct metagenomic functions. *Microbiome* **4**, 42 (2016).
123. Mitchell, C. M., Davy, B. M., Hulver, M. W., Neilson, A. P., Bennett, B. J. & Davy, K. P. Does exercise alter gut microbial composition? A systematic review. *Med. Sci. Sports Exerc.* **51**, 160–167 (2019).
124. Scheffer, D. D. L. & Latini, A. Exercise-induced immune system response: Anti-inflammatory status on peripheral and central organs. *Biochim. Biophys. Acta BBA - Mol. Basis Dis.* **1866**, 165823 (2020).
125. Odamaki, T., Kato, K., Sugahara, H., Hashikura, N., Takahashi, S., Xiao, J., Abe, F. & Osawa, R. Age-related changes in gut microbiota composition from newborn to centenarian: A cross-sectional study. *BMC Microbiol.* **16**, 90 (2016).
126. Barton, W., Penney, N. C., Cronin, O., Garcia-Perez, I., Molloy, M. G., Holmes, E., Shanahan, F., Cotter, P. D. & O'Sullivan, O. The microbiome of professional athletes differs from that of more sedentary subjects in composition and particularly at the functional metabolic level. *Gut gutjnl* 2016, 313627 (2017). doi:10.1136/gutjnl-2016-313627
127. Clark, A. & Mach, N. Exercise-induced stress behavior, gut-microbiota-brain axis and diet: A systematic review for athletes. *J. Int. Soc. Sports Nutr.* **13**, 43 (2016).
128. Araos, R., Andreatos, N., Ugalde, J., Mitchell, S., Mylonakis, E. & D'Agata, E. M. C. Fecal microbiome among nursing home residents with advanced dementia and clostridium difficile. *Dig. Dis. Sci.* **63**, 1525–1531 (2018).
129. Koblinsky, N. D., Power, K. A., Middleton, L., Ferland, G. & Anderson, N. D. The role of the gut microbiome in diet and exercise effects on cognition: A review of the intervention literature. *J. Gerontol. Ser. A* **78**, 195–205 (2023).
130. Kang, S. S., Jeraldo, P. R., Kurti, A., Miller, M. E. B., Cook, M. D., Whitlock, K., Goldenfeld, N., Woods, J. A., White, B. A., Chia, N. & Fryer, J. D. Diet and exercise orthogonally alter the gut microbiome and reveal independent associations with anxiety and cognition. *Mol. Neurodegener.* **9**, 36 (2014).
131. Den Besten, G., Van Eunen, K., Groen, A. K., Venema, K., Reijngoud, D.-J. & Bakker, B. M. The role of short-chain fatty acids in the interplay between diet, gut microbiota, and host energy metabolism. *J. Lipid Res.* **54**, 2325–2340 (2013).
132. Den Besten, G., Gerding, A., Van Dijk, T. H., Ciapaite, J., Bleeker, A., Van Eunen, K., Havinga, R., Groen, A. K., Reijngoud, D.-J. & Bakker, B. M. Protection against the metabolic syndrome by guar gum-derived short-chain fatty acids depends on peroxisome proliferator-activated receptor γ and glucagon-like peptide-1. *Plos One* **10**, e0136364 (2015).
133. Munukka, E., Rintala, A., Toivonen, R., Nylund, M., Yang, B., Takanen, A., Hänninen, A., Vuopio, J., Huovinen, P. & Jalkanen, S. Faecalibacterium prausnitzii treatment improves hepatic health and reduces adipose tissue inflammation in high-fat fed mice. *ISME J.* **11**, 1667–1679 (2017).
134. Aragón-Vela, J., Solis-Urra, P., Ruiz-Ojeda, F. J., Álvarez-Mercado, A. I., Olivares-Arancibia, J. & Plaza-Diaz, J. Impact of exercise on gut microbiota in obesity. *Nutrients* **13**, 3999 (2021).
135. Ramos, C., Gibson, G. R., Walton, G. E., Magistro, D., Kinnear, W. & Hunter, K. Systematic review of the effects of exercise and physical activity on the gut microbiome of older adults. *Nutrients* **14**, 674 (2022).

136. Mc Gettigan, N., O'Toole, A. & Boland, K. "Role of exercise in preventing and restoring gut dysbiosis in patients with inflammatory bowel disease": A letter to the editor. *World J. Gastroenterol.* **28**, 878–890 (2022).
137. Petersen, L. M., Bautista, E. J., Nguyen, H., Hanson, B. M., Chen, L., Lek, S. H., Sodergren, E. & Weinstock, G. M. Community characteristics of the gut microbiomes of competitive cyclists. *Microbiome* **5**, 98 (2017).
138. Morishima, S., Aoi, W., Kawamura, A., Kawase, T., Takagi, T., Naito, Y., Tsukahara, T. & Inoue, R. Intensive, prolonged exercise seemingly causes gut dysbiosis in female endurance runners. *J. Clin. Biochem. Nutr.* **68**, 253–258 (2021).
139. Roberts, J., Suckling, C., Peedle, G., Murphy, J., Dawkins, T. & Roberts, M. An exploratory investigation of endotoxin levels in novice long distance triathletes, and the effects of a multi-strain probiotic/prebiotic, Antioxidant Intervention. *Nutrients* **8**, 733 (2016).
140. Buchman, A. S. & Bennett, D. A. Cognitive frailty. *J. Nutr. Health Aging* **17**, 738–739 (2013).
141. Xu, R. & Wang, Q. Towards understanding brain-gut-microbiome connections in Alzheimer's disease. *BMC Syst. Biol.* **10**, 63 (2016).
142. Wei, Y. B., Melas, P. A., Wegener, G., Mathe, A. A. & Lavebratt, C. Antidepressant-like effect of sodium butyrate is associated with an increase in TET1 and in 5-Hydroxymethylation levels in the Bdnf gene. *Int. J. Neuropsychopharmacol.* **18**, pyu032–pyu032 (2015).
143. Hill, C., Guarner, F., Reid, G., Gibson, G. R., Merenstein, D. J., Pot, B., Morelli, L., Canani, R. B., Flint, H. J., Salminen, S., Calder, P. C. & Sanders, M. E. The international scientific association for probiotics and prebiotics consensus statement on the scope and appropriate use of the term probiotic. *Nat. Rev. Gastroenterol. Hepatol.* **11**, 506–514 (2014).
144. Bodke, H. & Jogdand, S. Role of probiotics in human health. *Cureus* **14**, e31313.
145. Office of Dietary Supplements – Probiotics. At https://ods.od.nih.gov/factsheets/Probiotics-HealthProfessional/
146. Bjarnason, I., Sission, G. & Hayee, B. A randomised, double-blind, placebo-controlled trial of a multi-strain probiotic in patients with asymptomatic ulcerative colitis and Crohn's disease. *Inflammopharmacology* **27**, 465–473 (2019).
147. Wang, W., Xing, W., Wei, S., Gao, Q., Wei, X., Shi, L., Kong, Y. & Su, Z. Semi-rational screening of probiotics from the fecal flora of healthy adults against DSS-induced colitis mice by enhancing anti-inflammatory activity and modulating the gut microbiota. *J. Microbiol. Biotechnol.* **29**, 1478–1487 (2019).
148. Rifkin, S. B., Giardiello, F. M., Zhu, X., Hylind, L. M., Ness, R. M., Drewes, J. L., Murff, H. J., Spence, E. H., Smalley, W. E., Gills, J. J., Mullin, G. E., Kafonek, D., Luna, L. L., Zheng, W., Sears, C. L. & Shrubsole, M. J. Yogurt consumption and colorectal polyps. *Br. J. Nutr.* **124**, 80–91 (2020).
149. Michels, K. B., Willett, W. C., Vaidya, R., Zhang, X. & Giovannucci, E. Yogurt consumption and colorectal cancer incidence and mortality in the Nurses' Health Study and the Health Professionals Follow-Up Study. *Am. J. Clin. Nutr.* **112**, 1566–1575 (2020).
150. Piwat, S., Sophatha, B. & Teanpaisan, R. An assessment of adhesion, aggregation and surface charges of Lactobacillus strains derived from the human oral cavity. *Lett. Appl. Microbiol.* **61**, 98–105 (2015).
151. Riccia, D. D., Bizzini, F., Perilli, M., Polimeni, A., Trinchieri, V., Amicosante, G. & Cifone, M. Anti-inflammatory effects of Lactobacillus brevis (CD2) on periodontal disease. *Oral Dis.* **13**, 376–385 (2007).
152. Spencer, C. N., McQuade, J. L., Gopalakrishnan, V., McCulloch, J. A., Vetizou, M., Cogdill, A. P., Khan, A. W., Zhang, X., White, M. G., Peterson, C. B., Wong, M. C., Morad, G., Rodgers, T., Badger, J. H., Helmink, B. A., Andrews, M. C., Rodrigues, R. R., Morgun, A., Kim, Y. S., Roszik, J., Hoffman, K. L., Zheng, J., Zhou, Y., Medik, Y. B., Kahn, L. M., Johnson, S., Hudgens, C. W., Wani, K., Gaudreau, P.-O., Harris, A. L., Jamal, M. A., Baruch, E. N., Perez-Guijarro, E., Day, C.-P., Merlino, G., Pazdrak, B., Lochmann, B. S., Szczepaniak-Sloane, R. A., Arora, R., Anderson, J., Zobniw, C. M., Posada, E., Sirmans, E., Simon, J., Haydu, L. E., Burton, E. M., Wang, L., Dang, M., Clise-Dwyer, K., Schneider, S., Chapman, T., Anang, N.-A. A. S., Duncan, S., Toker, J., Malke, J. C., Glitza, I. C., Amaria, R. N., Tawbi, H. A., Diab, A., Wong, M. K., Patel, S. P., Woodman, S. E., Davies, M. A., Ross, M. I., Gershenwald, J. E., Lee, J. E., Hwu, P., Jensen, V., Samuels, Y., Straussman, R., Ajami, N. J., Nelson, K. C., Nezi, L., Petrosino, J. F., Futreal, P. A., Lazar, A. J., Hu, J., Jenq, R. R., Tetzlaff, M. T., Yan, Y., Garrett, W. S., Huttenhower, C., Sharma, P., Watowich, S. S., Allison, J. P., Cohen, L., Trinchieri, G., Daniel, C. R. & Wargo, J. A. Dietary fiber and probiotics influence the gut microbiome and melanoma immunotherapy response. *Science* **374**, 1632–1640 (2021).

153. CFR - Code of Federal Regulations Title 21. at https://www.accessdata.fda.gov/scripts/cdrh/cfdocs/cfcfr/cfrsearch.cfm?fr=131.200
154. Nutrition, C. for F. S. and A. FDA Issues Draft Guidance on the Labeling of Dietary Supplements Containing Live Microbials. *FDA* (2018). https://www.fda.gov/food/hfp-constituent-updates/fda-issues-draft-guidance-labeling-dietary-supplements-containing-live-microbials
155. Degnan, F. H. The US food and drug administration and probiotics: Regulatory categorization. *Clin. Infect. Dis.* **46**, S133–S136 (2008).
156. Gibson, G. R., Hutkins, R., Sanders, M. E., Prescott, S. L., Reimer, R. A., Salminen, S. J., Scott, K., Stanton, C., Swanson, K. S., Cani, P. D., Verbeke, K. & Reid, G. Expert consensus document: The International Scientific Association for Probiotics and Prebiotics (ISAPP) consensus statement on the definition and scope of prebiotics. *Nat. Rev. Gastroenterol. Hepatol.* **14**, 491–502 (2017).
157. Davani-Davari, D., Negahdaripour, M., Karimzadeh, I., Seifan, M., Mohkam, M., Masoumi, S. J., Berenjian, A. & Ghasemi, Y. Prebiotics: Definition, types, sources, mechanisms, and clinical applications. *Foods* **8**(3), 92 (2019).
158. Walker, A. W., Duncan, S. H., McWilliam Leitch, E. C., Child, M. W. & Flint, H. J. pH and peptide supply can radically alter bacterial populations and short-chain fatty acid ratios within microbial communities from the human colon. *Appl. Environ. Microbiol.* **71**, 3692–3700 (2005).
159. Zafar, H. & Saier, M. H. Gut Bacteroides species in health and disease. *Gut Microbes* **13**, 1848158
160. Chen, R. Y., Mostafa, I., Hibberd, M. C., Das, S., Mahfuz, M., Naila, N. N., Islam, M. M., Huq, S., Alam, M. A., Zaman, M. U., Raman, A. S., Webber, D., Zhou, C., Sundaresan, V., Ahsan, K., Meier, M. F., Barratt, M. J., Ahmed, T. & Gordon, J. I. A microbiota-directed food intervention for undernourished children. *N. Engl. J. Med.* **384**, 1517–1528 (2021).
161. Markowiak, P. & Śliżewska, K. Effects of probiotics, prebiotics, and synbiotics on human health. *Nutrients* **9**, 1021 (2017).
162. Swanson, K. S., Gibson, G. R., Hutkins, R., Reimer, R. A., Reid, G., Verbeke, K., Scott, K. P., Holscher, H. D., Azad, M. B., Delzenne, N. M. & Sanders, M. E. The International Scientific Association for Probiotics and Prebiotics (ISAPP) consensus statement on the definition and scope of synbiotics. *Nat. Rev. Gastroenterol. Hepatol.* **17**, 687–701 (2020).
163. Eslamparast, T., Zamani, F., Hekmatdoost, A., Sharafkhah, M., Eghtesad, S., Malekzadeh, R. & Poustchi, H. Effects of synbiotic supplementation on insulin resistance in subjects with the metabolic syndrome: A randomised, double-blind, placebo-controlled pilot study. *Br. J. Nutr.* **112**, 438–445 (2014).
164. Lee, C., Kim, B. G., Kim, J. H., Chun, J., Im, J. P. & Kim, J. S. Sodium butyrate inhibits the NF-kappa B signaling pathway and histone deacetylation, and attenuates experimental colitis in an IL-10 independent manner. *Int. Immunopharmacol.* **51**, 47–56 (2017).
165. Chaiyasut, C., Sivamaruthi, B. S., Kesika, P., Khongtan, S., Khampithum, N., Thangaleela, S., Peerajan, S., Bumrungpert, A., Chaiyasut, K., Sirilun, S. & Sittiprapaporn, P. Synbiotic supplementation improves obesity index and metabolic biomarkers in Thai obese adults: A randomized clinical trial. *Foods* **10**, 1580 (2021).
166. Salminen, S., Collado, M. C., Endo, A., Hill, C., Lebeer, S., Quigley, E. M. M., Sanders, M. E., Shamir, R., Swann, J. R., Szajewska, H. & Vinderola, G. The International Scientific Association of Probiotics and Prebiotics (ISAPP) consensus statement on the definition and scope of postbiotics. *Nat. Rev. Gastroenterol. Hepatol.* **18**, 649–667 (2021).
167. Vinderola, G., Sanders, M. E. & Salminen, S. The concept of postbiotics. *Foods* **11**, 1077 (2022).
168. Martyniak, A., Medyńska-Przęczek, A., Wędrychowicz, A., Skoczeń, S. & Tomasik, P. J. Prebiotics, probiotics, synbiotics, paraprobiotics and postbiotic compounds in IBD. *Biomolecules* **11**, 1903 (2021).
169. Gill, P. A., van Zelm, M. C., Muir, J. G. & Gibson, P. R. Review article: short chain fatty acids as potential therapeutic agents in human gastrointestinal and inflammatory disorders. *Aliment. Pharmacol. Ther.* **48**, 15–34 (2018).
170. Melbye, P., Olsson, A., Hansen, T. H., Søndergaard, H. B. & Bang Oturai, A. Short-chain fatty acids and gut microbiota in multiple sclerosis. *Acta Neurol. Scand.* **139**, 208–219 (2019).
171. Thorburn, A. N., McKenzie, C. I., Shen, S., Stanley, D., Macia, L., Mason, L. J., Roberts, L. K., Wong, C. H. Y., Shim, R., Robert, R., Chevalier, N., Tan, J. K., Mariño, E., Moore, R. J., Wong, L., McConville, M. J., Tull, D. L., Wood, L. G., Murphy, V. E., Mattes, J., Gibson, P. G. & Mackay, C. R. Evidence that asthma is a developmental origin disease influenced by maternal diet and bacterial metabolites. *Nat. Commun.* **6**, 7320 (2015).

172. Andresen, V., Gschossmann, J. & Layer, P. Heat-inactivated Bifidobacterium bifidum MIMBb75 (SYN-HI-001) in the treatment of irritable bowel syndrome: A multicentre, randomised, double-blind, placebo-controlled clinical trial. *Lancet Gastroenterol. Hepatol.* **5**, 658–666 (2020).
173. Wegh, C. A. M., Geerlings, S. Y., Knol, J., Roeselers, G. & Belzer, C. Postbiotics and their potential applications in early life nutrition and beyond. *Int. J. Mol. Sci.* **20**, 4673 (2019).
174. Zagato, E., Mileti, E., Massimiliano, L., Fasano, F., Budelli, A., Penna, G. & Rescigno, M. Lactobacillus paracasei CBA L74 metabolic products and fermented milk for infant formula have anti-inflammatory activity on dendritic cells in vitro and protective effects against colitis and an enteric pathogen in vivo. *PloS One* **9**, e87615 (2014).
175. Stripling, J. & Rodriguez, M. Current evidence in delivery and therapeutic uses of fecal microbiota transplantation in human diseases—Clostridium difficile disease and beyond. *Am. J. Med. Sci.* **356**, 424–432 (2018).
176. Brandt, L. J. & Aroniadis, O. C. An overview of fecal microbiota transplantation: techniques, indications, and outcomes. *Gastrointest. Endosc.* **78**, 240–249 (2013).
177. Van Nood, E., Vrieze, A., Nieuwdorp, M., Fuentes, S., Zoetendal, E. G., De Vos, W. M., Visser, C. E., Kuijper, E. J., Bartelsman, J. F. W. M., Tijssen, J. G. P., Speelman, P., Dijkgraaf, M. G. W. & Keller, J. J. Duodenal infusion of donor feces for recurrent *Clostridium difficile*. *N. Engl. J. Med.* **368**, 407–415 (2013).
178. Moayyedi, P., Surette, M. G., Kim, P. T., Libertucci, J., Wolfe, M., Onischi, C., Armstrong, D., Marshall, J. K., Kassam, Z., Reinisch, W. & Lee, C. H. Fecal microbiota transplantation induces remission in patients with active ulcerative colitis in a randomized controlled trial. *Gastroenterology* **149**, 102–109. (2015).
179. Suskind, D. L., Brittnacher, M. J., Wahbeh, G., Shaffer, M. L., Hayden, H. S., Qin, X., Singh, N., Damman, C. J., Hager, K. R., Nielson, H. & Miller, S. I. Fecal microbial transplant effect on clinical outcomes and fecal microbiome in active Crohn's disease: *Inflamm. Bowel Dis.* **21**, 556–563 (2015).
180. Fehily, S. R., Basnayake, C., Wright, E. K. & Kamm, M. A. Fecal microbiota transplantation therapy in Crohn's disease: Systematic review. *J. Gastroenterol. Hepatol.* **36**, 2672–2686 (2021).
181. Paramsothy, S., Nielsen, S., Kamm, M. A., Deshpande, N. P., Faith, J. J., Clemente, J. C., Paramsothy, R., Walsh, A. J., Van Den Bogaerde, J., Samuel, D., Leong, R. W. L., Connor, S., Ng, W., Lin, E., Borody, T. J., Wilkins, M. R., Colombel, J.-F., Mitchell, H. M. & Kaakoush, N. O. Specific bacteria and metabolites associated with response to fecal microbiota transplantation in patients with ulcerative colitis. *Gastroenterology* **156**, 1440–1454. (2019).
182. Paramsothy, S., Kamm, M. A., Kaakoush, N. O., Walsh, A. J., Van Den Bogaerde, J., Samuel, D., Leong, R. W. L., Connor, S., Ng, W., Paramsothy, R., Xuan, W., Lin, E., Mitchell, H. M. & Borody, T. J. Multidonor intensive faecal microbiota transplantation for active ulcerative colitis: A randomised placebo-controlled trial. *The Lancet* **389**, 1218–1228 (2017).
183. Saint-Antoine IBD Network, Sokol, H., Landman, C., Seksik, P., Berard, L., Montil, M., Nion-Larmurier, I., Bourrier, A., Le Gall, G., Lalande, V., De Rougemont, A., Kirchgesner, J., Daguenel, A., Cachanado, M., Rousseau, A., Drouet, É., Rosenzwajg, M., Hagege, H., Dray, X., Klatzman, D., Marteau, P., Beaugerie, L. & Simon, T. Fecal microbiota transplantation to maintain remission in Crohn's disease: A pilot randomized controlled study. *Microbiome* **8**, 12 (2020).
184. Gopalakrishnan, V., Spencer, C. N., Nezi, L., Reuben, A., Andrews, M. C., Karpinets, T. V., Prieto, P. A., Vicente, D., Hoffman, K., Wei, S. C., Cogdill, A. P., Zhao, L., Hudgens, C. W., Hutchinson, D. S., Manzo, T., Petaccia De Macedo, M., Cotechini, T., Kumar, T., Chen, W. S., Reddy, S. M., Szczepaniak Sloane, R., Galloway-Pena, J., Jiang, H., Chen, P. L., Shpall, E. J., Rezvani, K., Alousi, A. M., Chemaly, R. F., Shelburne, S., Vence, L. M., Okhuysen, P. C., Jensen, V. B., Swennes, A. G., McAllister, F., Marcelo Riquelme Sanchez, E., Zhang, Y., Le Chatelier, E., Zitvogel, L., Pons, N., Austin-Breneman, J. L., Haydu, L. E., Burton, E. M., Gardner, J. M., Sirmans, E., Hu, J., Lazar, A. J., Tsujikawa, T., Diab, A., Tawbi, H., Glitza, I. C., Hwu, W. J., Patel, S. P., Woodman, S. E., Amaria, R. N., Davies, M. A., Gershenwald, J. E., Hwu, P., Lee, J. E., Zhang, J., Coussens, L. M., Cooper, Z. A., Futreal, P. A., Daniel, C. R., Ajami, N. J., Petrosino, J. F., Tetzlaff, M. T., Sharma, P., Allison, J. P., Jenq, R. R. & Wargo, J. A. Gut microbiome modulates response to anti–PD-1 immunotherapy in melanoma patients. *Science* **359**, 97–103 (2018).
185. Baruch, E. N., Youngster, I., Ben-Betzalel, G., Ortenberg, R., Lahat, A., Katz, L., Adler, K., Dick-Necula, D., Raskin, S., Bloch, N., Rotin, D., Anafi, L., Avivi, C., Melnichenko, J., Steinberg-Silman, Y.,

Mamtani, R., Harati, H., Asher, N., Shapira-Frommer, R., Brosh-Nissimov, T., Eshet, Y., Ben-Simon, S., Ziv, O., Khan, M. A. W., Amit, M., Ajami, N. J., Barshack, I., Schachter, J., Wargo, J. A., Koren, O., Markel, G. & Boursi, B. Fecal microbiota transplant promotes response in immunotherapy-refractory melanoma patients. *Science* **371**, 602–609 (2021).

186. Routy, B., Lenehan, J. G., Miller, W. H., Jamal, R., Messaoudene, M., Daisley, B. A., Hes, C., Al, K. F., Martinez-Gili, L., Punčochář, M., Ernst, S., Logan, D., Belanger, K., Esfahani, K., Richard, C., Ninkov, M., Piccinno, G., Armanini, F., Pinto, F., Krishnamoorthy, M., Figueredo, R., Thebault, P., Takis, P., Magrill, J., Ramsay, L., Derosa, L., Marchesi, J. R., Parvathy, S. N., Elkrief, A., Watson, I. R., Lapointe, R., Segata, N., Haeryfar, S. M. M., Mullish, B. H., Silverman, M. S., Burton, J. P. & Maleki Vareki, S. Fecal microbiota transplantation plus anti-PD-1 immunotherapy in advanced melanoma: A phase I trial. *Nat. Med.* **29**, 2121–2132 (2023).

187. Vangay, P., Johnson, A. J., Ward, T. L., Al-Ghalith, G. A., Shields-Cutler, R. R., Hillmann, B. M., Lucas, S. K., Beura, L. K., Thompson, E. A., Till, L. M., Batres, R., Paw, B., Pergament, S. L., Saenyakul, P., Xiong, M., Kim, A. D., Kim, G., Masopust, D., Martens, E. C., Angkurawaranon, C., McGready, R., Kashyap, P. C., Culhane-Pera, K. A. & Knights, D. US immigration westernizes the human gut microbiome. *Cell* **175**, 962–972. (2018).

7 Soil Health and Nutrient Density through a One Health Approach

Raiza Rezende, Cristina Laurenti, Eirini Tsirimokou, Sonia Blough, and Zuzanna Zielinska

INTRODUCTION

Soil, often taken for granted, is the foundation of life on Earth, with an astounding 95% of the food we consume being rooted in its depths. This uppermost layer of the Earth's crust is shaped by a complex interplay of physical, chemical, and biological processes, resulting in a dynamic ecosystem that includes mineral particles, organic matter, water, air, and a diverse community of living organisms. These components are structured into genetic soil horizons, which reflect the soil's capacity to sustain life. The health of the soil is intimately linked to the nutrient density of the crops it supports, directly influencing the nutritional quality of our food. This link has become more apparent in recent years as the nutrient density of common crops has continued to decline and food production outputs have suffered from soil degradation. The rise of industrial agriculture over the 21st century, deemed necessary to sustain the planet's growing population, has led to severe consequences for both ecosystems and human health.

Modern industrial agriculture often relies on intensive practices that degrade soil health. The widespread use of synthetic fertilizers, monocropping, and heavy machinery disrupts the natural balance of the soil, leading to erosion, loss of organic matter, and a decline in soil biodiversity. This degradation not only reduces the soil's ability to produce food but also diminishes the nutrient content of the crops, contributing to the global problem of nutrient deficiencies in human diets. Understanding the symbiotic relationship between soil health and crop nutrient density is crucial for developing sustainable agricultural practices that can ensure the long-term resilience of global food systems. Sustainable farming methods, such as agroecology and regenerative agriculture, emphasize soil regeneration, biodiversity, and the use of natural inputs. These practices not only help to restore and maintain soil health but also enhance the nutrient content of crops, ultimately improving food quality and human health.

In this chapter, we will explore the intricate connections between soil health and nutrient density in crops. Soil health is a challenging concept to quantify or define, given that soil itself is an incredibly complex ecosystem, with new insights about its composition and function continually emerging. However, what we do know is that soil health is essential to both human and planetary well-being, and it is imperative that we begin to prioritize the care of this living system that underpins all other forms of life. Finally, we will examine several sustainable agricultural models that are gaining increased attention as awareness of the importance of soil health continues to grow. By supporting a just transition to regenerative practices that restore and sustain soil health, we can safeguard our planet's health and the well-being of future generations.

 DOI: 10.1201/9781003469742-7

SOIL HEALTH DEFINITION

The concept of soil health had not been officially defined until recently (FAO, 2020). Early definitions of soil health were primarily anthropocentric, focusing on the soil's capacity to support human needs such as food and fiber production, and secondarily on maintaining other ecosystem services like climate regulation and biodiversity conservation.

Beginning in the 1950s, numerous authors have proposed various definitions of soil health. Dr. Rattan Lal, in his paper "Soil Health and Carbon Management" (Lal, 2016), makes an important distinction between soil health and soil quality. While these terms are related and share similarities, they should not be used interchangeably. A commonly used definition of soil health by (Pankhurst, Doube & Gupta, 1997) describes it as the "capacity of soil to function as a vital living system to sustain biological productivity, maintain environmental quality, and promote plant, animal, and human health."

Soil quality, on the other hand, relates to soil functions or what it does. These functions include supporting plant and animal productivity, regulating water, recycling nutrients, providing habitat for organisms, filtering pollutants, storing carbon, and sustaining biodiversity. However, soil functions are difficult to quantify and are usually measured through the assessment of soil quality indicators (SQIs). These indicators include a wide range of parameters, such as soil texture, bulk density, soil structure, porosity, soil depth, soil pH, electrical conductivity (EC), cation exchange capacity (CEC), nutrient levels, organic matter content, heavy metals, microbial biomass, soil respiration, enzyme activity, earthworm population, root health, soil fauna diversity, soil erosion, compaction, etc.

Two main elements distinguish soil health from soil quality: the time component, which reflects the soil's ability to continue functioning over time, and the recognition of soil as a vital living system, emphasizing the importance of soil biota to soil functioning (FAO, 2008)

"The question is not what is in the soil that can be measured, but what the soil does that must be quantified." (Lal, 2016). This means that to understand soil health, we should not simply measure a multitude of soil parameters. Instead, we should focus on understanding how soil functions (soil quality) affect soil health, and quantifying them by assessing relevant soil quality indicators (SQIs).

Soil health can also be defined in terms of soil properties: namely, physical health, chemical health, and biological health. Soil physical health can be characterized by aspects such as structure, texture, and density. Healthy physical properties of soil ensure good aeration, moisture retention, and root development, all of which are essential for robust plant growth and maintaining ecological balance. Soil biological health is the ability of soil to support large and diverse microbial communities, suppress pathogens, and support healthy crop development. Soil chemical health is defined by an optimum quantity of plant nutrients that are bioavailable and in balanced proportions, and which are available to plants without obstruction by other chemical compounds and properties. Soil chemical health also considers the presence or absence of harmful soil agrochemicals and pollutants (Shahane & Shivay, 2021).

The idea of the connection between soil health and human health is not new. Many scriptures throughout time bring the notion of this interconnectivity. As far back as approximately 1400BCE, bible passages from Moses describe his understanding that fertile soil was essential to the well-being of his people. In 400 BCE, the Greek philosopher Hippocrates, often referred to as the "Father of Medicine," provided a list of things that should be considered in a proper medical evaluation, and that list included soil.

Despite these early recognitions, the scientific investigation of soils and their impact on human health remains a relatively new field (Brevik and Sauer, 2012). Our separation from the land and the growth of our food systems have disconnected us from soil. We must reconnect with soil, if we want humans, animals and plants to be healthy.

SOIL DEGRADATION AND STATUS GLOBALLY

Degraded soils have impaired ability to sustain one health – soil, plant, animal and human health. Soil degradation takes many forms, encompassing damage in the physical, biological, and chemical quality of soil and resulting in erosion, poor nutrient content, and biodiversity loss, namely the bacteria and microorganisms in the soil that contribute to its ability to sustain ecosystems.

Soil erosion is broadly defined as the accelerated loss of topsoil from the land surface due to water, wind, or tillage. While soil erosion is a natural process occurring under all climatic conditions, it is significantly accelerated by human activities such as intensive agriculture, deforestation, overgrazing, and improper land use changes. These activities can increase erosion rates by up to 1,000 times (FAO, 2015).

On agricultural land, water erosion happens when rain or runoff loosens soil particles, and the resulting overland flow washes them away, often creating visible channels in the soil. Wind erosion takes place when dry, loose, bare soil is exposed to strong winds, which is particularly common in semi-arid areas where winds can easily mobilize soil particles during dry periods. Tillage erosion involves the direct downhill movement of soil, causing particles to be redistributed within a field. This leads to spatial variability of crop yields and lower overall yields (FAO, 2015).

Soil erosion decreases agricultural productivity, degrades ecosystem functions, increases hydrogeological risks, and leads to significant biodiversity losses and damage to infrastructure. As rapid soil loss outpaces soil formation, it is becoming more evident that protecting soil health is vital for sustaining life on Earth. The threat posed by soil erosion and land degradation extends to global food security and the achievement of Sustainable Development Goals (SDGs).

Currently, 33% of the Earth's soils are degraded, with projections indicating that over 90% could be degraded by 2050. This degradation is most likely to occur in areas with the largest remaining arable land, which could result in significant reductions in global crop yields. By 2050, it is estimated that over 90% of land areas will be substantially degraded, 4 billion people will live in drylands, and up to 700 million people may be forced to migrate. Annually, 75 billion tons of fertile soil are lost to erosion, and approximately 12 million hectares of land are degraded, reducing precious soil resources for future generations (FAO, 2020).

Assessing the case of the EU, a severe unified degradation of soils has been noticed. Approximately 115 million hectares, or 12% of Europe's land, are affected by water erosion, while wind erosion impacts 42 million hectares. About 45% of European soils have low organic matter content, especially in southern Europe, France, the UK, and Germany. Changes in land use, as observed in the Corine Land Cover database, have significantly affected soil, with 2.8% of Europe's land experiencing use changes between 1990 and 2000, including increased urbanization. The extent of surface sealing during this period varied widely among member states and regions, ranging from 0.3% to 10% (European Commission, 2006) _(Carla et al., 2012).

The European Mediterranean region has been identified as one of the most climate-vulnerable regions and a climate-change "hotspot". Soil degradation processes (physical, chemical, and biological) are linked with major soil threats in the Mediterranean countries, exhibiting the highest erosion rates, lowest levels of soil organic matter, severe salinization problems from inappropriate irrigation practices and seawater intrusion, and contamination, by heavy metals and pesticides (Ferreira et al., 2022). Southern Europe, as a semi-arid region, faces high risks of desertification, with 25% of its land identified as at high or very high risk (Ferreira et al., 2022).

Most soil degradation has been linked to modern anthropic causes, especially related to modern industrialized food production. Soil degradation can be attributed to a number of concomitant causes, from agricultural, industrial, and chemical pollution to urban expansion, unsustainable land management practices, and climatic changes (Rhodes C. J, 2014). However, we must recognize that soil degradation is not a modern issue.

In the book *Life in Syntropy*, authors Dayana Andrade and Felipe Pasini state "the lands where Jesus Christ walked were already degraded long before his birth". In the 10,000 years since the advent of agriculture – a mere 3% of our species' existence – many silvopastoral civilizations have risen and fallen. Despite their different belief systems, political structures, and social organizations, they all faced the same dilemma: the natural resources they relied on were consumed faster than they could regenerate. Soil degradation has sparked wars, invasions, and genocides throughout history, often attributed to political, ethnic, or religious causes.

From the Middle East to Europe, from China to the Pacific Islands, and from Australia to the Americas, the unsustainability of agriculture is not a modern problem, nor did it begin with the Industrial Revolution – though it was significantly exacerbated by it. The challenge of sustainable agriculture and the fate of civilizations that exploited their land is a recurring theme throughout history. In the book *Dirt: The Erosion of Civilizations*, David Montgomery invites us to examine the scars left in the soils of great civilizations like the Roman Empire. These agricultural societies followed a common pattern: population growth led to increased erosion due to intensified agriculture, overgrazing, and deforestation. The rapid depletion of natural resources inevitably resulted in instability, famine, disease, and conflict.

In the past, farmers were likely unaware of their impact on soil health and degradation, as changes occurred slowly and were noticed only across generations. Today, however, the negative effects of our agricultural practices are visible much more quickly, often within a single generation. As Felipe and Dayana reflect, while these times may feel more dramatic and intense, they also present opportunities for innovation. Crises are moments of opportunity, depending on how we choose to respond. If we can witness negative effects in one generation, we have the potential to see positive outcomes just as quickly – if we work in harmony with nature.

SOIL HEALTH & ONE HEALTH LINKAGE

> One Health is an integrated, unifying approach that aims to sustainably balance and optimize the health of people, animals, and ecosystems. It recognizes the health of humans, domestic and wild animals, plants, and the wider environment (including ecosystems) are closely linked and interdependent.
>
> *(One Health High-Level Expert Panel OHHLEP, 2022)*

In May 2021, four global organizations – the Food and Agriculture Organization (FAO), the World Organization for Animal Health (OIE), the United Nations Environment Programme (UNEP), and the World Health Organization (WHO) – created the One Health High-Level Expert Panel (OHHLEP) aiming to bring cooperation and efficiency across different sectors, industries, and communities, recognizing that a holistic approach is urgently needed to improve health across ecosystems.

There are five main principles defined under One Health, to ensure the holistic approach and applicability of the work done within it: (1) equity; (2) equal representation and inclusion across different cultures and social groups; (3) a balanced relationship between humans, animals, and the environment; (4) responsible care and the duty of humans to modify their actions; and (5) cross-disciplinary and multisector collaboration (One Health High-Level Expert Panel (OHHLEP, 2022).

Soil security is directly linked to food security and broadly defined as the maintenance and enhancement of the global soil resource to produce food, fiber, and fresh water; sustain biodiversity and ecosystem services; and support human health (Ohno & Hettiarachchi, 2018). The increasing global population and continued economic development are expected to intensify environmental stresses. Additionally, climate change impacts soil health and ecological processes, further affecting ecosystem health.

Soil health is pivotal to the concept of One Health, as soils are complex ecosystems that connect all life. Specifically, soil microbiomes perform over 40 functions that influence the health of soils, plants, animals, and humans and support ecosystem services (Figure 7.1). Hence, soil microbes play a crucial role in supporting One Health (Banerjee and van der Heijden, 2022). Focusing on soil

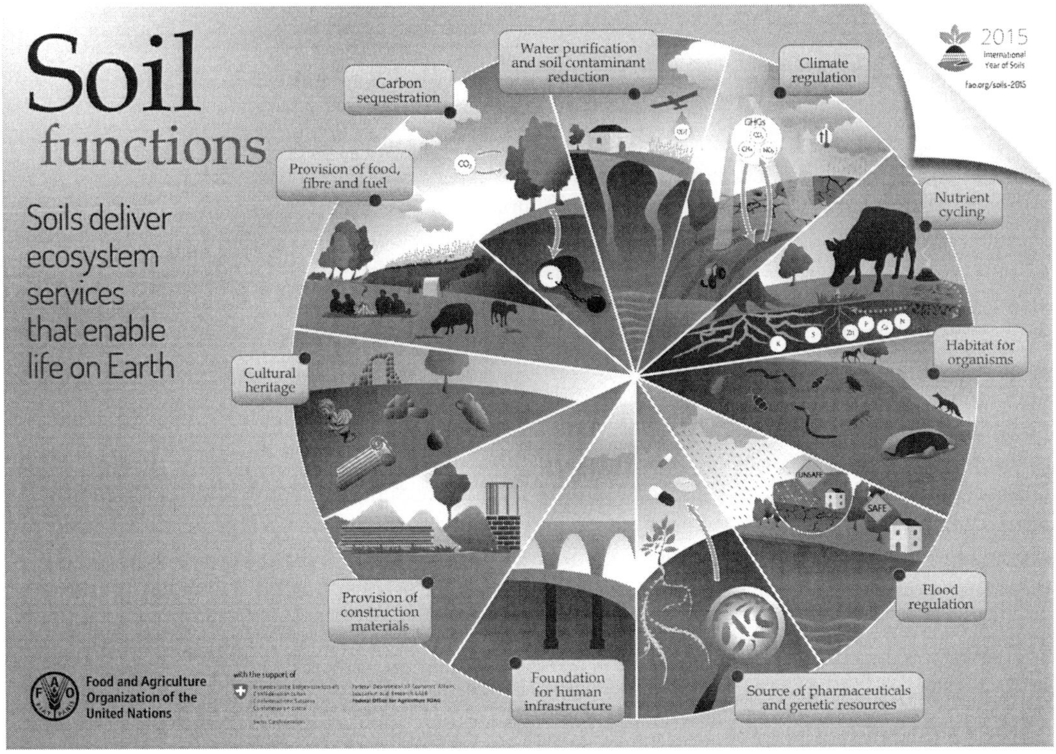

FIGURE 7.1 Soils deliver ecosystem services that enable life on Earth.
Source: © FAO 2015 "Soil Functions" https://www.fao.org/3/AX374E/ax374e.pdf. Date accessed 10/01/2022.

health provides solutions to critical One Health challenges, such as antibiotic resistance, zoonotic diseases, food insecurity and climate crisis.

Advocates of One Health recognize that our current agricultural system is harming soil, plant, animal, and human health, and a multistakeholder and cross-sector approach is necessary to tackle interconnected challenges like climate change, deforestation, zoonotic diseases, and antimicrobial resistance. However, the training of health professionals rarely covers soil health, and health professionals are rarely included in conversations about food systems. The contribution of the healthcare sector in this discussion is vital to be able to surmount these pressing global challenges in a more complete and efficient manner. Holistic education and interdisciplinary research are key to addressing One Health matters. Discussions prohealth in healthcare often focus on promoting plant-based diets, improving food safety, biosecurity, disease monitoring, and better use of antimicrobials. However, solutions must consider the relationship between soil and human health, the importance of microbiomes, the negative effects of chemical inputs, the relationship between biodiversity and the functioning of all living ecosystems. A global focus on soil health could lead to preventive strategies and stronger support and collaboration on One Health goals (Montgomery et al., 2024).

REGENERATIVE AGRICULTURE AND AGROECOLOGY AS TOOLS TO IMPROVE SOIL HEALTH

Regenerative agriculture and agroecology are well-known terms used to describe sustainable farming practices. They are both, at their core, principles that work with nature to improve soil health and human health. Although these terms are often used interchangeably, they have distinct definitions and applications by different stakeholders.

Regenerative agriculture has recently received significant attention not only from farmers and producers, but also from retailers, researchers, companies, consumers, governments, politicians, policymakers, and the mainstream media. Governments at various levels are exploring how regenerative agriculture can contribute to climate action plans and enhance food security. Meanwhile, private companies are integrating regenerative agriculture to build resilient supply chains (Newton et al., 2020). The authenticity of this shift has been questioned amongst scholars, as corporate plans to invest in regenerative agriculture are largely part of public relations campaigns, where the term "regenerative" takes on a very broad meaning.

Part of the problem with alternative movements gaining mainstream attention is that they can become watered down or appropriated to still serve the interest of those in power. Despite widespread interest, the term regenerative agriculture has no legal or regulatory definition, nor has a widely accepted definition emerged in common usage. Definitional ambiguity in regenerative agriculture means corporates and other actors can shape the discourse to fit their own goals, potentially resulting in greenwashing (Gordon et al., 2023). Some examples of major players investing in regenerative agriculture are Nestlé, Danone, and Bayer, the latter being one of the largest agrochemical producers in the world. This is one reason why it is imperative that a clear and shared definition of regenerative agriculture is adopted and enforced, to ensure that the growing momentum of this movement is not co-opted by the very actors responsible for creating many of the environmental problems in the first place.

Various authors, practitioners, and organizations have contributed to this ongoing discussion and generating various definitions. Definitions generally fall into two categories: process-based and outcome-based. Process-based definitions focus on specific practices such as integrating livestock, reducing or eliminating tillage, planting cover crops, incorporating perennials and trees, and promoting crop diversity (Newton et al., 2020). Francis and Harwood (1985) described the four principal tools used in regenerative agriculture: soil fertility practices, integrated pest management, advances in plant breeding, and integrated crop-animal systems.

Outcome-based definitions emphasize the results of these practices. Commonly mentioned outcomes include improving soil health, sequestering carbon, increasing biodiversity, enhancing watersheds, and boosting the social and economic well-being of communities (Newton et al., 2020). Grant (2017) defines regenerative agriculture as "any and all forms of agricultural practice that actively restore soil quality, biodiversity, ecosystem health, and water quality while producing sufficient food of high nutritional quality."

The absence of an agreed definition of regenerative agriculture has the potential to create several challenges. As discussed above, a broad term that does not have a commonly agreed upon definition can easily become diluted or corrupted over time, losing value and credibility. For farmers, uncertainty about which practices qualify as regenerative agriculture can make it difficult for them to implement effective strategies. It can be hard and expensive to access the right educational resources, funding, and technical support. Furthermore, current regulations (or lack thereof) pose additional challenges for farmers trying to access new markets, develop marketing strategies, and differentiate their products.

If you are a producer, practitioner, or proponent practicing regenerative agriculture, there may be a strong expectation that these practices will lead to particular outcomes. However, measuring outcomes is not so straightforward and it is highly possible that a particular process may not immediately, linearly, or directly lead to a particular outcome. However, making the shifts towards regenerative practices will also catalyze natural processes that can produce the desired outcomes (healthier soil, a biodiverse ecosystem, more nutrient dense crops, etc.) with time and patience. Generally, it is a combination of practices that yields the best results, and an understanding that every case is different, and adaptation and trial and error are a key part of any process of transformation.

When working with farmers transitioning to regenerative agriculture, it can be even more challenging to measure outcomes. The transition period from conventional to regenerative requires great efforts from the farmer and all the other actors involved in the supply chain. Transitioning means optimizing best practices by: reducing fertilizers, pesticide and fungicide applications; addressing the low-hanging fruits in the system; integrating regenerative practices; adapting machinery; generating internal knowledge, etc. A process of trialing, testing, failing and succeeding. Transitioning is a non-linear process that can include many ups and downs until the farm operation is better able to perform these regenerative practices. Therefore, markers for soil health, such as soil organic matter (SOM), soil organic carbon (SOC), soil life activity, nutrient density and others, can take a few years to show positive results. In this scenario, outcome-based definitions do not include transitioning farm operations, and process-based definitions seem to be more inclusive of all farmers on different journeys through regenerative agriculture.

The growing interest in regenerative agriculture from the industry without standardization has raised concern within the farming community and the non-profit sector regarding the potential for greenwashing. Companies can make vague, misleading claims or even exaggerated claims about their practices, presenting a false image of environmental responsibility. The absence of a standardized framework and regulatory mechanisms for regenerative agriculture enables companies to implement, define, and measure success according to their own criteria and key performance indicators (KPIs).

In the industry's use of regenerative agriculture, we see different approaches by different companies. Some supply chains incorporate not only agronomic and environmental aspects, but also social aspects, addressing farmer's livelihood and health, people, and communities. However, this is not necessarily considered by other companies whom only focus on agronomic environmental factors, omitting social aspects. The main difference between regenerative agriculture and agroecology, in our view, is agroecology's explicit inclusion of social aspects.

The term "agroecology" first emerged in the early 20th century, and its definition and scope have since evolved significantly. This evolution is primarily linked to the development of the two disciplines from which agroecology originates: agronomy and ecology. Initially recognized as a scientific discipline in the 1970s, agroecology gradually began to be also viewed as a movement and a set of practices, starting in the 1980s (Wezel et al., 2009). Agroecology is defined as a holistic approach to agriculture, which takes into account the ecological, social, political, and economic dimensions of producing food to build sustainable and resilient food systems that ensure food security and nutrition (Bezner Kerr et al., 2021). It is nowadays recognized as an alternative to the current food system, able to provide sustainable and healthy food for all, while preserving ecosystem health and guaranteeing a socially just food system.

If we focus on soil health, agroecological practices are able to maintain – and in many cases enhance – soil health, and many of them overlap with practices commonly used in regenerative agriculture (Dominguez et al., 2023). However, agroecology extends far beyond the mere care of the soil; it embraces a holistic approach that includes the rights of people to choose how and what to produce. This means recognizing and restoring the rights of Indigenous and local communities, as well as peasants, to their ancestral lands. It also involves ensuring that communities regain control over their seeds, and achieve food sovereignty. By doing so, agroecology not only promotes sustainable farming practices but also supports cultural heritage, biodiversity, and social justice, fostering a resilient and equitable food system for all.

In this chapter you will find the terms agroecology and regenerative agriculture being used interchangeably.

SOIL MICROBIOME AND NUTRIENT CYCLING

Soil is the foundation of our ecosystem as it links the different elements and provides sufficient nutrients to crops, grasses, and trees directly and indirectly to animals and humans. There are 16essential nutrients for plant growth, categorized into macronutrients and micronutrients. Macronutrients, required in larger quantities, include carbon, oxygen, hydrogen, nitrogen, phosphorus, potassium, calcium, magnesium, and sulfur. Micronutrients, needed in smaller amounts, include iron, zinc, manganese, boron, copper, molybdenum, and chlorine. Plants absorb all those macro and micronutrients from the soil. (FAO, 2015)

These essential nutrients for plants, animals, and microorganisms cycle in the soil between organic matter and inorganic nutrients, completing the so-called nutrient cycling. Nutrient cycling refers to the continuous, dynamic transfer of nutrients in the soil-plant system (Nair, Kumar & Nair, 2021).

Soil plays a critical role in plant growth and how the various nutrients and substances move along the ecosystem. Plants absorb the macro- and micronutrients required for their development and functions through their roots, such as nitrogen, phosphorus, potassium, iron, manganese, and more (Tucker, 2001). Water also plays a role in this nutrient uptake. As plant roots draw in water from the soil, they also absorb dissolved nutrients (Gitz & Meybeck, 2012). This symbiotic relationship with water ensures that plants receive a balanced diet of essential elements.

Additionally, plants engage in photosynthesis, a spectacular process where they convert carbon dioxide from the air and water from the soil into glucose and oxygen using sunlight. This process fuels plant growth but does not directly provide the nutrients plants need (Howard et al., 1999). Some plants enhance their nutrient absorption through symbiotic relationships with soil microorganisms. For example, mycorrhizal fungi form networks with plant roots, extending their reach and efficiency in nutrient uptake. Similarly, nitrogen-fixing bacteria convert atmospheric nitrogen into forms that plants can readily use, enriching the soil and boosting plant health (Thepbandit et al., 2024).

Another important element in soil nutrient cycling and soil health is soil organic matter (SOM), which also plays an important role in carbon sequestration. Crucially, SOM supports soil health, inactivates toxins, suppresses pathogens, and protects the ecosystem. It consists of plant and animal tissue in various decomposition stages, and includes three types: (1) plant residues and living microbial biomass, (2) active soil organic matter (detritus), and (3) stable soil organic matter (humus) (Slavich, 2001). Microorganisms decompose plant residues and detritus, releasing nutrients like nitrogen, phosphorus, and potassium, enhancing soil fertility. Humus, the final decomposition product, has less impact on fertility but improves soil structure, tilth, and cation exchange capacity (Wei, X et al., 2024).

Microorganisms in soil are crucial for soil ecosystem functions and services and all together form a community of microorganisms called the soil microbiome. This is why measuring the microbial population and their activity in soil is essential to assess the soil's potential to break down nutrients. Microorganisms can drive biological transformations, break down organic matter, cycle soil nutrients and support soil–plant–microbe interactions. By breaking down organic matter, they release nutrients that support soil structure, quality, and plant health. Negative changes in microbial populations are associated with soil degradation, while positive changes in microbial populations are associated with increased soil health. Overall, microbial activity is vital for plant health, soil health, and ecosystem services.

Soil microorganisms can consist of bacteria, protozoa, fungi, and algae and they are mostly found in the topsoil (up to 12 cm due to higher organic matter presence), and in the rhizosphere, the area near root surfaces. The rhizosphere is a biological active area where plants, soil, microorganisms, nutrients, and water interact (Mueller et al., 2024). It is a complex environment with bacteria, fungi, nematodes, and arthropods, and has more bacterial activity than regular soil but less variety in microbial population. The rhizosphere is also the place where various interactions between microbes and plants occur, for example microbes can produce enzymes or metabolites that inhibit plant pathogens acting as immune mediators, or produce phytohormones like auxins, gibberellins, and cytokinins essential for plant growth. Plants release a variety of organic compounds (sugars, amino acids, fatty acids), known as root exudates. Exudates influence carbon and nutrient cycles, shape physiochemical properties of soils, and act as nutrients and signals to root-associated microbial communities (McLaughlin et al., 2023) having direct impact on plant health and development. They can draw in both helpful and harmful microbes, showing the importance of plant-microbe interactions (Jamil et al., 2022).

Understanding how soil, plants, and microbes interact is crucial for maintaining soil health, quality, and ecosystem services which in return support human, plant, animal life, and improve the sustainability of crop production. Nevertheless, figuring out the soil microbiome and its impact on plant growth can be difficult due to the complex relationships between microorganisms, soil treatments, and environmental conditions.

Soil microorganisms play fundamental roles in agriculture by being essential to a number of processes that may lead to various direct and indirect benefits for crops, that is, nutrient cycling, improving soil quality and plant nutrition and health. Intensive conventional agriculture with focus on maximizing yields, has generated the degradation of biological and physicochemical properties of soil. Agricultural practices such as extensive tillage, the use of agrochemicals, monocropping among others have been found to influence negatively soil microbial community structure and soil health.

Tillage is one of the most recognized practices to alter soil's physical structure, moisture, temperature, aeration, and rate of mulch degradation, and to decrease soil macroaggregation. Minimizing soil disturbance with zero tillage or no-till practices is preferred over conventional tillage in order to maintain soil health and crop productivity. Studies show higher microbial activity in the surface layer of soils under no-tillage management with mulch when compared to conventional tillage without mulching.

Monoculture systems may change soil microbial parameters, particularly species richness, microbial activity, and community structure. Crop diversification is recommended, and to a great extent is being used, for enhancing land productivity. Crop diversification influences microbial diversity since different plants release varied root exudates which, in turn, determine the microbial community structure. Crop diversification also leads to diversity in carbon-rich compounds present in the rhizosphere, influencing the diversity of native microbial species.

Commonly used organic and inorganic agricultural inputs have a significant impact on soil microflora and have a great role in determining the microbial community structure in the soil. Pesticides can affect microbial diversity in two ways: first, by immediately harming microbial communities due to their toxicity, and second, by causing long-term changes as new microbial communities gradually take over. In both cases, the changes in the microbial communities will affect soil health.

Enhancing soil health can help restore and improve soil's physical, chemical, and biological properties (Bhattacharyya, S.S, & Furtak, K., 2022), which, in turn, will result in better nutrient decomposition, uptake, and increase crop nutrient-density. To do so, there are a number of evidenced-based practices to explore and implement. Namely, agroecology and regenerative agriculture come as answers to our agricultural crisis, where practices improve soil health, increase microbial activity and increase nutrient-density in crops, ultimately affecting animal and human health. Other techniques such as nutrient uptake enhancement, and biocontrol of pests and diseases using microbial inoculants may replace/reduce agrochemicals in agricultural production systems. Combined with agroecological practices, beneficial microorganisms seem to have great potential to provide an alternative opportunity to overcome the negative effects of various components of conventional intensive agriculture.

To meet the demand of a growing population, especially in developing countries, the intensification of agriculture has become inevitable. Jeopardizing crop quality and soil health to maximize yields has become the norm. However, awareness among the population about healthy food has grown and consequently a growing demand for different types and varieties of food, as well as for better quality produce. The demand for a high-quality diet and the rising popularity of organic produce are requiring an urgent need for ecologically sound agricultural advancement, which will produce nutrient dense crops and generate ecosystem resilience to be able to feed the increasing population, in the face of ever-changing climatic conditions (Gupta et al., 2022).

NUTRIENT DENSITY

Nutrient Decline in Crops Over the Years

Research has shown that today's fruits, vegetables, and grains have lower levels of protein, calcium, phosphorus, iron, riboflavin, and vitamin C than those grown decades ago. Even those who avoid processed foods are consuming less nutritious produce. The decline in nutrient content of crops can be attributed to several factors, including agricultural practices, land and soil management, and climate change. This, coupled with changes in modern diets and lifestyles, means we are now living in a world where many people are overfed but undernourished, while hunger remains a pressing issue in many parts of the world. This is a remarkable paradox given that food production is at an all-time high. Perhaps part of the problem is that, for the past century, policy and dominant agricultural practices have been prioritizing quantity (yield) over quality (nutrient density).

Agricultural practices that increase yields but harm soil health, including irrigation, fertilization, monocropping, and harvesting methods that disrupt plant–soil–fungi interactions, have been linked to reduced nutrient absorption by plants (Bhardwaj et al., 2024). The intensive use of chemicals (pesticides, herbicides, insecticides, and fungicides) is another element of modern industrialized agriculture that plays a role in nutrient decline. The word pesticide derives from the English (pest) and Latin -cide (killer): they are by definition used to make an environment uninhabitable for living organisms deemed unwanted. The problem is all these organisms play an important role in soil ecosystems; they are what keep our soil alive and well. Healthy soil is critical for nutrient uptake by plants. Thus, without healthy soil, we cannot produce healthy plants, and without healthy plants, we cannot sustain healthy humans.

Research from the University of Texas at Austin, using USDA nutrient data from 1950 and 1999, analyzed 13 nutrients in 43 garden crops, including asparagus, snap beans, strawberries, and watermelon. The results of this study are striking, as they paint a worrisome picture of the real decline we have seen in a matter of decades. They observed notable reductions in six nutrients: protein, calcium, phosphorus, iron, riboflavin, and vitamin C, with decreases ranging from 6% for protein to 38% for riboflavin (Davis, Epp and Riordan, 2004). In a similar study, protein content in wheat saw a 23% reduction from 1955 to 2016, accompanied by notable decreases in manganese, iron, zinc, and magnesium (Mariem et al., 2020). The list could go on and on, for these are not isolated incidents. Even more troubling are the available nutrition data that suggest 80% of nutritional dilution has happened during the last 30–40 years (Bhardwaj RL et al., 2024), indicating that we are facing an exponentially growing trend of decline.

Aside from practices that degrade soil health, this decline can also be attributed to the introduction and widespread adoption of novel strains/varieties of crops which produce more yield, have higher growth rate and are more resistant to pest and disease. High-yielding uniform crops have failed to produce nutrient-density and monocropping has also reduced availability of diverse nutrients in the soil and in the average diet (Bhardwaj et al., 2024).

The alarming patterns of broad-spectrum nutrient decline across all major crops globally are clearly linked to the reality that more than two billion people are suffering from micronutrient insufficiency and malnutrition. Micronutrients, vitamins and minerals, are crucial for human health as the core regulators of fundamental body processes such as the immune system, energy functions, growth, bone health, fluid balance, and other biological processes. Nutritional deficiencies have been linked to a wide range of physiological impairments, from metabolic imbalances, inflammation, and compromised cardiovascular health, to cognitive function and immunity (Clemente-Suárez et al., 2023).

While narratives of malnutrition often center on "low-income" or "developing" countries, "hidden hunger" or micronutrient deficiency is becoming a more common problem worldwide, even in "rich" or "developed" countries. Micronutrient deficiency is considered a global health issue. Factors such as dietary patterns and the prevalence of a Western-style diet consisting of more processed foods and fewer phytonutrient-rich fruits and vegetables, further contribute to micronutrient deficiencies in populations (Cena & Calder, 2020).

Globally, we have been witnessing this troubling pattern over the past half century. A simultaneous rise has occurred in non-communicable diseases, such as obesity, diabetes, and cardiovascular diseases, alongside a steadily increasing number of people who are malnourished and world hunger on the rise.

We can no longer ignore the connection between soil health, our agricultural practices, and human health. It's important to highlight the different ways that nutrient deficiency affects human health and why we need to take action. When reflecting on government, agricultural, and health policies, the consumption of sustainably, organically, regeneratively, or agroecologically grown foods should be used as a strategy to promote public health and combat diseases prevalent worldwide. Little has been done to address the systemic problems underlying public health concerns. Slowly but steadily, our plates were filled with fewer nutrients, less variety, more chemicals, more processed foods, more additives and more sugar, so we are now harvesting the consequences: a pandemic of chronic health problems.

As a public health strategy, we must promote nutrient-dense, plant forward, diverse diets as a holistic approach to health in preventive care and management of chronic conditions. Understanding that conventional agricultural practices can lead to soil degradation, biodiversity loss, nutrient loss, and food contamination, is crucial to engage various stakeholders within the food chains and the public health sector in this discussion. In this way, it will be possible to strengthen organic, agroecological, and regenerative practices that benefit the environment while also enhancing the nutritional density of foods.

Defining Nutrient Density with Nutrient Variation

The term "nutrient density" is commonly defined as the amount of selected nutrients per reference amount of food (e.g., 100kcal, 100g, or a serving size) (Drewnowski et al., 2019). For instance, 100g of kale typically contains 1.5 to 1.9g of iron, or one orange (130g), contains 70mg of vitamin C. This conventional definition which creates averages for micronutrient concentration in foods, assumes that all kale or all oranges are nutritionally uniform and have the same amount of iron or vitamin C, which we now know is completely inaccurate. The nutrient content of kale, oranges and other crops vary significantly based on agricultural practices, growing conditions, crop variety and many other aspects.

Despite this conventional definition being widely adopted by health professionals, the public sector, food companies, and the entire food supply chain, it has recently started to be questioned. The concept of nutrient-dense foods was first included in the 2005 Dietary Guidelines for Americans (DGA) without a formal definition. The DGA, which guides much of the nutrition policy and education in the United States, provided a subjective list of nutrient-dense food groups based on nutrients to limit, rather than nutrients to encourage. This list included whole grains, fruits, vegetables, fat-free or low-fat milk and milk products, seafood, lean meats, poultry, eggs, beans, peas, nuts, and seeds, all prepared without added fats or sugars (Drewnowski et al., 2019).

In recent years, organizations like the Bionutrient Food Association (BFA), led by Dan Kittredge, have been at the forefront of redefining nutrient density. In attempting to define nutrient density, in the first years of research (BFA, 2018; BFA, 2019), the BFA discovered significant nutrient variation within our food systems. By analyzing hundreds of samples, they found substantial differences in nutrient concentrations. Crops analyzed included carrots, spinach, kale, lettuce, cherry tomatoes, and grapes, all showing wide-ranging nutritional values. These variations in nutritional values were correlated to many factors, such as soil health, agricultural practices, and crop variety. This recognition of nutrient variation makes it challenging to define nutrient density consistently. Since the same type of food can have vastly different nutrient profiles depending on how and where it is grown, a single definition of nutrient density may not accurately reflect the true nutritional value of foods. The BFA's research underscores the importance of considering these variations when discussing nutrient density, as it moves beyond a one-size-fits-all approach to a more nuanced understanding of food nutrition.

Defining nutrient density, as approached by the BFA, is not about establishing a static written definition. Instead, it involves analyzing a diverse array of samples to establish a spectrum that captures the full range of nutritional variation for each crop. Nutrient density can only be defined once the full variability is understood – when the outliers, representing the lowest and highest nutrient concentrations, have been identified. This enables the creation of thresholds to establish what are nutrient-dense crops and what is lower nutritional quality.

Along with crop nutritional analysis, data on soil health, farm management practices, fertilization programs, soil type, climate and other information are also collected to draw meaningful conclusions. This holistic process helps to understand variation within the food system, establish benchmarks for nutrient density, and evaluate how factors such as management practices influence nutritional outcomes.

For example, the BFA found a significant variation (up to 200:1) in antioxidants, polyphenols, and minerals in crops like carrots, spinach, kale, lettuce, grapes, and cherry tomatoes. An interesting result from the 2020 study was the large variation in antioxidant content in blueberries. Blueberries are known for their high antioxidant values, which was confirmed by the average antioxidant content of blueberries being higher than any of the other crops in the study. However, some low-quality blueberries had no more antioxidants than the average oat or tomato sample. This finding implies that merely consuming blueberries does not guarantee a specific amount of antioxidants. Instead we must go beyond and analyze how these blueberry trees were grown and if the soil was healthy to achieve higher nutrient content.

This is especially important when considering the potential of bioactive compounds in mitigating chronic diseases like obesity, diabetes, cancer, and cardiovascular disorders. Polyphenols, for example, combat oxidative stress, modulate gene expression, and regulate glucose metabolism, benefiting gastrointestinal health, gut microbiota, pancreatic–endocrine functions, and weight management (Trujillo and Ramírez, 2024). The 2020 research (BFA, 2020) continued to explore this variation more in-depth, trying to understand how that variation can affect consumers in more practical terms. Results continued to see a significant variation in nutrient and mineral content in almost all of the 20 crops analyzed.

Given the great nutritional variation in crops within food supply chains, redefining nutrient density requires a deeper understanding of the One Health framework, which emphasizes the interconnectedness of soil health, plant health, animal health, environmental health, and human health. Only in this way, we will be able to acknowledge that the quality of our food is intricately linked to the health of the soil in which it is grown, and, therefore, cultivating healthy soils is essential for growing nutrient-dense crops. As healthy soils provide the necessary nutrients and support for robust plant growth, we must recognize the need to adopt agricultural practices that promote soil health, such as regenerative agriculture and agroecology, to enhance the overall well-being of ecosystems and communities.

Soil Health and Nutrient Density – Case Studies

In order to highlight the connection between soil health and nutrient density we will present two very important case studies.

The first is the study "Soil Health and Nutrient Density: Preliminary Comparison of Regenerative and Conventional Farming" (Montgomery et al., 2022).

The study compares soil health and nutrient density in crops from both regenerative and conventional farming systems. The farms included in the research either used conventional methods, relying on synthetic fertilizers and herbicides, or followed regenerative practices for five to ten years. Preliminary findings suggest that regenerative farming, which emphasizes soil health through practices like no-till farming, cover cropping, and crop rotation, generally results in crops with higher nutrient density compared with conventional methods.

The comparison included crops such as peas, sorghum, corn, and soybeans from eight paired farms in the US, using either conventional or regenerative methods. Additionally, the same variety of cabbage was compared between regenerative no-till and organic till operations. Soil health was assessed using tests like soil organic matter analysis and the Haney test, which measures microbial activity. For the crops, vitamins (B1, B2, B6, C, E, K), minerals (calcium, phosphorus, copper, iron, potassium, magnesium, sodium, and others), and phytochemicals (total phenols, phytosterols, and carotenoids) were analyzed.

Results varied across farms and crops, but on average, regenerative practices produced crops higher in vitamins B1, B2, E, and K, as well as carotenoids, phenolics, and phytosterols. Crops from regenerative farms also showed higher levels of calcium, phosphorus, and copper. However, regenerative crops tended to have lower amounts of vitamin B6 and manganese. For example, regenerative soybeans were lower in vitamins B1, B3, B6, and vitamin C compared to their conventional counterparts.

In addition, the study examined the fatty acid composition of beef and pork. Beef samples came from three types of operations: 100% grass-fed and finished on regeneratively managed land, grass-fed beef, and conventional beef fed on a corn and soy diet. Pork samples came from regenerative pasture-raised pork on mixed grains, pork raised with outdoor access and non-GMO feed, and conventionally raised pork on GMO corn and soy diets. In both beef and pork, the regenerative grass-fed and finished meat had superior unsaturated fatty acid profiles, with a healthier omega-3 to omega-6 ratio.

These findings suggest that regenerative farming significantly improves the nutritional quality of both plant and animal foods. However, further research is needed to understand why some micronutrients were found in lower levels with regenerative methods, and how to address these gaps for better results.

The second study to be showcased is the 2020 BFA report. This study is interesting for the nuance that it adds to the conversation. In the report they state:

> Within the regenerative agriculture community there is a belief that 'regenerative practices' will improve soil health, which will in turn improve the nutrient density of the crops grown in that soil. However, given the number of factors that influence nutrient density – climate, soil type, crop variety, management practices, and so on – it is difficult to test this hypothesis.
>
> (BFA, 2020)

Let's understand why. The study compared soil health and nutrient density across regenerative and conventional farming methods, with many positive correlations between soil health indicators and nutrient outcomes. For example, regenerative practices like no-till, cover cropping, and organic amendments significantly improved the nutrient density of crops such as beets, zucchini, and grains. Antioxidant levels in beets increased by 22% with the use of mulch, while zucchini saw a 10% rise in polyphenols. Regenerative practices also resulted in higher levels of vitamins B1, B2, K, and E, as well as phytochemicals like carotenoids and phenolics. Overall, crops grown with regenerative methods tended to show enhanced nutritional profiles, reinforcing the benefits of regenerative farming techniques.

The study also examined the impact of tillage intensity on soil carbon and respiration, showing that no-till practices generally improve soil structure and organic matter retention. In smaller, diversified farms where organic amendments are common, light tillage even led to increases in soil carbon. Conversely, heavy tillage in larger grain farms reduced soil carbon by breaking down organic matter too quickly. These findings highlight the complex relationship between farm management practices, soil health, and crop nutrient density, which are influenced by factors like crop type, scale of operation, and soil conditions.

Despite these positive and promising findings, some results were mixed. For example, while mulch increased antioxidant levels in beets, it led to a 63% decrease in antioxidant content in carrots. Regenerative soy crops showed lower levels of vitamins B1, B3, B6, and C compared to conventional methods. These mixed results can be disappointing if we expect a clear and consistent advantage from regenerative practices. However, we must look at the results from a scientific perspective where such outcomes are an opportunity for deeper exploration.

In research, we should not have a bias, and instead be open to any result that might come. Negative or inconsistent findings don't mean failure – they are actually essential to refining our understanding. When regenerative practices result in lower nutrient density, the question isn't whether the approach is flawed, but rather *why* these outcomes occurred. What specific variables in regenerative agriculture may be causing these results? For example, if a farmer is planting cover crops, what species are they using? Are they the right mixture for the local environment and soil? How and when are these cover crops being planted, and are they optimizing soil health effectively? The execution of regenerative practices is crucial and very context-specific. There are both effective and ineffective strategies for implementing regenerative techniques like cover cropping, and those nuances may influence nutrient outcomes. We must also recognize the work of many conventional farmers who might have a bad reputation in the regenerative community, but are nonetheless extremely knowledgeable about their crops and very precise about their fertilization programs, which might reflect in better nutrient density outcomes for their crops.

To truly understand and optimize regenerative agriculture, we need to embrace these and many other questions, examining the context and specifics of how each practice applied. Researchers at BFA emphasize we need to have larger sample sizes and more years of data to better understand these dynamics and draw stronger conclusions. For research to advance, we need more than just a commitment from individual researchers and farmers. We must promote, create, and fund research. The nutrient-density space is very new and significantly underfunded. We must support the pioneer organizations on this work, as well as create new multistakeholder consortiums with universities farmers, companies, and others to create robust data and studies in the field.

In conclusion, while the BFA study demonstrates a compelling link between soil health and nutrient density, it also highlights the complexity and nuances involved in the results. The variations seen in nutrient density across different crops and practices demonstrate the need for more research in the field, with better methodologies and larger data samples, to understand the complexity of why some regenerative practices generate lower levels of certain nutrients. Through sustained research and multistakeholder collaboration we will be able to unlock many pathways in the nutrient density field.

How to Measure Nutrient Density

There are different ways to measure nutrient variation and density in crops. We discuss two tools: the Brix meter and the Bionutrient meter.

Brix is a scale used to measure the density of plant juices by quantifying how much light bends as it passes through a liquid. For example, if one were to place a hand in a pond and measure the apparent bending, then do the same in the ocean, the difference in light bending would be observed. This "light-bending" measurement is known as refraction and is captured by a refractometer. Different types of refractometers are available for various applications. The Brix scale was pioneered by German chemist Adolf Ferdinand Wenceslaus Brix in the 19th century and was further developed by Dr. Carey A. Reams, an agricultural engineer, who popularized its use in food production. Reams also created a Brix chart to evaluate crop quality, highlighting the connection between Brix readings and the mineral content of crops, which led to improvements in farming practices (BFA, n.d.-b).

Traditionally used by winemakers and the agriculture industry, the Brix meter provides an indication of plant composition and is commonly used to assess the overall health and quality of fruits, vegetables, and other crops. The Brix scale accounts for various components in plant juice, including sugars, minerals, amino acids, and other solids, and classifies food as poor, average, good, or excellent in terms of nutrient content. Generally, higher Brix readings indicate higher nutritional value and better taste.

In practice, for plants, this process involves extracting juice from leaves and measuring the light refraction, which will vary according to the concentration of dissolved solids (sugars, vitamins, amino acids, and minerals) in the juice. The simplicity and accessibility of the Brix meter make it an essential tool for farmers and consumers in evaluating the nutritional density of foods, promoting better agricultural practices, and improving food quality.

Another gadget under development is the Bionutrient meter, a handheld spectrometer created by the Bionutrient Institute lab in Michigan to measure the nutrient density in food and crops, along with soil carbon. It utilizes spectroscopy technology to analyze materials noninvasively using light flashes. Different light reflectance at specific wavelengths is correlated to different amounts of compounds in food such as vitamins, antioxidants, and aromatic compounds, as well as in organic carbon in soil and chlorophyll levels in plants. The biometer could show a consumer on their phone screen an indication value of antioxidants and polyphenols, based on the light bouncing of the examined crop compared against the lab analyzeddataset. The aim is to create an inexpensive handheld device for consumers to test crop quality before purchase, potentially shifting economic incentives in crop production (BFA, n.d.-a).

Since 2018, the institute has collected thousands of soil, food, and crop samples to calibrate the meter, which is part of a larger strategy including an open-source data platform accessible to all. By making high-quality crop production principles accessible, the institute hopes to raise consumer awareness of food quality disparities and empower them to make informed purchasing decisions. The ongoing development of the Bionutrient meter seeks to address declining nutrient concentrations in fresh produce and related health issues, while promoting ecologically regenerative agricultural practices (BFA, n.d.-a).

Both the Brix meter and the Bionutrient meter offer farmers and consumers an accessible way to assess nutrient density without needing complex laboratory analysis. Traditionally, evaluating the nutrient density of crops involves detailed lab tests that measure both micro and macronutrients, including minerals, vitamins, and phytochemicals. Laboratory analysis are effective, however are not accessible to the public nor affordable for many farmers. They are also complex as they depend on machinery availability in labs.

For farmers looking to gather data on their crops via labs without incurring high costs, a practical approach is to focus on specific key nutrients rather than analyzing a wide range of compounds. For instance, when testing tomatoes, the focus could be on vitamin C and polyphenols like flavonoids. If testing spinach, the focus might be on iron, magnesium, and potassium. By narrowing down the nutrients of interest for each crop, farmers can reduce analysis costs while still obtaining valuable scientific data about their produce.

From Farm to Gut – Factors Influencing Nutrient Density/Availability

The concept of nutrient density in crops is crucial to understanding the quality of the food we consume. To fully grasp this, it's essential to recognize the elements and phases that influence nutrient density, from the moment a seed is planted to the final stages of digestion. This understanding requires us to embark on a journey that begins deep within the soil and finishes on our large intestine.

Soil, water, and climate are just a few of the key factors that influence nutrient density in crops in the field (Lal, 2006;). Nutrient density can be affected by other external factors such as processing and storage. It's important to recognize that most of our food doesn't come directly from a farm; it undergoes various processes and stages where ingredients are mixed, transformed, and stored, which can alter their nutritional composition (Rickman, Barrett, & Bruhn, 2007). Additionally, understanding how our bodies absorb these nutrients, and the role consumers have in preserving foods they purchase is crucial. Addressing all of these factors in detail would require far more than a single chapter. However, in this section, we will briefly outline them – not to delve into each one, but to provide an overview of the complexity involved in this topic.

At this point, it should be clear that the foundation of nutrient-dense crops lies in healthy soils. Rich, organic soils, balanced in pH and nutrient content, are the bedrock of nutritious plants (Lal, 2006). Practices like crop rotation, cover cropping, and organic farming enhance soil health and, consequently, the nutrient content of crops and are also environmentally sound.

Moving from the soil to the food system, the first factor that influences the nutrient density of a crop is the time of the harvest. Crops harvested at their peak ripeness generally have higher nutrient levels compared to those picked too early or too late. For instance, tomatoes ripened on the plant are richer in flavour and nutrients than those harvested green and ripened afterwards (Rickman, Barrett, & Bruhn, 2007).

Once harvested, crops must be handled carefully to preserve their nutrient content. Exposure to air, light, and heat can degrade sensitive nutrients like vitamin C (Rickman, Barrett & Bruhn, 2007). Proper washing, cooling, and storage are essential to maintain their nutritional integrity.

Minimal processing is key to retaining nutrients. While some processes, like freezing, can preserve nutrients, others, such as extensive milling or refining, can lead to significant nutrient loss. Whole foods, close to their natural state, are typically the most nutrient-dense (Howard et al., 1999).

Appropriate storage conditions are also crucial. Fresh produce should be kept at optimal temperatures and humidity levels and handled gently to prevent nutrient loss and physical damage The journey from farm to retailer involves transportation, which also plays a role in nutrient density. Prolonged transit times and exposure to fluctuating temperatures can degrade nutrients (Managa et al., 2018). Thus, locally sourced produce often retains more nutrients due to shorter transportation times (Gitz & Meybeck, 2012). Finally, how food is displayed and handled in retail settings can also impact its nutrient content.

And last, but not least, let's not forget the final actors in this supply chain: consumers – all of us. We play a crucial role in preserving the nutrient density of our food. How we choose to store or cook our food can either help maintain its nutritional value or lead to a significant loss in its composition For example, leafy greens are best stored in cool, moist environments to retain their vitamins and minerals (Kader, 2005) and root vegetables like potatoes should be kept in a cool, dry place. Freezing fruits and vegetables shortly after purchase can also help retain their nutritional content, especially if you won't be using them right away.

When it comes to cooking, methods like steaming, or sautéing with minimal water can help preserve vitamins and minerals that are easily lost in water-based cooking methods like boiling. Avoiding overcooking is important, as excessive heat can destroy nutrients like vitamin C and vitamin B9 (folate). Using the cooking water or broths (vegetable and/or meat) in soups, sauces, grains and beans, to retain food's original nutrients increase nutrient concentration in new preparations.

As we move from the dining table, the focus shifts to the human digestive system. Humans cannot synthesize all required micronutrients for proper biological function, which therefore need to be acquired from the three major resources of (1) dietary components, (2) oral supplements, or (3) synthesis by gut bacteria (Hadadi et al., 2021). Consuming nutrient-dense crops is just the beginning; our bodies must efficiently absorb these nutrients to benefit from them.

Our digestion process starts in the mouth. The food that we consume is broken down physically by chewing, and chemically by enzymes which break down complex food molecules. Food then moves to the stomach, where we also have both physical (stomach contractions) and chemical breakdown (hydrochloric acid) which digests fats and proteins. After the stomach, food moves to the intestines, first the small intestine, and then the large intestine. Absorption occurs mainly in the small intestine, where the inner walls are lined with tiny, finger-like projections called villi. These villi increase the surface area for nutrient absorption, allowing vitamins, minerals, and other essential nutrients to pass into the bloodstream. From there, they are transported to cells throughout the body, supporting various physiological functions, including energy production, growth, and repair.

After the small intestine, we move to the large intestine.Previously, scientists thought that not much happened in the large intestine. It was believed that food was mostly digested in the small intestine and the large intestine was mainly producing the stool. However, with the advances in research, we now know that the large intestine is where most of our microbes are located, and it's what we call our gut microbiome. Recent clinical evidence suggests an increasingly important contribution from the gut microbiome in nutrient production and absorption. Commensalism is a long-term biological interaction in which members of one species gain benefits while those of the other species neither benefit nor are harmed. Commensal microorganisms can, in fact, regulate micronutrient levels, both by influencing how they are produced in the body and by controlling how they are absorbed (Barone et al., 2022). Micronutrients affect the quantity and variety of bacteria in the gut; therefore, micronutrients can lead to either positive or negative effects on the microbiome and consequently on the body (Hadadi et al., 2021).

The relationship between micronutrients and the gut microbiome, the micronutrient–microbiome axis, is bidirectional. Gut microbes consume micronutrients, from food or supplements, for their growth and function, meaning what you eat affects the composition and health of your microbiome. But also gut microbes produce important vitamins, such as vitamin K and B vitamins, and help the absorption of minerals such as iron and calcium (Hadadi et al., 2021).

There's growing evidence that the composition of the human gut microbiome has a great contribution to the micronutrient status in the body. Studies show that several micronutrient deficiencies could be positively or negatively associated with gut microbiota. When we consider healthcare and tools for healthcare practitioners, personalized strategies targeting the gut microbiome could be a new way to address micronutrient deficiencies. By adjusting the microbiome, we address the underlying cause of the problem rather than its symptoms, meeting our body's needs and supporting overall health.

An example of such holistic research is currently an ongoing study led by Leiden University called "Soils2Gut: Steering Soil Microbiomes for Better Crop Quality: The Holy Grail to Improve Human Health?" (Leiden University, 2023). This study explores the critical link between soil health, the gut microbiome, and the prevention of non-communicable diseases (NCDs). The research integrates experimental, clinical, and social sciences to examine how various farming practices affect soil and crop microbiomes, crop nutritional quality, and human gut health. The project also assesses consumer preferences for microbiome-optimized crops, aiming to improve both human and soil health, while enhancing agricultural sustainability and farmer livelihoods. The interdisciplinary consortium driving this work brings together academic, farming, and industry partners to create holistic solutions for sustainable health and food systems.

Understanding the journey of nutrients from soil to plate highlights the interconnectedness of One Health. From soil microbiome to human microbiome, from agroecological farming practices to conscious consumption, every step on this journey matters. Taking care of soil health ensures that we eat food that nourishes both our bodies and the planet.

WAYS FORWARD/CONCLUSION

Humans have evolved with soil: from our ape ancestors to nomadic hunter gatherers, and all the way to the development of modern agricultural systems, it's clear that our existence depends on soil. The complex and beautiful relationship between soil health and human health is highlighted by the One Health perspective, which acknowledges the interconnectivity between human, animal, and environmental health. Soil health directly affects the quality of food we consume, and the anthropogenic degradation of soil throughout our agricultural history has dramatically changed the nutritional value of crops we consume. For this reason, understanding soil health is not just an environmental matter – it is a critical component of human health.

Regenerative agriculture and agroecology are tools to improve soil health, and as a consequence, these practices can produce more nutrient dense food. Agroecology and regenerative agriculture movements have started to gain momentum in the past 10 years. These practices not only sustain the environment but also contribute to a food system that supports human well-being at the most fundamental level. Farmers, consumers, companies and governments have started to recognize the importance of improving ecosystem resilience and shift their mentalities towards more holistic priorities – like that of soil health. Despite these positive trends, conventional agricultural practices still dominate the global food system, indicating that deeper and systemic transformations are urgently needed. Systemic changes require both bottom-up and top-down action: while we must push for our governments to increase regulation on agrichemicals, and promote agroecology, we also have a role to play as consumers and community members to support this transition.

The negative impacts of our agricultural systems are seen not only in the fields, but also in our bodies. As humans grow detached from their land and their soils, they also disconnect from their food and traditions. The modern processed diet which has become the norm is making us sick, as evidenced by epidemics of obesity and other non-communicable diseases. Micronutrient deficiencies, malnutrition, and hidden hunger are pressing health issues all over the world. The growing burden of hidden hunger – where many suffer from a lack of essential nutrients despite having enough calories – poses a serious public health challenge, highlighting the urgent need to view soil health as a fundamental tool for human health within the framework of public health. We need policy to promote soil health and human health by prioritizing long-term solutions to regenerate soil. Information on how agricultural practices can affect soil health, the deep implications for nutrient-density and in human health are necessary to be able to fight our agricultural and health crises.

The need for healthcare practitioners to be trained in One Health is clear. By having a more holistic approach to health, and understanding how agricultural practices impact human health, healthcare providers can make more informed recommendations for theirpatients. This cross-sector knowledge is an opportunity for innovative healthcare strategies, and especially in the field of Regenerative Healthcare. Regenerative Healthcare is a term coined by the Rodale Institute, and it highlights the connection between agriculture and healthcare working together for a prevention-based approach to human and environmental health. Rather than relying on chemicals to solve agricultural issues and pharmaceutical interventions to manage the disease, regenerative healthcare aims to prevent disease through organic, whole foods, recommending a plant-forward diet that begins on farms which work in harmony with nature.

The expansion of multidisciplinary research and funding in the nutrient density and One Health fields are crucial for the development of systemic solutions and influence policies. The intersection of soil health and human health represents a critical frontier in both agricultural and nutritional sciences. Many scientific gaps still remain, particularly in proving the direct link between soil health and human health.

When taking a socio-economic perspective, there is an additional challenge to ensure that nutrient-rich foods are accessible and affordable to a wider population, so that the benefits of agroecological agricultural practices are not confined to a privileged few. This highlights the necessity for broader policy changes and consumer education to drive meaningful change in the food system. Governments, policymakers, corporations, and civil society must embrace the One Health framework and work towards transforming food systems in ways that promote health for all. As research increasingly shows, soil ecosystems play a key role in either mitigating or exacerbating climate change. We can no longer afford to overlook or underestimate the potential of healthy soils to solve some of the world's most pressing problems. Restoring soil health isn't just a matter of human survival – it's a pathway toward planetary resilience and the future of life on Earth.

BIBLIOGRAPHY

Andrade, D. and Pasini, F. (2022). *Vida em Sintropia*. Editor Labrador.

Banerjee, S. and van der Heijden, M. G. A. (2022). Soil microbiomes and one health. *Nature Reviews Microbiology*, 21, https://www.nature.com/articles/s41579-022-00779-w

Barone, M., D'Amico, F., Brigidi, P. and Turroni, S. (2022). Gut microbiome–micronutrient interaction: The key to controlling the bioavailability of minerals and vitamins? *BioFactors*, 48(2), DOI: 10.1002/biof.1835

Bezemer, M. (2023). *Soil2Guts*. Leiden University, The Netherlands. https://www.universiteitleiden.nl/en/research/research-projects/science/ibl-soils2guts

Bezner Kerr, R., Madsen, S., Stüber, M., Liebert, J., Enloe, S., Borghino, N., Parros, P., Mutyambai, D. M., Prudhon, M. and Wezel, A. (2021). Can agroecology improve food security and nutrition? A review. *Global Food Security*, 29(29), https://www.sciencedirect.com/science/article/abs/pii/S221191242100050X

Bhardwaj, R. L., Parashar, A., Parewa, H. P. and Vyas, L.. (2024). An alarming decline in the nutritional quality of foods: The biggest challenge for future generations' health. *Foods*. https://pubmed.ncbi.nlm.nih.gov/38540869/

Bhattacharyya, S. S. and Furtak, K. (2022). Soil–plant–microbe interactions determine soil biological fertility by altering rhizospheric nutrient cycling and biocrust formation. *Sustainability*, 15(1), https://www.mdpi.com/2071-1050/15/1/625

Bionutrient Food Association. (2018). 2018 data report. https://our-sci.gitlab.io/bionutrient-institute/bi-docs/2018_report/

Bionutrient Food Association. (2019). 2019 data report. https://our-sci.gitlab.io/bionutrient-institute/bi-docs/2019_report/

Bionutrient Food Association. (2020). 2020 data report. https://www.bionutrientinstitute.org/2020datareport

Bionutrient Food Association. (n.d.-a). The bionutrient meter. https://www.bionutrient.org/bionutrientmeter

Bionutrient Food Association. (n.d.-b). BRIX. https://www.bionutrient.org/bionutrientmeter

Brevik, Eric and Sauer, Thomas. (2012). The Past, Present, and Future of Soils and Human Health Studies. In *SOIL Discussions. Paper presented at the European Geosciences Union General Assemble in Vienna*, Austria.

Cena, H. and Calder, P. C. (2020). Defining a healthy diet: Evidence for the role of contemporary dietary patterns in health and disease. *Nutrients*, 12(2), https://www.ncbi.nlm.nih.gov/pmc/articles/PMC7071223/

Clemente-Suárez, V. J., Beltrán-Velasco, A. I., Redondo-Flórez, L., Martín-Rodríguez, A. and Tornero-Aguilera, J. F. (2023). Global impacts of western diet and its effects on metabolism and health: A narrative review. *Nutrients*, 15(12), https://www.mdpi.com/2072-6643/15/12/2749

Davis, D. R., Epp, M. D. and Riordan, H. D. (2004). Changes in USDA food composition data for 43 garden crops, 1950 to 1999. *Journal of the American College of Nutrition*, 23(6), https://pubmed.ncbi.nlm.nih.gov/15637215/

Dollinger, J. and Jose, S. (2018). Agroforestry for soil health. *Agroforestry Systems*, 92(2). DOI: 10.1007/s10457-018-0223-9.

Domínguez, A., Escudero, H. J., Rodríguez, M. P., Ortiz, C. E., Arolfo, R. V. and Bedano, J. C. (2023). Agroecology and organic farming foster soil health by promoting soil fauna. *Environment, Development and Sustainability*. DOI: 10.1007/s10668-022-02885-4

Drewnowski, A., Dwyer, J., King, J. C. and Weaver, C. M. (2019). A proposed nutrient density score that includes food groups and nutrients to better align with dietary guidance. *Nutrition Reviews*, 77(6), https://www.ncbi.nlm.nih.gov/pmc/articles/PMC6489166/

European Commission. (2006). *Communication from the Commission to the Council, the European Parliament, the European Economic and Social Committee and the Committee of the Regions: Thematic Strategy for Soil Protection (COM(2006) 231 Final)*. European Commission, Brussels. https://esdac.jrc.ec.europa.eu/ESDB_Archive/Policies/Directive/com_2006_0231_en.pdf?utm_source=chatgpt.com

FAO. (2008). NSP – What is a healthy soil? https://www.fao.org/agriculture/crops/thematic-sitemap/theme/spi/soil-biodiversity/the-nature-of-soil/what-is-a-healthy-soil/en/#:~:text=Healthy%20soils%20maintain%20a%20diverse,and%20nutrient%20holding%20capacity%2C%20and

FAO (2015). Status of the world's soil resources. https://www.fao.org/fileadmin/user_upload/newsroom/docs/FAO-world-soils-report-SUMMARY.pdf

FAO. (2020). Towards a definition of soil health. ITPS – Intergovernmental Technical Panels on Soils. chrome-extension://efaidnbmnnnibpcajpcglclefindmkaj /https://openknowledge.fao.org/server/api/core/bitstreams/ffb5feaf-8388-4e2f-b319-2260a9a6f5a2/content

Ferreira, C. S. S., Seifollahi-Aghmiuni, S., Destouni, G., Ghajarnia, N. and Kalantari, Z. (2022). Soil degradation in the European Mediterranean region: Processes, status and consequences. *Science of The Total Environment*, 805, https://www.sciencedirect.com/science/article/pii/S0048969721051810

Francis, C. A. (1985). Rationality of new technology for small farmers in the tropics. *Agriculture and Human Values*, 2(2), DOI: 10.1007/BF01530551

Gitz, V. and Meybeck, A. (2012). Risks, vulnerabilities and resilience in a context of climate change. *Agriculture and Consumer protection departament, FAO*, 8(1), 34–37.

Gordon, E., Davila, F. and Riedy, C. (2023). Regenerative agriculture: A potentially transformative storyline shared by nine discourses. *Sustainability Science*, 18(4), DOI: 10.1007/s11625-022-01281-1

Grant, S. (2017). Organizing alternative food futures in the peripheries of the industrial food system. *Journal of Sustainability Education*. https://www.fao.org/4/i3084e/i3084e03.pdf

Gupta, A., Singh, U. B., Sahu, P. K., Paul, S., Kumar, A., Malviya, D., Singh, S., Kuppusamy, P., Singh, P., Paul, D., Rai, J. P., Singh, H. V., Manna, M. C., Crusberg, T. C., Kumar, A. and Saxena, A. K. (2022). Linking soil microbial diversity to modern agriculture practices: A review. *International Journal of Environmental Research and Public Health*, 19(5), https://www.ncbi.nlm.nih.gov/pmc/articles/PMC8910389/

Hadadi, N., Berweiler, V., Wang, H. and Trajkovski, M. (2021). Intestinal microbiota as a route for micronutrient bioavailability. *Current Opinion in Endocrine and Metabolic Research*, DOI: 10.1002/biof.1835

Howard, L. A., Wong, A. D., Perry, A. K. and Klein, B. P. (1999). β-Carotene and ascorbic acid retention in fresh and processed vegetables. *Journal of Food Science*, DOI: 10.1111/j.1365-2621.1999.tb15943.x

Jamil, F., Mukhtar, H., Fouillaud, M. and Dufossé, L. (2022). Rhizosphere signaling: Insights into plant-rhizomicrobiome interactions for sustainable agronomy. *Microorganisms*, https://www.mdpi.com/2076-2607/10/5/899

Kader, A. A. (2005). Increasing food availability by reducing postharvest losses of fresh produce. *Acta Horticulturae*, 682, 2169–2176, https://www.actahort.org/books/682/682_296.htm

Lal, R. (2006). Enhancing crop yields in the developing countries through restoration of the soil organic carbon pool in agricultural lands. *Land Degradation & Development*, DOI: 10.1002/ldr.696

Lal, R. (2016). Soil health and carbon management. *Food and Energy Security*, DOI: 10.1002/fes3.96

Leiden University (n.d.) *Soils2Guts Research Project*. Leiden University, Institute of Biology Leiden. https://www.universiteitleiden.nl/en/research/research-projects/science/ibl-soils2guts

Managa, M. G., Tinyani, P. P., Senyolo, G. M., Soundy, P., Sultanbawa, Y. & Sivakumar, D. (2018). Impact of transportation, storage, and retail shelf conditions on lettuce quality and phytonutrient losses in the supply chain. *Food Science & Nutrition*, 6(6), 1527–1536, DOI: 10.1002/fsn3.685

Mariem, S. B., Gámez, A. L., Larraya, L., Fuertes-Mendizabal, T., Cañameras, N., Araus, J. L., McGrath, S. P., Hawkesford, M. J., Murua, C. G., Gaudeul, M., Medina, L., Paton, A., Cattivelli, L., Fangmeier, A., Bunce, J., Tausz-Posch, S., Macdonald, A. J. and Aranjuelo, I. (2020). Assessing the evolution of wheat grain traits during the last 166 years using archived samples. *Scientific Reports*, https://www.nature.com/articles/s41598-020-78504-x

McLaughlin, S., Zhalnina, K., Kosina, S., Northen, T. R. and Sasse, J. (2023). The core metabolome and root exudation dynamics of three phylogenetically distinct plant species. *Nature Communications*, https://www.nature.com/articles/s41467-023-37164-x

Montgomery, D. R., Biklé, A., Archuleta, R., Brown, P. and Jordan, J. (2022). Soil health and nutrient density: Preliminary comparison of regenerative and conventional farming. *PeerJ*. https://peerj.com/articles/12848/

Montgomery, D. R., Rabinowitz, P., Sipos, Y. and Wheat, E. E. (2024). Soil health: A common focus for one health and planetary health interventions, *One Health*, https://www.sciencedirect.com/science/article/pii/S2352771423001933

Mueller, C., Baumert, V., Carminati, A., Germon, A., Holz, M., Kögel-Knabner, I., Peth, S., Schlüter, S., Uteau, D., Vetterlein, D., Teixeira, P. P. and Vidal, A.. (2024). From rhizosphere to detritusphere – Soil structure formation driven by plant roots and the interactions with soil biota. *Soil Biology and Biochemistry*. https://www.sciencedirect.com/science/article/pii/S0038071724000853

Nair, P. K. R., Kumar, B. M. and Nair, V. D. 2021. Soil organic matter (SOM) and nutrient cycling. In: *An Introduction to Agroforestry*. Springer, Cham. DOI: 10.1007/978-3-030-75358-0_16

Newton, P., Civita, N., Bartel, K. and Johns, C. (2020). What is regenerative agriculture? A review of scholar and practitioner definitions based on processes and outcomes. *Frontiers in Sustainable Food Systems*, DOI: 10.3389/fsufs.2020.577723

Ohno, T. and Hettiarachchi, G. M. (2018), Soil chemistry and the one health initiative: Introduction to the special section. *Journal of Environmental Quality*, DOI: 10.2134/jeq2018.08.0290

One Health High-Level Expert Panel (OHHLEP), Adisasmito, W. B., Almuhairi, S., Behravesh, C. B., Bilivogui, P., Bukachi, S. A. et al. (2022) One health: A new definition for a sustainable and healthy future. *PLoS Pathog*. DOI: 10.1371/journal.ppat.1010537

Pankhurst, C. E., Doube, B. M. and Gupta, V. (1997). Biological indicators of soil health, https://www.researchgate.net/publication/267266422_Biological_Indicators_of_Soil_Health

Rhodes, C. J. (2014) Science progress (2014). *Soil Erosion, Climate Change and Global Food Security: Challenges and Strategies*, 97(2), 97–153. DOI: 10.3184/003685014X13994567941465

Rickman, J. C., Barrett, D. M., & Bruhn, C. M. (2007). *Nutritional comparison of fresh, frozen and canned fruits and vegetables. Part I. Vitamins C and B and phenolic compounds. Journal of the Science of Food and Agriculture*, 87(7), 930–944. https://doi.org/10.1002/jsfa.2825

Shahane, A. A., & Shivay, Y. S. (2021). Soil health and its improvement through novel agronomic and innovative approaches. *Frontiers in Agronomy*. DOI: 10.3389/fagro.2021.680456

Slavich, P. (2001). *Soil Health: A Systems Approach to Soils*. Southern Cross University.

Thepbandit, W. and Athinuwat, D. (2024 Mar 11). Rhizosphere microorganisms supply availability of soil nutrients and induce plant defense. *Microorganisms.*, 12(3), 558. DOI: 10.3390/microorganisms12030558. PMID: 38543610; PMCID: PMC10975764.

Trujillo, J. and Ramírez, V. (2024). Polyphenols and health benefits: Volume I. *Foods*, https://www.mdpi.com/2304-8158/13/8/1221

Tucker, K. L. (2001). Eat a variety of healthful foods: old advice with new support. *Nutrition Reviews*, https://pubmed.ncbi.nlm.nih.gov/11396697/

Wei, X., Xie, B., Wan, C., Song, R., Zhong, W., Xin, S. and Song, K. (2024). Enhancing soil health and plant growth through microbial fertilizers: Mechanisms, benefits, and sustainable agricultural practices. *Agronomy*. https://www.mdpi.com/2073-4395/14/3/609

Wezel, A., Bellon, S., Doré, T., Francis, C., Vallod, D. and David, C.. (2009). Agroecology as a science, a movement and a practice. *Agronomy for Sustainable Development*. DOI: 10.1051/agro/2009004

8 Sustaining Soil Health for Human Health

Mansonia Pulido-Moncada and Bruce C. Ball

INTRODUCTION

Human health, as defined in the constitution of the World Health Organization, is a "state of complete physical, mental, and social well-being and not merely the absence of disease or infirmity". This implies that humans need to develop in an environment in which there is a positive balance between factors and processes in health development. Among all possible influencing factors, soils are one of the central foundations of sustainability in human development (Bouma et al. 2019). Soil is connected not only to human health but also to culture, civilization and livelihood (Minami, 2009). From a historical perspective, soils have provided services to humans for housing (caves and houses made with rocks, daub structures, mud or fired bricks), agriculture (food and fiber), archaeological evidence, cooking pottery, food preparation and storage, preservation of cadavers, painting materials, water extraction, medicine, mineral mining, construction materials for civilization development, and waste disposal, among others (Baveye et al., 2016). In recent centuries, soil services to humans have been influenced by socio-economic and industrial needs, and more recently by ecological concerns.

The preservation of soil services to humans relates to preserving soil health. The term soil health is widely used by farmers and other stakeholders as a measure of the overall suitability of the soil to provide soil services. Moreover, soil health was defined by the Intergovernmental Technical Panel on Soils of the Food and Agriculture Organization of the United Nations as "the ability of the soil to sustain the productivity, diversity, and environmental services of terrestrial ecosystems" (ITPS-FAO, 2020). Based on this, soil health scope extends beyond human health to planetary health, and its sustainability objectives are broad (Lehmann et al., 2020). This is reflected in the approach of the One Health concept, in which soil health plays a key part in connecting the health of humans, animals, and ecosystems (Keith et al., 2016). Additionally, soil health has a significant role in achieving part of the targets of the United Nations' Sustainable Development Goals (Tóth et al., 2018) and the European Green Deal (Montanarella and Panagos, 2021).

Society, hence, faces the challenge of maintaining soil health to secure an adequate food supply and to preserve natural resources (Bampa et al., 2019), and human health. Maintaining healthy soils encompasses promoting sustainable ecosystems, sustainable agriculture, and environmental well-being, which are in part achieved through avoiding adverse soil challenges and adopting sustainable soil management practices.

Soil resources have been overexploited in modern society and are currently on the verge of collapsing, but this may be averted by development of an ethical attitude toward soil through increased awareness of the links between soil and people (Minami, 2009).

The concept of soil security can give a useful framework for relating soil health and human health. It broadens the scope of soil health and quality to embrace interdisciplinary and transdisciplinary approaches (Steffan et al., 2018). One of these that we discuss is connectivity, specifically the connectivity between public health, food, agriculture, and soil health. Important aspects are an adequate and continuous supply of nutritious food, lack of contaminants in the soil, good quality water for drinking and crop irrigation, and sustained good mental health. Integral to

DOI: 10.1201/9781003469742-8

these linkages are soil aspects of fertility, pollution control, and mitigating climate change. Another important aspect of soil health for human health is the soil microbiome. This connects to human health through its effects on the productivity and nutrient density of crops and its potential role as an important source of components for human disease (Suman et al., 2022). Nevertheless, it is important to be aware that soil structure and soil chemistry are vital for providing the conditions for the soil microbial community and that these help form the foundations of soil health (Coyne et al., 2022).

Although soil health is still seen as an evolving concept with some academic criticism of its validity, both public and private sectors are increasingly adopting the concept of soil health to address current challenges such as food security, environmental degradation, and climate changes (Wander, 2023).

As recognized earlier by Sir Albert Howard (1873–1947) "the health of soil, plant, animal, and man is one and indivisible." Therefore, it is fundamental to increase the awareness of the relationship between soil and human health, to influence behavioral change for more sustainable soil management that promotes resilience in food production and disease mediation in our socio-ecological landscapes (Keith et al., 2016).

The objective of this chapter is to highlight the importance of recognizing and promoting the connection between soil health and human health, identifying key aspects of soil health that are relevant to human health.

THE CONNECTION OF SOIL HEALTH TO HUMAN HEALTH: COMMUNICATING SOIL–HUMAN HEALTH KNOWLEDGE AVAILABILITY

The connection between humans and soil has been widely known, as mentioned above, with the most general historical connection being soil as the foundation for both food production and human habitation. However, society needs to become more aware of its dependence on soil by developing conscious connections with soil. Ball et al., (2018) recognized three types of connections between humans and soils that are fundamental in building awareness of the soil as an essential resource: direct, indirect, and temporal connections. The authors identified a direct connection with the soil during visual and tactile examination of it, for instance through gardening, farming, and research. The second, indirect type of connection involves a broader understanding of the environment and its links to human health and food, ecosystem services, climate change, biodiversity, and related decisions. Temporal connections are made by the awareness of the dynamics of soil processes and soil management, via observation and experience of changes such as soil fertility loss and structural degradation. These connections are fostered through knowledge availability and transfer across all age groups, enabling engagement in experiences that nurture respect for the soil (Ball et al., 2018).

Another way to classify the connections between humans and soils includes considering the nature of the connection or examining the impact of soils on human health. These connections could be biological, chemical, physical, or social (Table 8.1). Biological connections refer to when soils help in prevention of allergies and immunity-related disorders through direct ingestion and serve as an important reservoir of antibacterial agents, or a negative connection when individuals are exposed to soil pathogens (Brevik et al., 2019). Chemical connections may be soils providing nutrients through plant and animal consumption (Brevik et al., 2019), or on the contrary not contributing to human nutrition when foods are cultivated in nutrient-deficient soils (lacking iodine, zinc, and iron) (Silver et al., 2021). Chemical connections may also occur when individuals are exposed to toxic materials (heavy metals, radioactive materials, organic chemicals, and nitrogen pollutants) (Zhang et al., 2012; Katz, 2020a; Alengebawy et al., 2021), and when soils are used as reservoirs of medicinal compounds (Carretero and Pozo, 2010; Hamamoto et al., 2021). Examples of physical connections could be soils acting as a natural filter for clean water sources (Pierzynski, 2015), facilitating carbon sequestration, or off-site effects of erosion and landslides. Social connections could be soil's capability to hold recreational vegetation or use soil mechanisms as a beneficial

TABLE 8.1
Instances of Soil Health and Human Health Connections

Type of Connection	Soil Health	Human Health	Reference
Biological	Terroir	Nutritious food and drink	Brevik et al. (2019)
	Soil microbiome (microbial biodiversity)	Direct effects on human health and indirect effects through type of agricultural system	Suman et al. (2022)
	Presence of certain active bacteria and fungi	Beneficial effects (e.g., serotonin) and negative effects (e.g., tetanus, mycotoxins)	Lal (2020b)
Chemical	Content of nutrients and other compounds for crops, animals and humans	Decline of soil nutrients in soils or use of poor soils can lead to nutrient deficiencies. High-quality nutrient-rich organic soils may improve health	Jones et al. (2013); Silver et al. (2021); Vigar et al. (2020)
	Presence of pollutants, wastes and micro- (nano-) materials	Possible toxicity or diseases	Lal (2020b); Katz (2020a); Loyal et al. (2023); Surendran et al. (2023)
	Reservoir of medicinal components	Provide prevention of or recovery from disease	Carretero and Pozo (2010); Hamamoto et al. (2021)
Physical	Soil fertility and structure	Handling healthy soil produces benefits to physical and mental health Healthy soil safeguards water quality and facilitates carbon management	Chalmin-Pui et al. (2021); Penniman (2018); Lewandowski and Cates (2023); Neary et al. (2009); Lal (2016)
	Soil degradation and landslides	Off-site effects of soil erosion pollute water sources Deforested areas or climate events can cause landslides and land degradation with risk to health	Issaka and Ashraf (2017); Bowyer et al. (2009)
Social	Soil mechanisms and processes	Using soil mechanisms as a beneficial model of mental and social activity; recognizing soil as a unifying entity; developing cultural understanding of soil to improve human–soil relationships	Ball et al. (2018); Ball (2022); Goss (2020); Patzel et al. (2023)
	Capability to hold recreational vegetation	Beneficial to social activities and mental health	Nisbet et al. (2020)

model of mental and social activity (Nisbet et al., 2020; Ball, 2022). Overall connections of humans and soil can also be improved by developing a cultural understanding of soils from many cultures (Patzel et al., 2023).

Across these groupings, examples of how to achieve "connecting people and soil" could be by promoting visual and tactile inspection, gardening, open-air activities, and buying food from systems with mindful soil management, such as organic farming, or agroecology, among others. Soil visual examination methods (Emmet-Booth et al., 2016; Pulido-Moncada et al., 2021), for instance, could be used to link people to soil by the of monitoring soil health by farmers, advisers, or gardeners. Such an assessment of soil health brings focus to the physical and chemical interactions in soil that are important for soil functions (Coyne et al., 2022). Soil health assessed in this way is also important for gauging soil degradation, successful tillage methods, as well as adaptation to and mitigation of climate change.

Brevik et al. (2019) also discussed the crucial importance of enhancing people's connection with the soil. The authors suggested that a good "social marketing campaign" of already existing

Sustaining soil health for human health

Key factors	Good health	Fragile health
Awareness and communication	Conscious connections with soils Soil awareness General conscious sustainable management Socioeconomic and ecological driven management decisions	No conscious connections with soils Lack of soil governance and poor compliance Mainly economically driven management decisions
Advanced technology	High productivity for soil security Good food quality _nutritious Sustainable inputs _precision agriculture	Soil degradation _heavy machinery Toxicity and unknown effects of chemicals in food chain Hidden effects of micro (nano) materials (food chain)
Type of agriculture	Conservation biodiversity and good microbiome Maintaining soil security, soil and water quality Good food quality _ nutritious and gut microbiome helper Gardening and urban agriculture	Soil degradation _heavy machinery Mobility and bioavailability of pollutants Poor biodiversity and gut microbiome Poor food and water quality
Fauna and microbe diversity	Increase/maintenance of biodiversity Immunity, source of medicinal compounds	Development of anaerobic processes (greenhouse gases) Pathogens in soil and water
Chemical components	Drivers of biologically related processes Medicines and supplements Good food, soil and water quality	Nutrients deficiency Toxics: heavy metals, radioactive materials, and organic chemicals _food chain Presence of fertilisers, pesticides, medicines in food chain
Weather and climate	Mitigation climate change practices Long term land use	Soil degradation, floods, droughts Human migration
Conscious use	Healthy soils for fresh air and water Exercise and recreation Useful place in society Mental health _good overall health All generations engagement	Disposal of pollutants Overexploitation Agricultural land expansion

FIGURE 8.1 Factors affecting soil health for sustaining human health.

concepts such as terroir, soil health, and soil security, could establish better connections between soil and people and help to change society's behavior. A clear understanding of concepts in sustainable soil management is the basis for a common language and increases the sense of belonging to the environment (Weninger et al., 2024).

In the context of this chapter, the challenge is to better communicate the available knowledge on human health dependency on soil health to promote awareness of connections with soils from daily single activities to research focus programs (Figure 8.1). In the literature, soils are acknowledged for their dual impact on human health; both positive and negative effects. Although extensive effort has been dedicated to identifying soil impacts on human health, as summarized by Brevik et al. (2020) and Brevik et al. (2015), many potential negative effects may remain unknown.

There is a pressing need for further communication of soil knowledge for environmental decision-making by stakeholders and people in general. One of the mechanisms for knowledge communication is knowledge co-production engagement activities, bringing researchers and stakeholders together to address epistemological differences – or different knowledge cultures (Gerlak et al., 2023). Through collaborative projects and activities, participation establishes a collective sense of ownership and common goals to interact in development of soil–human health harmony. General education of the public on the food-production stages including harvesting, storage, processing, packaging, sales and consumption, may enhance, for example, the field-to-fork perspective and create consciousness of soil as the root of human health and well-being (El-Ramady et al., 2014). The illustration of the intertwined roots of human and flower by Ball (2022) is a good visual impression of a sense of rootedness of life in soil (Figure 8.2).

Additionally, awakening consciousness of soil health and human health connection may be achieved by understanding the wide scope of social principles that resemble soil mechanisms (Table 8.1). In society, treating others well is based on fostering united communities, promoting mental health, increasing awareness of the need for mutual assistance, and advancing toward a peaceful, balanced, and well-established society (e.g., Bregman, 2020). This principle applies to maintaining any ecosystem or resource. When people are educated to recognize that kindness and respectful behavior extend beyond family and society to all the components of the planet's life, there is an awakening consciousness of the connection between aspects such as soil health and human health. This realization does not just come from the transfer of scientific knowledge but also through

FIGURE 8.2 All life has its roots in soil: the soil brings beauty to life.
Acrylic on paper, first published in Ball, 2022.

recognizing that human health relies on other resources of the planet. When these resources are degraded, human life itself is endangered.

Communicating soil health to society is also the aim of establishing effective soil governance resources. In their study of the global standard bearers of soil governance, Peake and Robb (2022) highlighted that many wealthy industrialized countries have established soil governance frameworks and legal instruments to protect their environment, however, rarely do they achieve complete compliance. Conversely, some poorer countries may achieve effectiveness in environmental protection due to their local context. Additionally, while individual country commitments to developing soil governance are fundamental, there is a need for enhanced international integration of environmental, land, and soil governance with agendas related to climate and biodiversity, among others (Montanarella and Panagos, 2021).

The authors of this chapter are in favor of policies and instruments that help in soil protection and, hence human health protection. However, we raise awareness of blanket application of regulations that overlook site-specific conditions. Aligning with Peake and Robb (2022), we also stress the need for appropriate regulations, with both national and international scope, though calling attention to knowledge co-production strategies to avoid unsuccessful projects and wrong policies (Prager and McKee, 2015; Gerlak et al., 2023).

CURRENT CHALLENGES TO SOIL HEALTH AND HUMAN HEALTH HARMONY

While we have long known about the relationship between soils and human health, the way soils have been managed has traditionally been driven by socio-economic priorities. It is only in recent decades that soil-related ecological and human health concerns have gained widespread attention.

This may be linked to the increased concern about climate change and its influence on both local and global economies. However, regardless of the trigger for increased awareness or interest in the soil and human health harmony, there are currently numerous challenges to address (Figure 8.1). We have selected five soil challenges to be discussed in this chapter, without diminishing the significance of other existing challenges that also warrant attention.

Soil Degradation: Impact on Human Health

The decline in land quality caused by human activities, known as land degradation, has been a concern since the 20th century and still is today. It has a harmful impact on the environment, world food security, and quality of life (Eswaran et al., 2019). Among the various processes contributing to land degradation, soil degradation is a world problem that is significant, particularly in agricultural regions, as it directly influences food security and people's needs by reducing yield, forcing farmers to use more inputs, and eventually leading to soil abandonment (Gomiero, 2016).

Among the soil degradation impacts on human health, the following are mentioned as examples considered to be currently poorly documented.

Soil Pollution through Chemicals

Of the various chemicals polluting the soils, there are currently knowledge gaps on the impacts of heavy metals, pesticides, fertilizers, and medicines.

- Toxicity of heavy metals and pesticides is well-documented to contribute to the deterioration of human health, causing diseases such as neurodegenerative disorders, musculoskeletal diseases, reproductive hormonal imbalance, sarcoma, cancer, asthma, genetic damage, and endocrine disruption (Alengebawy et al., 2021). However, little is known about the co-occurrence of toxic mixtures, their interactions, and combined toxicity. There is also a need for better knowledge of approaches for phytoremediation and bioremediation of environmental toxicants (Alengebawy et al., 2021).
- Amongst the impacts of inadequate fertilization practices, the contribution of soil to nitrogen pollution is of high relevance for its impact on soil health and human health. As summarized by Katz (2020b), efforts have been made to assess how excess nitrogen input has caused environmental degradation (e.g., greenhouse gases, terrestrial biodiversity, surface water, and groundwater contamination), adverse human health effects (e.g., respiratory infections, heart diseases, cancer, blue-baby syndrome), and economic costs associated with environmental and human health impact (drinking-water treatment costs, losses to commercial fishing, recreational activities, decrease in real estate values). Collaborative efforts are recommended for assessing sustainable practices to reduce human nitrogen footprint with local and global challenges (Castner et al., 2017). There is also a need for studies on the effects of nitrogen pollution on less well-known effects on human health.
- Regarding medicine usage, extensive knowledge exists when it comes to their side-effects on humans and animals. Yet, uncertainties persist regarding their potential impact on different organisms in the environment and the implications for environmental health, which inevitably intertwines with human health in some way or another. Boxall (2004) critically analyzed how human and veterinary medicines that follow distinct pathways into soils and water bodies are affecting human and environmental health. For instance, veterinary antibiotics and resistant bacteria reach agricultural fields with manure application, which undergoes transportation and is distributed to air, water, soil, or sediment with potential implications for human health (Jechalke et al., 2014). Importantly, as reflected by Boxall (2004), though risk studies have been conducted, there are still many uncertainties such as: (1) the impact of mixtures of veterinary and human medicines in soils and surface

waters on both the environment and human health (2) risks of potential uptake from soils into crops and biomagnification through the food chain; (3) substances that have not yet been thoroughly studied and the need for improved assessment of ecotoxicity; (4) understanding the real-world implications of ecotoxicity data from laboratory studies; (5) determining the uptake of pharmaceuticals into organisms and through the food chain; and lastly (6) a pressing question about whether environmental exposure contributes to antibacterial resistance, which could pose a serious threat to both human and animal health.

- Interdisciplinary and comprehensive studies are needed to close these knowledge gaps. Radu et al. (2021) gave an example. They observed that antibiotic resistance gene concentrations affected by fertilizer and pesticide amendments that "return to background levels after manure-induced exposure within a crop growing season, highlighting the resilience capacity of soil".

Subsoil Compaction

Subsoil compaction is a long persisting condition, due to limited biological activity and physical processes when compared with topsoil (Håkansson and Reeder, 1994; Berisso et al., 2013), and limited lasting recovery-management practices (Olesen and Munkholm, 2007). Apart from potential negative effects on root growth, crop yield, and fluid movement, subsoil compaction may influence the preferential flow of pollutants (Etana et al., 2013), and production and transport of greenhouse gases such as nitrous oxide (Clough et al., 2005; Schjønning et al., 2019). However, there is a paucity of knowledge about the contribution of subsoil compaction to the fate of greenhouse gases (Pulido-Moncada et al., 2022). Plowing and subsoiling (biological or mechanical) are options to mitigate soil compaction, whereas controlled traffic farming (Tamirat et al., 2022) and compaction-intelligent traffic, such as robots' light machinery (Calleja-Huerta et al., 2023), are soil-compaction avoidance strategies. While these strategies become successful in avoiding subsoil compaction, the low recovery of the already existing compacted subsoil remains a threat to soil health and a hidden contributor to emissions and climate change.

Impact of Type of Agriculture on Food Quality and Choice and on Human Health

The food we eat, both in terms of its quality and quantity, is directly affected by soil health and intricately connected to the surrounding environment; human health, in turn, relies on the food we eat (Reeve et al., 2016).

It has been shown that the type of agriculture helps determine soil management and thus soil health (Albaladejo et al., 2021). Sustainable agricultural practices that focus on soil health such as agroecology or conservation agriculture (also known as regenerative agriculture) reduce the likelihood of soil degradation, thereby conserving soil health. Long-term cropping of older soils in less-developed countries such as India can lead to stripping of trace elements from soils that reduces the content of nutrients for crops, animals, and humans (Table 8.1) thereby causing some impoverishment of the quality of food (Jones et al., 2013). However, there is controversy about whether depletion in agricultural soil minerals results in a decline in mineral nutrient composition of vegetables, fruits, and grains (Marles, 2017).

Trace elements (or micronutrients) and phytochemicals (antioxidants and anti-inflammatory compounds) have been shown to be higher in organically farmed produce than in conventional practice produce (Montgomery and Biklé, 2021). Nevertheless, factors such as the overlap in management practices among farming systems and environmental and crop species, and/or cultivar interactions may have stronger effects than management (Reeve et al., 2016). These authors also point out that the controversy over differences between organic and conventional crops needs to be broadened to include other types of agronomy and to include those compounds beneficial to the maintenance of health and the prevention of disease.

As pointed out by Halonen et al. (2021) the Farm to Fork strategy as part of the Green Deal of the European Commission (Veerman et al., 2020) aims to build a sustainable food chain that works for consumers, producers, climate, and the environment. The authors reflected that this strategy entails a shift to healthier diets, requiring corresponding changes in the agricultural and food sectors, such as sustainable and environmentally friendly food production, nutritionally adequate and safe food, as well as socially and culturally acceptable food.

Hirt (2020) suggested that the concept of a healthy diet, traditionally focused on fiber, minerals, and vitamins from fruits and vegetables, should expand to include plant microbes. This is because the human gut microbiome and soil microbiome influence human metabolism, health, and overall well-being. However, conventional agricultural practices, characterized by the widespread use of antibiotics, herbicides, fungicides, and pesticides (including treatments during food storage and shipping), pose a threat to both human gut (through food consumption) and soil/plant microbiomes (Hirt, 2020). It is, therefore, important to prioritize the establishment and maintenance of sustainable soil practices for a healthy microbiome in agriculture to contribute to sustainably providing healthy food.

Organic farming can improve beneficial soil microbe and invertebrate activities and increase plant health, though the presence of a link to improved human health requires more study (Reeve et al., 2016). Although the role of organic farming in improving nutritional quality of food is debatable, the absence of synthetic inputs is considered to make organic produce safer and healthier (Aulakh et al., 2022). The benefits of organic farming in small-scale urban horticulture and agriculture are increasingly evident in enhancing human health. This includes access to fresh, organically grown food, as highlighted by Ball (2022), along with fostering a greener urban environment devoid of conventional chemicals. Additionally, organic farms have been recognized for their role in mental recovery programs. Initiatives like the Growing Veterans organic farm in Washington, United States, offer individuals with post-combat traumatic stress disorder an opportunity to connect with soil, fostering a sense of rootedness while producing healthy food to support life (Detjens, 2014).

Organic farming and other agroecological practices, however, may not produce sufficient food if practiced widely but there is evidence for productive opportunities in northern Europe (Röös et al., 2018), Africa (Neuhoff and Kwesiga, 2021), and India (Das et al., 2020). The need to strengthen local food production through initiatives such as home gardening and urban agriculture should be perceived as a strategy for enhancing food quality, choice and availability at household and community levels (Lal, 2020a). However, there is a need for studies to assess and tackle contaminants in urban soils to ensure the production of safe food.

Hidden Risk for Soil–Water–Plant–Human Health: Micro- and Nanomaterials in the Food Chain

The long and complicated food chains across the world have a heavy impact on various dimensions of sustainability, including but not limited to food security, poverty, and land use (Halonen et al., 2021). In food chains, attention should also be paid to hidden risks for human health such as transported micro- and nanomaterials. While the application of micro- and nanotechnologies in agriculture is still in development at the time of this publication, they contribute to more efficient agricultural systems that address yield gaps. However, concerns persist about potential harmful impacts.

As summarized by Loyal et al. (2023), negative effects of nanosized pesticides and fertilizers, as well as nano-formulated pesticides and fertilizers, have been found on crops, soil, environment, and human health. For instance, genotoxic and phytotoxic effects on different plant parts, harmful effects on non-targeted soil microorganisms, and disturbance in the ecosystems and their components such as pollinators and aquatic life have not yet been studied. Regarding human health, there is a risk of genotoxic effects due to direct contact with nanochemicals (e.g., occupational exposure by farmers), and due to consumption of contaminated sources (nanomaterials that enter the food chain) (Loyal et al., 2023).

In recent years, concern has grown about the excessive use and disposal of plastic. However, individuals need to recognize the significance of micro- and nano-plastics on soil health and their potential transmission into the food chain. Given their small size, these particles can be ingested by a wide range of organisms, spanning in size from soil microfauna to humans, consequently posing a threat to human health (Kumar et al., 2020). The work by Surendran et al. (2023) summarized the main sources and transport of micro- and nano-plastics that have become pervasive in the environment and affect soil properties, impact soil fauna, and influence soil functions. Some of the sources are sewage sludge, agricultural mulching films, coated fertilizers, and abrasion from vehicle tires. Though some governing agencies are proposing policy measures to address plastic pollution in different parts of the world, mitigation of micro- (nano)plastic pollution, especially in soils, is challenging and there remains a need for further studies focusing on potential implications on soil ecosystems and human health (Surendran et al., 2023).

The micro- (nano)particle pollution might also have an ecological influence on plant-available water in the soil (green water), water bodies (blue water), and, hence, on drinking water, which consequently harm human health. An ecosystem service of soils is the capacity to filter water, chemical filtering (bio/chemical reactions) and physical filtering (infiltration potential, water-storage capacity and residence time of soil water), which determine the potential of soil to remove contaminants (Keesstra et al., 2021). However, limited knowledge is available on the potential capacity of the soil to filter micro- (nano)particles or the hazard posed by interactions of these particles with soil's potentially toxic elements.

Unknown negative effects of micro and nanoparticles originating from the agriculture sector on the food chain and environment could potentially counterbalance their contributions to enhancing sustainability. Therefore, interdisciplinary and comprehensive studies of the trade-offs of using micro and nanotechnologies in farming are needed to better understand environmental risks and to ensure human safety regarding micro- and nanoparticle fate, toxicity, and disease links (Rajput et al., 2020).

Climate Change Adaptation Related to Soil and Human Health

Climate change can affect soil functions directly through changes in temperature, precipitation and moisture regimes; it can also have indirect effects through adaptations such as irrigation, crop rotation changes, and tillage practices (Hamidov et al., 2018). Climate change also causes perturbation of soil cycles (Amundson, 2021). Projections suggest desertification due to water scarcity in certain areas worldwide poses further threats to soil health and food security, and thereby to human health (Hatten and Liles, 2019). Areas where climate change makes weather so extreme that adequate crop production can no longer be supported, are considered to be outside of the "safe climatic space" (Kummu et al., 2021). Studies in Europe have shown that adaptation options of conservation tillage, irrigation, improved drainage, and modified crop rotations reduced erosion threats and increased soil organic carbon; they also improved the soil functions of food and biomass production, soil organic-carbon storage and storing, filtering, transforming, and recycling capacities (Hamidov et al., 2018). Thus, changes in cropping and soil management, including agroecological options, offer possibilities for adaptation that conserve soil function while increasing resilience to climate change (Altieri et al., 2015), reducing the need for communities to migrate to more favorable areas to survive climate change. Although climate-change mitigation is on the global agenda, Halonen et al. (2021) stress that "measures to adapt to ongoing climate change are urgently needed to sustain Planetary Health and that their relative efficacy needs to be evaluated".

Population Movements: Migrations of People from Poor Soil/Economies

"Food security exists when all people, at all times, have physical and economic access to sufficient, safe and nutritious food that meets their dietary needs and food preferences for an active and healthy life" (World Food Summit, 1996). To maintain food security, soil must be protected and preserved from degradation. However, other factors directly affect soil security and favor human migration.

Environmental changes such as rainfall patterns or unsustainable practices such as water diversions, deforestation, overgrazing, and inappropriate irrigation can affect agricultural productivity and lead to economic problems and migration (Mianabadi et al., 2023). Land problems are a major motivator for migration with particular importance in developing countries (Albaladejo et al., 2021; Majumder and Rahman, 2023). The multidimensional and complex impact of migration and relocation of people can lead to environmental degradation of other vulnerable areas with poor soil quality, and problems such as sanitation or hygiene due to lack of infrastructure, and therefore have adverse effects on human health (Majumder and Rahman, 2023).

Addressing environmental human migration is a key concern for policymakers. The varying patterns of migration resulting from gradual and sudden environmental changes, along with economic, social, demographic, and political influences, pose challenges for policymakers in devising effective plans and allocating resources for data collection and research (Mianabadi et al., 2023). By promoting sustainable practices including those discussed here, the negative consequences on soil health can be minimized, potentially reducing the need for migration driven by economic challenges related to agriculture.

CURRENT AND FUTURE ACTIONS FOR CLOSING KNOWLEDGE GAPS ON SOIL MANAGEMENT BASED ON HUMAN HEALTH NEEDS

Soil Management

The management of soil in the context here depends on the type and intensity of agriculture. In general, low-intensity systems such as agroecology or organic farming tend to have sustainable soil management. In contrast, more intensive systems tend to have more risky soil management such as using heavy machinery and poorly timed operations – for example, working the soil when too wet which is prone to cause compaction and possible runoff leading to soil erosion. Thus, good quality soil management is essential for the sustainability of soil health. Although the type of agriculture may mainly determine the soil management, it is often possible to adopt a favorable management system for most types of agriculture. The common type of beneficial soil management is conservation agriculture or conservation tillage practices that operate at the farm or field scale. Management techniques for conservation agriculture include conservation tillage such as no-tillage, contour farming, strip tillage, also use of cover crops, crop rotation, buffer strips, and terracing (Hobbs et al., 2008). Although all these conservation or sustainable practices are well known, their adoption is rather limited across agricultural areas worldwide.

From the soil conservation point of view, Albaladejo et al. (2021) proposed the main reasons that explain the failure of soil conservation initiatives. The first is the improper design of soil conservation projects, as most lack integration assessments on the interactions of the soil with other components of the socio-ecosystem: land users and farmers, social reality, and socio-economic and cultural context. Additionally, knowledge generated from projects lacks effort to ensure implementation of conservation measures and commitment to dealing with human pressures (health, poverty, migration, and land rights). The second reason mentioned by the authors was the failure to attract social and political awareness, and lastly, the lack of adaptation of soil conservation measures to global change. These were considered by the authors as the basis for the development of an Eco-Holistic Soil Conservation framework, which also includes practical recommendations for its implementation with the main focus on transdisciplinary enhancement of social awareness and soil ethic.

In line with the perspectives above, tackling soil-health challenges requires coordinated effort. Effective solutions to soil-health risks require broad approaches that integrate multidisciplinary research, address the impacts of climate change, engage with social factors, prioritize investments in knowledge and innovative knowledge transfer (Amundson et al., 2015) and adopt a perspective that recognizes the potential risks to human health posed by soil problems.

The One Health initiative and Planetary Health paradigm are ongoing integrating initiatives. The One Health initiative seeks to create an environment of interdisciplinary collaboration between medical professionals and other relevant scientific disciplines to promote human, animal, and environmental health (World Health Organization, 2017). Accordingly, the Planetary Health paradigm – the health of human civilization and the state of natural systems on which it depends – is a framework aiming for a shift from fragmented approaches to policy and practice toward systematic actions to promote human health and the health of the planet by conserving nature and biodiversity, and to halt climate change (Halonen et al., 2021).

Policies and Legislation to Improve Sustainable Soil Management

Strategies used to attempt to assure soil sustainability involve legislation and policies for soil conservation and management that operate at a range of scales from global to national. The Food and Agriculture Organization (FAO) established a Global Soil Partnership in 2012 with the mission to position soils in the global agenda to improve soil governance and to promote sustainable soil management. There are now 194 partners engaged.

One of the few regional systems involving soil legislation is the EU soil strategy for 2030. It is a framework of practical measures to protect and restore soils and to ensure their sustainable use. The legislation proposed is to achieve good soil health by 2050. Typical objectives include limiting land takes, restoring degraded soils and promoting sustainable soil management. Other country initiatives include the Australian National Soil Strategy and the USDA National Resources Conservation Service, though strategies vary between states. Nevertheless, they are considered to provide an opportunity for broader collaboration with public health (Delmendo et al., 2021). Other initiatives which form part of farm and land management include the Soil Initiative for Africa, Living Soils of the Americas, Soil Action Plan for England, and Managing Scotland Soils. Many of these approaches are voluntary and based mainly on conservation agriculture with some attracting funding support at the farm level, making agricultural policies key to the protection of our soils (Calvo, 2022).

Nevertheless, public awareness of the importance of the link between farming and soil health is also important. The UK Soil Association shows that this can be achieved by describing five ways to save your soil at home.

Regarding policy implementation and target setting in agriculture, it is important to consider flexible approaches that allow for site-specific variations. This approach can offer adaptable prioritization and decision-making (Wander, 2023).

Strategies to Improve the Linkage between Soil Health and Human Health

In the context of this chapter, knowledge needs of the linkage between soil health and human health were highlighted above, other authors have also summarized identified needs (e.g., Brevik et al., 2020). Research needs in less traditional areas have also been highlighted such as possible human health benefits (physical and mental) of contact with healthy soil and organic farming (Brevik et al., 2015). There is also a need for studies on human health benefits from maintaining soil health provided by home gardening and urban agriculture, as well as benefits from cultural connections and children's education on healthy soils (Ball, 2022).

The need for soil protection and its connection with humans have been known since the beginning of civilization, as reflected in hieroglyphs and holy books (e.g., the Torah, Bible, Koran). However, there remains a gap in human awareness regarding the importance of internalizing the fragility of both human health and its dependency on soil health. Positively, there appears to be a growing interest in integrating soil knowledge availability with present and future environmental challenges, as well as soil-related human threats. Examples are the targets of the United Nations' Sustainable Development Goals (Tóth et al., 2018). An example at a national level, is the Ecological Civilization development strategy in China (in the constitution since 2018), which aims to improve humanity's relationship with the environment, and emphasizes political and cultural aspects (Wei et al., 2021). This approach involves ongoing research for effective strategies like eco-compensation mechanisms and social engagement strategies.

Worldwide, there is also a rising interest in knowledge co-production efforts and effective knowledge communication and implementation to foster a multidisciplinary approach to a sustainable relationship between soil and humanity. Despite acknowledging the inherent complexity of these approaches, the recent interest shown by decision-makers and all stakeholders involved suggests a promising trajectory in the years ahead.

CONCLUSION

Efforts have been made to assess human dependency on soil health; however, there is a need to communicate this knowledge to promote conscious awareness and develop a sense of rootedness of life in the soil among people. This understanding should impact both daily activities and research programs, as well as food-chain processes. Current challenges for sustaining soil health for human health include knowledge gaps, identified as: the threat from soil degradation (e.g., chemical pollution and subsoil compaction); type of agriculture as influencing human gut microbiome and nutritional quality of food; unknown effects of micro- (nano)materials mixture and interactions with other components of soil and environment; climate change scenarios and their impact on soil health and food security and thereby on human health; the potential of sustainable agricultural practices to reduce human health risks; and the complex links between climate change, soil health and human migration. Closing knowledge gaps in relation to the links between soil health and human health calls for expanding educational interventions to improve producer and consumer knowledge of the connections between soil and public health, especially in urban populations. Fundamental changes in communication, within both local and international communities, are needed to reduce irrational use of resources and consumption of food and to fight poverty by developing an environmental ethic to lead to more consciousness of soil (and land). Multidisciplinary collaborations are required at a range of scales, for example, the One Health initiative and the Planetary Health paradigm, to ensure that the health of human civilization and the state of our ecosystems, including soil, are major policy drivers. Other strategies to sustain and develop the connection between soil health and human health are a focus on farm policies and funding support for soil and human health by use of sustainable soil management techniques, preferably through targeting improved farmer profit margins rather than more regulations (Delmendo et al., 2021). Maintaining and improving the current focus on environmental approaches is vital for planetary health, as is the need to increase the emphasis on food quality in agricultural production.

ACKNOWLEDGEMENTS

This work was funded by the EJP SOIL project under the European Union's Horizon 2020 research and innovation program: Grant agreement No 862695.

REFERENCES

Albaladejo, J., Díaz-Pereira, E., de Vente, J., 2021. Eco-holistic soil conservation to support land degradation neutrality and the sustainable development goals. *Catena* 196, 104823.

Alengebawy, A., Abdelkhalek, S.T., Qureshi, S.R., Wang, M.Q., 2021. Heavy metals and pesticides toxicity in agricultural soil and plants: Ecological risks and human health implications. *Toxics* 9(3), 42.

Altieri, M.A., Nicholls, C.I., Henao, A., Lana, M.A., 2015. Agroecology and the design of climate change-resilient farming systems. *Agronomy for Sustainable Development* 35, 869–890.

Amundson, R., 2021. Factors of soil formation in the 21st century. *Geoderma* 391, 114960.

Amundson, R., Berhe, A.A., Hopmans, J.W., Olson, C., Sztein, A.E., Sparks, D.L., 2015. Soil science. Soil and human security in the 21st century. *Science* 348, 1261071.

Aulakh, C.S., Sharma, S., Thakur, M., Kaur, P., 2022. A review of the influences of organic farming on soil quality, crop productivity and produce quality. *Journal of Plant Nutrition* 45, 1884–1905.

Ball, B.C., 2022. *Healing soil.* Kindle Direct Publishing Services.

Ball, B.C., Hargreaves, P.R., Watson, C.A., 2018. A framework of connections between soil and people can help improve sustainability of the food system and soil functions. *Ambio* 47, 269–283.

Bampa, F., O'Sullivan, L., Madena, K., Sandén, T., Spiegel, H., Henriksen, C.B., Ghaley, B.B., Jones, A., Staes, J., Sturel, S., 2019. Harvesting European knowledge on soil functions and land management using multi-criteria decision analysis. *Soil Use and Management* 35, 6–20.

Baveye, P.C., Baveye, J., Gowdy, J., 2016. Soil "ecosystem" services and natural capital: Critical appraisal of research on uncertain ground. *Frontiers in Environmental Science* 4, 41.

Berisso, F.E., Schjønning, P., Keller, T., Lamandé, M., Simojoki, A., Iversen, B.V., Alakukku, L., Forkman, J., 2013. Gas transport and subsoil pore characteristics: Anisotropy and long-term effects of compaction. *Geoderma* 195, 184–191.

Bouma, J., Montanarella, L., Evanylo, G., 2019. The challenge for the soil science community to contribute to the implementation of the UN Sustainable Development Goals. *Soil Use and Management* 35, 538–546.

Bowyer, C., Withana, S., Fenn, I., Bassi, S., Lewis, M., Cooper, T., Benito, P., Mudgal, S., 2009. Land degradation and desertification; European Parliament; Policy Department A, Economic and Scientific Policy: Brussels, Belgium.

Boxall, A.B., 2004. The environmental side effects of medication. *EMBO Reports* 5, 1110–1116.

Bregman, R., 2020. *Humankind: A hopeful history.* Bloomsbury Publishing.

Brevik, E.C., Cerdà, A., Mataix-Solera, J., Pereg, L., Quinton, J.N., Six, J., Van Oost, K., 2015. The interdisciplinary nature of *SOIL. The Soil* 1, 117–129.

Brevik, E.C., Slaughter, L., Singh, B.R., Steffan, J.J., Collier, D., Barnhart, P., Pereira, P., 2020. *Soil and human health: Current status and future needs* (p. 13). Air, Soil and Water Research.

Brevik, E.C., Steffan, J.J., Rodrigo-Comino, J., Neubert, D., Burgess, L.C., Cerdà, A., 2019. Connecting the public with soil to improve human health. *European Journal of Soil Science* 70, 898–910.

Calleja-Huerta, A., Lamandé, M., Green, O., Munkholm, L., 2023. Impacts of load and repeated wheeling from a lightweight autonomous field robot on the physical properties of a loamy sand soil. *Soil and Tillage Research* 233, 105791.

Calvo, F., 2022. *How Farm policies can help protect our soils.* International Institute for Sustainable Development.

Carretero, M.I., Pozo, M., 2010. Clay and non-clay minerals in the pharmaceutical and cosmetic industries Part II. Active ingredients. *Applied Clay Science* 47, 171–181.

Castner, E.A., Leach, A.M., Leary, N., Baron, J., Compton, J.E., Galloway, J.N., Hastings, M.G., Kimiecik, J., Lantz-Trissel, J., Reguera, E.D.L., 2017. The nitrogen footprint tool network: A multi-institution program to reduce nitrogen pollution. *Sustainability: The Journal of Record* 10, 79–88.

Chalmin-Pui, L.S., Griffiths, A., Roe, J., Heaton, T., Cameron, R., 2021. Why garden?–Attitudes and the perceived health benefits of home gardening. *Cities* 112, 103118.

Clough, T.J., Sherlock, R.R., Rolston, D.E., 2005. A Review of the Movement and Fate of N2O in the Subsoil. *Nutrient Cycling in Agroecosystems* 72, 3–11.

Coyne, M.S., Pena-Yewtukhiw, E.M., Grove, J.H., Sant'Anna, A.C., Mata-Padrino, D., 2022. Soil health – It's not all biology. *Soil Security* 6, 100051.

Das, S., Chatterjee, A., Pal, T.K., 2020. Organic farming in India: A vision towards a healthy nation. *Food Quality and Safety* 4, 69–76.

Delmendo, M., Sipos, Y., Montgomery, D., Cole, R., Otten, J., 2021. The scope of US state soil health legislation: A mixed-methods policy analysis. *Journal of Agriculture, Food Systems, and Community Development* 10, 69–90.

Detjens, A., 2014. *The healing power of soil.* Center for Sustaining Agriculture and Natural Resources.

El-Ramady, H., Abdalla, N., Alshaal, T., Domokos-Szabolcsy, É., Elhawat, N., Prokisch, J., Sztrik, A., Fári, M., El-Marsafawy, S., Shams, M.S., 2014. Selenium in soils under climate change, implication for human health. *Environmental Chemistry Letters* 13, 1–19.

Emmet-Booth, J., Forristal, P., Fenton, O., Ball, B., Holden, N., 2016. A review of visual soil evaluation techniques for soil structure. *Soil Use and Management* 32, 623–634.

Eswaran, H., Lal, R., Reich, P., 2019. Land degradation: An overview. In: Bridges, E., Hannam, I., Oldeman, L., Penning de Vries, F., Scherr, S., Sombatpanit, S. (eds.), *Response to land degradation* (pp. 20–35). CRC Press/Taylor & Francis.

Etana, A., Larsbo, M., Keller, T., Arvidsson, J., Schjønning, P., Forkman, J., Jarvis, N., 2013. Persistent subsoil compaction and its effects on preferential flow patterns in a loamy till soil. *Geoderma* 192, 430–436.

Gerlak, A.K., Guido, Z., Owen, G., McGoffin, M.S.R., Louder, E., Davies, J., Smith, K.J., Zimmer, A., Murveit, A.M., Meadow, A., 2023. Stakeholder engagement in the co-production of knowledge for environmental decision-making. *World Development* 170, 106336.

Gomiero, T., 2016. Soil degradation, land scarcity and food security: Reviewing a complex challenge. *Sustainability* 8(3): 281.

Goss, S., 2020. An eco-system view of change and a different role for the state. *Compass* 1, 22.

Håkansson, I., Reeder, R.C., 1994. Subsoil compaction by vehicles with high axle load—extent, persistence and crop response. *Soil and Tillage Research* 29, 277–304.

Halonen, J.I., Erhola, M., Furman, E., Haahtela, T., Jousilahti, P., Barouki, R., Bergman, Å., Billo, N.E., Fuller, R., Haines, A., 2021. A call for urgent action to safeguard our planet and our health in line with the Helsinki declaration. *Environmental Research* 193, 110600.

Hamamoto, H., Panthee, S., Hashimoto, K., Tsuchida, T., Sekimizu, K., 2021. An efficient method to screen for the soil bacteria producing therapeutically effective antibiotics. *The Journal of Antibiotics* 74, 850–855.

Hamidov, A., Helming, K., Bellocchi, G., Bojar, W., Dalgaard, T., Ghaley, B.B., Hoffmann, C., Holman, I., Holzkämper, A., Krzeminska, D., 2018. Impacts of climate change adaptation options on soil functions: A review of European case-studies. *Land Degradation and Development* 29, 2378–2389.

Hatten, J., Liles, G., 2019. Chapter 15 – A 'healthy' balance – The role of physical and chemical properties in maintaining forest soil function in a changing world. *Developments in soil science*, Vol. 36 (pp. 373–396). Elsevier.

Hirt, H., 2020. Healthy soils for healthy plants for healthy humans: How beneficial microbes in the soil, food and gut are interconnected and how agriculture can contribute to human health. *EMBO Reports* 21, e51069.

Hobbs, P.R., Sayre, K., Gupta, R., 2008. The role of conservation agriculture in sustainable agriculture. *Philosophical Transactions of the Royal Society, B: Biological Sciences* 363, 543–555.

Issaka, S., Ashraf, M.A., 2017. Impact of soil erosion and degradation on water quality: A review. *Geology, Ecology, and Landscapes* 1, 1–11.

ITPS-FAO, 2020. *Towards a definition of soil health* (p. 3). Soil Leters.

Jechalke, S., Heuer, H., Siemens, J., Amelung, W., Smalla, K., 2014. Fate and effects of veterinary antibiotics in soil. *Trends in Microbiology* 22, 536–545.

Jones, D.L., Cross, P., Withers, P.J., DeLuca, T.H., Robinson, D.A., Quilliam, R.S., Harris, I.M., Chadwick, D.R., Edwards-Jones, G., 2013. Nutrient stripping: The global disparity between food security and soil nutrient stocks. *Journal of Applied Ecology* 50, 851–862.

Katz, B., 2020a. Exploring the widespread impacts of ongoing nitrogen pollution. *Eos* 101, e10654.

Katz, B.G., 2020b. *Nitrogen overload: Environmental degradation, ramifications, and economic costs*. John Wiley & Sons.

Keesstra, S., Sannigrahi, S., López-Vicente, M., Pulido, M., Novara, A., Visser, S., Kalantari, Z., 2021. The role of soils in regulation and provision of blue and green water. *Philosophical Transactions of the Royal Society B* 376, 20200175.

Keith, A.M., Schmidt, O., McMahon, B.J., 2016. Soil stewardship as a nexus between Ecosystem Services and One Health. *Ecosystem Services* 17, 40–42.

Kumar, M., Xiong, X., He, M., Tsang, D.C.W., Gupta, J., Khan, E., Harrad, S., Hou, D., Ok, Y.S., Bolan, N.S., 2020. Microplastics as pollutants in agricultural soils. *Environmental Pollution* 265, 114980.

Kummu, M., Heino, M., Taka, M., Varis, O., Viviroli, D., 2021. Climate change risks pushing one-third of global food production outside the safe climatic space. *One Earth* 4, 720–729.

Lal, R., 2016. Soil health and carbon management. *Food and Energy Security* 5, 212–222.

Lal, R., 2020a. Home gardening and urban agriculture for advancing food and nutritional security in response to the COVID-19 pandemic. *Food Security* 12, 871–876.

Lal, R., 2020b. *The soil-human health-nexus*. CRC Press.

Lehmann, J., Bossio, D.A., Kögel-Knabner, I., Rillig, M.C., 2020. The concept and future prospects of soil health. *Nature Reviews Earth and Environment* 1, 544–553.
Lewandowski, A.M., Cates, A., 2023. Connecting soil health and water quality in agricultural landscapes. *Journal of Environmental Quality* 52, 412–421.
Loyal, A., Pahuja, S., Sharma, P., Malik, A., Srivastava, R.K., Mehta, S., 2023. Potential environmental and human health implications of nanomaterials used in sustainable agriculture and soil improvement. *Engineered nanomaterials for sustainable agricultural production, soil improvement and stress management* (pp. 387–412). Elsevier.
Majumder, S.C., Rahman, M.H., 2023. Rural–urban migration and its impact on environment and health: Evidence from Cumilla City Corporation, Bangladesh. *GeoJournal* 88, 3419–3437.
Marles, R.J., 2017. Mineral nutrient composition of vegetables, fruits and grains: The context of reports of apparent historical declines. *Journal of Food Composition and Analysis* 56, 93–103.
Mianabadi, A., Davary, K., Mianabadi, H., Kolahi, M., Mostert, E., 2023. Toward the development of a conceptual framework for the complex interaction between environmental changes and rural-urban migration. *Frontiers in Water* 5, 1142307.
Minami, K., 2009. Soil and humanity: Culture, civilization, livelihood and health. *Soil Science and Plant Nutrition* 55, 603–615.
Montanarella, L., Panagos, P., 2021. The relevance of sustainable soil management within the European Green Deal. *Land Use Policy* 100, 104950.
Montgomery, D.R., Biklé, A., 2021. Soil health and nutrient density: Beyond organic vs. conventional farming. *Frontiers in Sustainable Food Systems* 5, 417.
Neary, D.G., Ice, G.G., Jackson, C.R., 2009. Linkages between forest soils and water quality and quantity. *Forest Ecology and Management* 258, 2269–2281.
Neuhoff, D., Kwesiga, J., 2021. Para-organic intensification of future farming as alternative concept to reactor-based staple food production in Africa. *Organic Agriculture* 11, 209–215.
Nisbet, E.K., Shaw, D.W., Lachance, D.G., 2020. Connectedness with nearby nature and well-being. *Frontiers in Sustainable Cities* 2, 18.
Olesen, J.E., Munkholm, L.J., 2007. Subsoil loosening in a crop rotation for organic farming eliminated plough pan with mixed effects on crop yield. *Soil and Tillage Research* 94, 376–385.
Patzel, N., Grunwald, S., Brevik, E.C., Feller, C., 2023. Summary and conclusions: Understanding soil-cultural relations. *Cultural understanding of soils: The importance of cultural diversity and of the inner world* (pp. 521–540). Springer.
Peake, L.R., Robb, C., 2022. The global standard bearers of soil governance. *Soil Security* 6, 100055.
Penniman, L., 2018. *Farming while black: Soul fire farm's practical guide to liberation on the land.* Chelsea Green Publishing.
Pierzynski, G.M., 2015. *Soil: Earth's largest natural water filter.* Kansas State University.
Prager, K., McKee, A., 2015. Co-production of knowledge in soils governance. *International Journal of Regional, Rural and Remote Law and Policy*1, 81–97.
Pulido-Moncada, M., Ball, B.C., Cornelis, W.M., 2021. *Advances in visual soil evaluation techniques. Advances in measuring soil health* (pp. 71–110). Burleigh Dodds Science Publishing.
Pulido-Moncada, M., Petersen, S.O., Munkholm, L.J., 2022. Soil compaction raises nitrous oxide emissions in managed agroecosystems. A review. *Agronomy for Sustainable Development* 42, 38.
Radu, E., Woegerbauer, M., Rab, G., Oismuller, M., Strauss, P., Hufnagl, P., Gottsberger, R.A., Krampe, J., Weyermair, K., Kreuzinger, N., 2021. Resilience of agricultural soils to antibiotic resistance genes introduced by agricultural management practices. *Science of The Total Environment* 756, 143699.
Rajput, V., Minkina, T., Mazarji, M., Shende, S., Sushkova, S., Mandzhieva, S., Burachevskaya, M., Chaplygin, V., Singh, A., Jatav, H., 2020. Accumulation of nanoparticles in the soil-plant systems and their effects on human health. *Annals of Agricultural Sciences* 65, 137–143.
Reeve, J.R., Hoagland, L.A., Villalba, J.J., Carr, P.M., Atucha, A., Cambardella, C., Davis, D.R., Delate, K., 2016. Organic farming, soil health, and food quality: Considering possible links. *Advances in Agronomy* 137, 319–367.
Röös, E., Mie, A., Wivstad, M., Salomon, E., Johansson, B., Gunnarsson, S., Wallenbeck, A., Hoffmann, R., Nilsson, U., Sundberg, C., 2018. Risks and opportunities of increasing yields in organic farming. A review. *Agronomy for Sustainable Development* 38, 1–21.
Schjønning, P., Lamandé, M., Thorsøe, M., 2019. *Soil compaction-drivers, pressures, state, impacts and responses: Vidensgrundlag inden for jordpakning.* DCA - Nationalt Center for Fødevarer og Jordbrug, 40.

Silver, W., Perez, T., Mayer, A., Jones, A., 2021. The role of soil in the contribution of food and feed. *Philosophical Transactions of the Royal Society B* 376, 20200181.

Steffan, J., Brevik, E., Burgess, L., Cerdà, A., 2018. The effect of soil on human health: An overview. *European Journal of Soil Science* 69, 159–171.

Suman, J., Rakshit, A., Ogireddy, S.D., Singh, S., Gupta, C., Chandrakala, J., 2022. Microbiome as a key player in sustainable agriculture and human health. *Frontiers in Soil Science* 2, 821589.

Surendran, U., Jayakumar, M., Raja, P., Gopinath, G., Chellam, P.V., 2023. Microplastics in terrestrial ecosystem: Sources and migration in soil environment. *Chemosphere*, 318, 137946.

Tamirat, T.W., Pedersen, S.M., Farquharson, R.J., de Bruin, S., Forristal, P.D., Sørensen, C.G., Nuyttens, D., Pedersen, H.H., Thomsen, M.N., 2022. Controlled traffic farming and field traffic management: Perceptions of farmers groups from Northern and Western European countries. *Soil and Tillage Research* 217, 105288.

Tóth, G., Hermann, T., da Silva, M.R., Montanarella, L., 2018. Monitoring soil for sustainable development and land degradation neutrality. *Environmental Monitoring and Assessment* 190, 1–4.

Veerman, C., Correia, T.P., Bastioli, C., Biro, B., Bouma, J., Cienciala, E., Emmett, B., Frison, E.A., Grand, A., Filchev, L.H., 2020. *Caring for soil is caring for life: Ensure 75% of soils are healthy by 2030 for healthy food, people, nature and climate*. Luxembourg: European Commission.

Vigar, V., Myers, S., Oliver, C., Arellano, J., Robinson, S., Leifert, C., 2020. A systematic review of organic versus conventional food consumption: Is there a measurable benefit on human health? *Nutrients* 12, 7.

Wander, M.M., 2023. Soil health: Definitions, history, key concepts and hurdles. In: W. Horwath (Ed.) *Improving soil health* (pp. 2–21). Burleigh Dodds Science Publishing.

Wei, F., Cui, S., Liu, N., Chang, J., Ping, X., Ma, T., Xu, J., Swaisgood, R.R., Locke, H., 2021. Ecological civilization: China's effort to build a shared future for all life on earth. *National Science Review* 8, nwaa279.

Weninger, T., Ramler, D., Bondi, G., Asins, S., O'Sullivan, L., Assennato, F., Astover, A., Bispo, A., Borůvka, L., Buttafuoco, G., 2024. Do we speak one language on the way to sustainable soil management in Europe? A terminology check via an EU-wide survey. *European Journal of Soil Science* 75, e13476.

World Food Summit, 1996. Report of the World Food Summit, Part 1. Rome: Food and agriculture organization of the United Nations (FAO). Rome, Italy. https://openknowledge.fao.org/handle/20.500.14283/w3548e

World Health Organization, 2017. *One health*. World Health Organization. https://www.who.int/health-topics/one-health#tab=tab_1

Zhang, X., Yang, L., Li, Y., Li, H., Wang, W., Ye, B., 2012. Impacts of lead/zinc mining and smelting on the environment and human health in China. *Environmental Monitoring and Assessment* 184, 2261–2273.

9 Case Study of Lifestyle Medicine Intervention
Promoting Healthy Lifestyle and Nutrition

Bilal Walker and Saul J. Bautista

INTRODUCTION

This chapter presents a detailed case study of a lifestyle medicine intervention conducted by Al-Munir LLC, in collaboration with Dr. Saul J. Bautista, aimed at enhancing the health and nutrition of participants from La Casa De Don Pedro.

The intervention spanned nine weeks, with sessions conducted four hours per month, focusing on nutrition education, food preparation, health assessments, physical exercise, and general lifestyle medicine practices. The program engaged 40 participants aged between 21 and 65, with over half of them being immigrants from Central and South America. This diverse demographic provided a unique context for exploring the impact of lifestyle interventions on low-income residents experiencing food insecurity. The thematic analysis of this intervention reveals significant insights into promoting healthy lifestyles and addresses the challenges and successes encountered during the program.

DR. SAUL J. BAUTISTA AND HIS CONTRIBUTIONS TO LIFESTYLE MEDICINE

Dr. Saul J. Bautista is a physician in the field of lifestyle medicine, with a background in both medical education and community health.

- **Investigator for Food is Medicine vs Lifestyle Medicine**: A Community Based Pragmatic Randomized Control Trial for Patients With Cardiovascular Kidney Metabolic (CKM) Syndrome (NCT06461273)
- **Former Marketing Committee Chair**, Lifestyle Medicine Pre-Professional Working Group for the American College of Lifestyle Medicine (ACLM): In this role, Dr. Bautista played a crucial part in promoting lifestyle medicine to a broad audience, enhancing the visibility and impact of the field.
- **Founder and President**, Lifestyle Medicine Interest Group at Rutgers New Jersey Medical School: He established this group to foster interest and advance education in lifestyle medicine among medical students, laying the groundwork for future health professionals committed to preventative care and healthy living.
- **Awards and Recognition**: Dr. Bautista received the Donald A. Pegg Lifestyle Medicine Student Leadership Award from ACLM in 2018 and the Trailblazer Award from the Student National Medical Association in 2019. These honors recognize his leadership, innovation, and dedication to advancing lifestyle medicine.

DOI: 10.1201/9781003469742-9

INTRODUCTION: FOOD SYSTEM INEQUITIES AND THEIR ROOTS IN BLACK AND INDIGENOUS LAND LOSS AND URBAN GENTRIFICATION

The food system in the United States is deeply entangled with historical and contemporary issues of land loss and urban development, particularly impacting Black and Indigenous communities. The displacement and disenfranchisement of these communities have contributed significantly to the food and health inequities they face today.

Historical Context of Land Loss

The roots of the current inequities can be traced back to systemic land dispossession. For Indigenous peoples, this began with European colonization, which resulted in the seizure of vast tracts of ancestral lands. The federal government enforced policies like the Indian Removal Act of 1830 and the General Allotment Act (Dawes Act) of 1887, which fragmented Indigenous land holdings and transferred millions of acres to non-Indigenous settlers and corporations.

Similarly, Black land ownership has been systematically undermined. After the Civil War, Black Americans were initially able to acquire land, particularly during the Reconstruction era. However, through mechanisms like discriminatory lending practices, violence, and legal manipulation, much of this land was gradually lost. The Great Migration saw many Black families move to urban areas, seeking better opportunities but often encountering new forms of systemic racism and exclusion from economic opportunities.

Urban Gentrification and Displacement

In urban areas, gentrification has exacerbated food and health inequities. Historically marginalized neighborhoods, often populated by Black and Indigenous people, have been targeted for redevelopment. While gentrification can bring economic investment, it frequently leads to the displacement of long-term residents due to rising property values and living costs. This displacement disrupts community networks and forces residents into areas with poorer access to healthy food options and healthcare services.

Gentrification also alters the landscape of food accessibility. As wealthier populations move into formerly disinvested neighborhoods, the demand for upscale grocery stores and dining establishments grows. While this can improve food availability for new residents, it often leads to the closure of smaller, community-focused food outlets that served the original residents, effectively reducing their access to affordable, healthy food options.

Impact on Food and Health Inequities

These historical and ongoing processes have created significant food and health inequities. Black and Indigenous communities disproportionately reside in food deserts – areas with limited access to affordable and nutritious food. The lack of grocery stores, farmers' markets, and healthy food options forces reliance on convenience stores and fast food, contributing to poor diet and associated health problems such as obesity, diabetes, and cardiovascular disease.

Moreover, these communities often face greater barriers to healthcare access. Historical mistrust of the medical system, driven by a legacy of exploitation and discrimination, further complicates health outcomes. The combined effect of poor diet and inadequate healthcare access perpetuates a cycle of health disparities.

Urban Farming as an Intervention

Urban agriculture offers a promising intervention to mitigate these inequities. Organizations like Newark Science and Sustainability (NSS) and the Urban Agriculture Cooperative (UAC) in

Newark, New Jersey, are pioneering efforts to address food insecurity and promote health equity through urban farming. These organizations work collaboratively to develop transformative projects that serve residents with the greatest need for fresh, healthy food in their neighborhoods.

NSS and UAC cultivate land and support community gardening activities at various sites in Newark. They are also building their own markets and urban agriculture supply chains in partnership with urban and rural growers and other local food advocacy organizations. This shift from a solely charity-driven model to building a customer base and markets for urban farm produce exemplifies how small-scale urban agriculture can be an economic development tool.

NSS has expanded from developing health and wellness community workshops to cultivating international partnerships, blending global and local efforts to build a sustainable and just food system. They lead a network of Newark-based urban farmers, and partner with rural producers to supply food for a community Farm-to-Table Co-op program, which feeds more than two dozen families in the community.

NSS is also recognized for its signature events, such as the Sustainable Living Empowerment Conference and Citywide Garden Tour, which attract hundreds of participants. During a period of rapid redevelopment and gentrification in Newark, NSS took an active stance to protect productive urban gardens and farms, acquiring ownership of key sites like the Garden of Hope. This garden is the centerpiece of a planned food hub, a shared-use space where fresh produce is aggregated and marketed to local consumers. This hub, with classroom and kitchen space for community workshops, aims to increase the availability of fresh, locally grown produce.

The UAC focuses on perfecting an urban model for the aggregation, distribution, and marketing of produce in urban communities. They operate urban farm sites and farmers markets, providing programs and services for growers in their network. UAC's efforts to transition some urban agriculture from a charity model to small business support the viability of local food economies. They address challenges such as food costs by partnering with rural growers to supply larger amounts of fresh, organic produce.

The work of NSS and UAC in Newark exemplifies how urban farming can build community food systems, create new local businesses and jobs, and support community-led economic development. These efforts highlight the potential of urban agriculture to address food and health inequities by providing fresh, affordable food to underserved communities.

Conclusion

Addressing the food and health inequities faced by Black and Indigenous communities requires acknowledging and addressing the historical and ongoing processes of land loss and gentrification. Sustainable solutions must involve equitable urban planning, protection of community land rights, and the development of food systems that prioritize accessibility and affordability for marginalized populations. Urban agriculture initiatives like those led by NSS and UAC demonstrate the potential of community-driven solutions to create healthier, more equitable food systems. Only through such comprehensive approaches can we begin to rectify the deep-seated inequities that shape the current food and health landscapes in the United States.

AL-MUNIR LLC: URBAN AGRICULTURE AND FOOD JUSTICE

Al-Munir LLC is a pioneering organization committed to advancing urban agriculture and promoting food justice. The company's mission is to create sustainable, equitable food systems in urban communities, ensuring that all residents have access to nutritious, affordable food. Al-Munir LLC focuses on community-driven projects that empower local residents through education and hands-on involvement in urban farming.

The organization collaborates with community groups, schools, and local governments to establish urban gardens and food production systems. These initiatives not only provide fresh produce but also serve as educational platforms, teaching participants about sustainable agriculture, nutrition, and healthy living. Al-Munir LLC's efforts are geared towards addressing food deserts and improving food security in underserved urban areas, making nutritious food accessible to all.

NOURISHING NEWARK: FUNDING COMMUNITY HEALTH INITIATIVES

Nourishing Newark is a funding initiative designed to support projects that improve food access and promote healthy eating in Newark. The initiative provides grants to community organizations, enabling them to implement innovative programs that address food insecurity and enhance nutritional education. Nourishing Newark aims to create a sustainable food system in the city by investing in projects that empower residents and build community capacity.

The lifestyle medicine course taught by Dr. Saul J. Bautista was one such project funded by Nourishing Newark. This program leveraged the grant to offer a comprehensive intervention focused on nutrition education, physical exercise, and overall lifestyle improvements. The funding allowed for the development of tailored educational materials, hands-on cooking sessions, and health assessments, ensuring that participants received practical, actionable guidance to improve their health.

LA CASA DE DON PEDRO: A LEGACY OF COMMUNITY SUPPORT

La Casa De Don Pedro is a prominent community-based organization in Newark, New Jersey, with a rich history of serving the local community. Founded in 1972, La Casa has been a pillar of support for residents, providing a wide range of services including education, housing assistance, immigration services, and economic development programs. The organization's mission is to foster self-sufficiency, empowerment, and community well-being through comprehensive programs that address the diverse needs of Newark's residents.

Over the years, La Casa has developed numerous initiatives aimed at improving the quality of life for low-income families and individuals. Their programs are designed to promote social, economic, and environmental justice, ensuring that all community members have the resources and opportunities to thrive. La Casa's commitment to community health and wellness is evident in their support for projects like the lifestyle medicine course, which aligns with their broader goals of fostering healthy, resilient communities.

MAJOR THEME 1: PROMOTING HEALTHY LIFESTYLE AND NUTRITION

The primary theme of "Promoting Healthy Lifestyle and Nutrition" encompasses various facets of the intervention aimed at improving the participants' overall well-being through education and practical lifestyle changes.

Importance of Healthy Lifestyle

Participants emphasized the critical role of individual behavior in maintaining a healthy lifestyle. They recognized the need for well-designed diet plans, regular exercise, and adequate sleep to sustain good health. Awareness about health tips and a structured timetable for meals were seen as essential components. Self-control was identified as a crucial factor, highlighting the importance of avoiding unhealthy food behaviors.

Key Insights:

- Healthy lifestyle is behavior-dependent.
- Importance of diet, exercise, and sleep.
- Need for awareness and self-control in nutrition.

Nutrition and Diet

The influence of family and upbringing on nutrition was a significant sub-theme. Participants stressed the importance of consuming fruits, vegetables, and grains while avoiding salty foods, sugar, and red meat. They acknowledged that balanced nutrition contributes to overall well-being and the absence of medical issues.

Key Insights:

- Family and upbringing influence nutrition.
- Balanced diet includes fruits, vegetables, and grains.
- Avoidance of unhealthy foods like salt, sugar, and red meat.

Challenges in Maintaining Healthy Habits

Participants identified several challenges, including anxiety-related behaviors, irregular eating habits, and busy routines that hindered their ability to maintain healthy habits. Strategies to manage anxiety, such as drinking water, were discussed, along with the importance of regular meals.

Key Insights:

- Anxiety affects eating habits.
- Importance of regular meals.
- Strategies like drinking water to manage anxiety.

Food Safety and Awareness

Concerns about fast food and its negative impact on health were prevalent. Participants highlighted the need for increased awareness of food safety, particularly regarding chemical substances and hormones in food. Cultural differences and availability of diverse food options in the USA also posed challenges.

Key Insights:

- Fast food has adverse health effects.
- Need for awareness about food safety and chemicals.
- Cultural differences impact dietary choices.

MAJOR THEME 2: LEARNING IMPACT AND PARTICIPANTS' REFLECTIONS

The second major theme focuses on the participants' experiences, learning outcomes, and reflections on the intervention.

Group Dynamics and Communication

Participants valued strong group communication and suggested the need for regular class sessions and group chats. They proposed restructuring the class to focus on different aspects such as cooking, exercise, and theoretical knowledge, indicating a desire for a holistic learning experience.

Key Insights:

- Importance of strong group communication.
- Need for regular and structured class sessions.

Learning and Education

Participants expressed a keen interest in learning about exercise, emotional therapy, and practical skills like cooking. They highlighted the importance of addressing specific health issues and integrating emotional support into the program.

Key Insights:

- Desire to learn about exercise and emotional therapy.
- Importance of practical skills and addressing specific health issues.

Positive Changes and Progress

The intervention led to positive changes in participants' behaviors, such as exercising at home, implementing learned strategies, and becoming more mindful about food choices. Participants reported efforts to eliminate soda and reduce junk food consumption, reflecting their commitment to healthier lifestyles.

Key Insights:

- Positive behavioral changes and commitment to healthy habits.
- Increased mindfulness about food choices.

Obstacles and Mindset

Participants identified obstacles like age-related weight-loss difficulties and the need for a positive mindset and self-love. These reflections underscore the importance of addressing psychological barriers in promoting healthy lifestyles.

Key Insights:

- Age and mindset are significant obstacles.
- Importance of self-love and mental strength.

Class Experience and Appreciation

Participants appreciated the efforts of the instructor and the positive class environment. They reported acquiring new strategies to improve their lifestyle and expressed satisfaction with the tangible results achieved.

Key Insights:

- Positive class experience and instructor efforts.
- Tangible improvements in lifestyle and health.

THE IMPORTANCE OF CULTURALLY RESPONSIVE HEALTH INTERVENTIONS

Introduction

The growing diversity in many urban communities necessitates culturally responsive health interventions that address the unique needs and experiences of different populations. Culturally responsive health interventions recognize and respect the cultural contexts of participants, enhancing the effectiveness of health education and behavior change initiatives. This chapter emphasizes the importance of such interventions, supported by relevant research and references.

The Necessity of Cultural Responsiveness in Health Interventions

Culturally responsive health interventions are crucial for several reasons:

1. **Enhanced Engagement and Participation**: Interventions that are culturally relevant are more likely to engage participants effectively. When individuals see their cultural practices and values reflected in health programs, they are more likely to participate and adhere to the recommended changes.
2. **Improved Health Outcomes**: Cultural responsiveness in health interventions can lead to better health outcomes. Programs that consider cultural dietary preferences, communication styles, and health beliefs are more likely to resonate with participants, leading to sustained behavioral changes and improved health metrics.
3. **Reduction of Health Disparities**: Minority populations often face significant health disparities. Culturally responsive interventions can help address these gaps by providing tailored health education and resources that consider the specific barriers and facilitators experienced by different cultural groups.
4. **Building Trust and Rapport**: Trust is a critical component in the success of health interventions. Culturally responsive programs help build trust between health providers and participants by showing respect and understanding for cultural differences.

Research Supporting Culturally Responsive Health Interventions

Research consistently supports the effectiveness of culturally responsive health interventions. Several studies highlight the positive impact of culturally tailored programs on health behaviors and outcomes.

Engagement and Participation

A study by Resnicow et al. (2000) demonstrated that culturally tailored dietary interventions significantly increased fruit and vegetable consumption among African American participants compared to non-tailored interventions. The study found that incorporating cultural elements such as traditional foods and culturally relevant messages enhanced participant engagement and adherence to dietary recommendations (Resnicow et al., 2000).

Improved Health Outcomes

Research by Jernigan et al. (2010) showed that culturally adapted diabetes management programs for Native American populations resulted in significant improvements in glycemic control and diabetes-related knowledge. The study emphasized the importance of incorporating cultural practices and traditional foods into the intervention, which contributed to its success (Jernigan et al., 2010).

Lifestyle medicine interventions, which focus on modifying behaviors related to diet, physical activity, and stress management, have been shown to significantly improve health outcomes across various populations. Here, we provide an overview of the data demonstrating the efficacy of these interventions.

Cardiovascular Health

1. **Hypertension**: A study published in the *Journal of the American College of Cardiology* found that lifestyle interventions, including dietary changes and increased physical activity, resulted in a significant reduction in systolic and diastolic blood pressure. Participants experienced an average reduction of 11 mmHg in systolic blood pressure and 6 mmHg in diastolic blood pressure after 12 weeks of intervention.

2. **Cholesterol Levels**: Research published in the *American Journal of Lifestyle Medicine* indicated that a plant-based diet and regular exercise significantly reduced LDL cholesterol levels by an average of 20–30% within six months.

Diabetes Management

1. **Glycemic Control**: The *Diabetes Prevention Program (DPP)* demonstrated that intensive lifestyle interventions could reduce the incidence of type 2 diabetes by 58% over three years compared to a control group. Participants who maintained healthy diets and regular physical activity experienced significant improvements in HbA1c levels.
2. **Weight Loss**: Lifestyle interventions focusing on diet and exercise led to an average weight loss of 5–7% of body weight, which is clinically significant for diabetes management and prevention.

Obesity

1. **Weight Reduction**: A comprehensive review in the *Journal of Obesity* highlighted that structured lifestyle interventions resulted in an average weight loss of 5–10% of initial body weight over 6–12 months. This weight loss was associated with improvements in obesity-related conditions such as hypertension, dyslipidemia, and insulin resistance.
2. **Sustained Weight Loss**: Long-term follow-up studies indicate that individuals who continue to engage in lifestyle interventions maintain an average weight loss of 4–6% after two years.

Mental Health

1. **Depression and Anxiety**: A meta-analysis in the *Journal of Psychiatric Research* found that lifestyle interventions, including regular physical activity and dietary changes, significantly reduced symptoms of depression and anxiety. Participants reported a 25–30% reduction in depressive symptoms and a 20–25% reduction in anxiety symptoms after 12 weeks of intervention.
2. **Quality of Life**: Lifestyle interventions have been associated with significant improvements in overall quality of life, including better mental health, physical functioning, and social interactions. These improvements were sustained over long-term follow-up periods.

Cancer Prevention

1. **Risk Reduction**: The *World Cancer Research Fund* reported that lifestyle interventions, such as maintaining a healthy weight, engaging in regular physical activity, and consuming a diet rich in fruits, vegetables, and whole grains, could reduce the risk of several types of cancer by up to 40%.
2. **Survival Rates**: Cancer survivors who adopt lifestyle changes, including diet and exercise, have shown improved survival rates and quality of life. A study in the *Journal of Clinical Oncology* indicated that breast cancer survivors who engaged in regular physical activity and maintained a healthy diet had a 50% lower risk of recurrence and a 40% lower risk of mortality compared to those who did not adopt these changes.

Conclusion

The data strongly supports the efficacy of lifestyle medicine interventions in improving a wide range of health outcomes. These interventions not only help manage and prevent chronic diseases but also enhance overall well-being and quality of life. By incorporating lifestyle changes, individuals can achieve significant health benefits, underscoring the importance of lifestyle medicine in public health strategies.

REDUCTION OF HEALTH DISPARITIES

The importance of culturally responsive interventions in reducing health disparities is highlighted in a review by Goode et al. (2006). The review found that culturally competent healthcare services, including tailored health education and communication strategies, significantly improved health outcomes for minority populations, thereby reducing disparities (Goode et al., 2006).

Building Trust and Rapport

A study by Saha et al. (2008) explored the impact of cultural competence training for healthcare providers on patient–provider relationships. The findings indicated that patients who perceived their providers as culturally competent were more likely to trust them and adhere to medical advice, underscoring the importance of cultural responsiveness in building trust and rapport (Saha et al., 2008).

CONCLUSION

The findings of this study emphasize the importance of promoting healthy lifestyles and nutrition among low-income residents experiencing food insecurity. The intervention successfully addressed key factors influencing healthy habits, including individual behavior, nutrition knowledge, and food safety awareness. Participants' reflections highlight the positive impact of the intervention, the significance of group dynamics, and the need for addressing psychological barriers. These insights can guide the development of targeted interventions to foster healthy eating habits and overall well-being in similar populations.

The evidence strongly supports the importance of culturally responsive health interventions. These interventions not only enhance engagement and participation but also improve health outcomes, reduce health disparities, and build trust between providers and participants. As urban communities continue to diversify, incorporating cultural responsiveness into health programs becomes increasingly essential for promoting health equity and improving overall public health.

Considerations for Community-Led Interventions

It has been suggested that engaging members of disadvantaged communities in public health initiatives can reduce health inequities (Agonafer et al. 2021). Collaborations between health systems and community-based organizations (CBOs) are becoming increasingly common mechanisms to address the unmet health-related social needs of high-risk populations. Direct funding of community-based organizations and individuals to implement health promotion programs is a strategy being utilized by some organizations and local municipalities (City of Newark (n.d.); Community Foundation of New Jersey (n.d.).

The rationale for this strategy lies with the notion that community members themselves know their needs and best know how to execute public health programming to meet their needs.

There is evidence that suggests that community engagement in public health interventions leads to improvements in social capital, social cohesion, but its impact on mortality, morbidity, health behaviors, or impact on inequalities is variable (O'Mara-Eves et al., 2015; Popay et al., 2007). A significant challenge is that community partners typically lack the academic skills, credentials, and experience to develop and structure evidence-based public health interventions. The development of effective health promotion interventions often requires reviews of the relevant literature, application of theories, collection of new data, involvement of experts, and other stakeholders in addition to community members in the planning process (Fernandez et al., 2019).

Community members and community-based organizations would benefit from using systematic frameworks to design and implement health promotion interventions. This intervention as originally planned was developed by the La Casa administrative staff.

The staff available did not have public health, nutrition, or clinical experience. The development of programming was not guided by a theoretical health-behavior change framework nor informed by existing literature. A physician was recruited as a consultant at the start of the intervention to deliver health education including nutrition education. It is well known that physicians receive very little nutrition education within medical school (Adams et al., 2010) which raises the question whether a physician is the appropriate health professional to deliver nutrition education relative to a registered dietitian.

Lifestyle-medicine physicians are uniquely trained and achieve competency in nutrition science, assessment, and dietary prescriptions (Lianov et al., 2022). Prior to the recruitment of the physician, the consultants from Almunir LLC were going to utilize the USDA My Plate toolkit and curricular materials. The use of existing national toolkits is reasonable, however a conflict exists as national food-based dietary guidelines may differ from the rigor and specificity of emergent dietary clinical practice guidelines, For the treatment, reversal, and prevention of lifestyle-related chronic disease, ACLM recommends an eating plan based predominantly on a variety of minimally processed vegetables, fruits, whole grains, legumes, nuts, and seeds which is consistent with aggregated findings from global food based dietary guidelines and clinical practice guidelines (Cara et al., 2023 and ACLM 2024).

If the goal of public-health interventions is health restoration and improvement of health outcomes than allowing community-based organizations to implement and communicate curricula that do not meet the rigor of clinical practice guidelines raises an ethical concern as this could lead to the dissemination of inaccurate or substandard health information. This can potentially harm individuals by not providing them with evidence-based recommendations, leading to ineffective or even harmful interventions, ultimately compromising the goal of health restoration and the improvement of health outcomes.

To mitigate these risks the authors recommend community-based organizations form community-academic partnerships to design and deliver these interventions. This allows community-based organizations to plan and develop interventions with the scientific rigor to demonstrate efficacy and effectiveness, and to increase the legitimacy of their work among funders, policy-makers, and other stakeholders, while retaining the value added in cultural competency and trust from community members engaged in the research/intervention process.

BIBLIOGRAPHY

Adams, K. M., Kohlmeier, M., & Zeisel, S. H. (2010 Sep). Nutrition education in U.S. medical schools: latest update of a national survey. *Academic Medicine*, 85(9), 1537–1542. doi:10.1097/ACM.0b013e3181eab71b. PMID:20736683; PMCID: PMC4042309.

Agonafer, E. P., Carson, S. L., & Nunez, V. et al. (2021). Community-based organizations' perspectives on improving health and social service integration. *BMC Public Health* 21, 452. doi:10.1186/s12889-021-10449-w

American College of Lifestyle Medicine. (2024). Positions. Retrieved June 14, 2024, from https://lifestylemedicine.org/positions/

Cara, K. C., Goldman, D. M., Kollman, B. K., Amato, S. S., Tull, M. D., & Karlsen, M. C. (2023 May). Commonalities among dietary recommendations from 2010 to 2021 clinical practice guidelines: A meta-epidemiological study from the American college of lifestyle medicine. *Advances in Nutrition*, 14(3), 500–515. doi:10.1016/j.advnut.2023.03.007. Epub 2023 Mar 20. PMID: 36940903; PMCID: PMC10201822.

City of Newark. (n.d.). Nourishing Newark community grants program. Retrieved June 14, 2024, from https://www.newarknj.gov/card/nourishing-newark

Community Foundation of New Jersey. (n.d.). Prudential community grants program. Retrieved June 14,2024, from https://cfnj.org/prudential/

Ergönül, B. (2013). Consumer awareness and perception to food safety: A consumer analysis. *Food Control*, 32(2), 461–471.

Fernandez, M. E., Ruiter, R. A., Markham, C. M., & Kok, G. (2019). Intervention mapping: theory-and evidence-based health promotion program planning: Perspective and examples. *Frontiers in Public Health*, 7, 436767.

Goode, T. D., Dunne, M. C., & Bronheim, S. M. (2006). *The Evidence Base for Cultural and Linguistic Competency in Health Care*. The Commonwealth Fund.

Jernigan, V. B. B., Salvatore, A. L., Styne, D. M., & Winkleby, M. (2010). Addressing food insecurity in a Native American reservation using community-based participatory research. *Health Education Research*, 25(4), 652–660.

Lee, J., Jeong, S., Ko, G., Park, H., & Ko, Y. (2016). Development of a food safety and nutrition education program for adolescents by applying social cognitive theory. *Osong Public Health and Research Perspectives*, 7(4), 248–260.

Lianov, L. S., Adamson, K., Kelly, J. H., Matthews, S., Palma, M., Rea, B. L. (2022 Aug 31). Lifestyle medicine core competencies: 2022 update. *American Journal of Lifestyle Medicine*, 16(6), 734–739. doi:10.1177/15598276221121580. PMID:36389051; PMCID: PMC9644148.

O'Mara-Eves, A., Brunton, G., Oliver, S., Kavanagh, J., Jamal, F., & Thomas, J. (2015 Feb 12). The effectiveness of community engagement in public health interventions for disadvantaged groups: a meta-analysis. *BMC Public Health*, 15, 129. doi:10.1186/s12889-015-1352-y. PMID: 25885588; PMCID: PMC4374501.

Popay, J., Attree, P., Hornby, D., Milton, B., Whitehead, M., French, B., et al. *Community Engagement in Initiatives Addressing the Wider Social Determinants of Health: A Rapid Review of Evidence on Impact, Experience and Process*. Lancaster: University of Lancaster; 2007.

Resnicow, K., Baranowski, T., Ahluwalia, J. S., & Braithwaite, R. L. (2000). Cultural sensitivity in public health: defined and demystified. *Ethnicity & Disease*, 9(1), 10–21.

Saha, S., Beach, M. C., & Cooper, L. A. (2008). Patient centeredness, cultural competence and healthcare quality. *Journal of the National Medical Association*, 100(11), 1275–1285.

10 Mitigation of Arsenic Poisoning Through Soil and Water Management in Bangladesh

Md Nayem Hasan Munna, Nargis Akter Shapna, and Rattan Lal

INTRODUCTION

Arsenic (As) contamination presents a severe global environmental and public health challenge, affecting millions of people in various regions around the world. According to the World Health Organization (WHO), over 140 million people across more than 70 countries are exposed to As-contaminated drinking water, with the most heavily impacted areas located in South Asia, Latin America, and Southeast Asia. Nations such as India, Bangladesh, Vietnam, and Chile face some of the highest levels of contamination, primarily from groundwater sources used for both drinking and agricultural irrigation (Smith et al., 2000). Prolonged exposure to As has been linked to numerous serious health conditions, including cancers of the skin, lungs, and bladder, as well as cardiovascular diseases and developmental disorders (Rahman et al., 2009). The widespread nature of this issue highlights the urgent need for comprehensive mitigation strategies, especially for populations most at risk.

Globally, efforts to mitigate As contamination have had mixed success. In regions such as Latin America, natural As deposits in the Andean region have led to severe groundwater contamination, especially in rural communities (Bundschuh et al., 2012). Similarly, Southeast Asian countries, including Cambodia and Vietnam, are struggling with As contamination in groundwater used for both drinking and agricultural purposes (Berg et al., 2001). Although technologies such as As-removal filters, deep tube wells, and improved water management practices have been implemented, their long-term sustainability is limited by cost, infrastructure challenges, and access issues, particularly in remote areas. These persistent challenges highlight the need for innovative and context-specific solutions for As mitigation.

Among the hardest-hit countries is Bangladesh, where As contamination has escalated into a major environmental and public health crisis. It is now considered one of the most pressing health concerns of the 21st century for the country (Islam et al., 2023). The issue gained international attention when it was discovered that groundwater – on which millions of Bangladeshis rely for drinking and irrigation – was severely contaminated with As (Ahmed et al., 2019). The country, situated in the Ganges-Brahmaputra delta, has long benefitted from a vast network of rivers and water bodies that support its agricultural economy and human settlements.However, this same geographical advantage has also contributed to one of the largest As contamination crises in the world, as naturally occurring As deposits in the earth's crust have leached into the groundwater. This largely unnoticed crisis has drastically impacted millions, particularly in rural communities where groundwater is a critical resource (Smith et al., 2000). Table 10.1 provides an overview of

 DOI: 10.1201/9781003469742-10

TABLE 10.1
Arsenic (As) Concentration Ranges and Health Critical Limits for Soil, Water, and Different Foods

Source	As Concentration Range (mg/kg)	Critical Limit for Human Health (mg/kg) by WHO or EFSA	References
Soil	0.1–50	>20	Baker and Chesnin (1975); Vinogradov (1959); Bowen (1979); Kabir et al. (2016); UNICEF (2010)
Groundwater	0.000003–5 mg/L	>0.01 mg/L	Chen and Costa (2021); Gupta and Chatterjee (2017); Ahmad and Bhattacharya (2019)
Cow milk	0.0046–0.0053 mg/L	>0.01 mg/L	EFSA (2009)
Human breast milk	0.0032–0.0054 mg/L	0.00064 mg/L	Ulman et al., (1998)
Fish	0.097–1.318	0.1	Adeloju et al. (2021)
Rice[a]	0.01–2.00	>0.2	Zavala and Duxbury (2008)
Wheat[b]	<0.235	0.1–0.2	Mondal et al. (2021)
Potato[c]	0.005–0.176	0.1	Mondel et al. (2021)
Tomato[d]	0.017–0.023	0.1	Matos-Reyes et al. (2010)
Spinach[e]	0.051–0.063	0.1	Matos-Reyes et al. (2010)
Garlic[f]	0.001–0.310	0.1	Song et al. (2009)
Capsicum[g]	0.001–0.116	0.1	Song et al. (2009)
Kidney bean[h]	0.001–0.202	0.1	Song et al. (2009)
Eggplant[i]	0.004–0.075	0.1	Song et al. (2009)
Chickpea[j]	0.059–0.063	0.1	Matos-Reyes et al. (2010)
Onion[k]	0.034–0.044	0.1	Matos-Reyes et al. (2010)
Carrot[l]	0.48–0.70	0.1	Adeloju et al. (2021)
Lettuce[m]	0.042–0.054	0.1	Matos-Reyes et al. (2010)
Apple[n]	1.95–2.55	0.1	Shandana et al. (2024)

[a] *Oryza sativa*
[b] *Triticum aestivum*
[c] *Solanum tuberosum*
[d] *Solanum lycopersicum*
[e] *Spinacia oleracea*
[f] *Allium sativum*
[g] *Capsicum annuum*
[h] *Phaseolus vulgaris*
[i] *Solanum melongena*
[j] *Cicer arietinum*
[k] *Allium cepa*
[l] *Daucus carota*
[m] *Lactuca sativa*
[n] *Malus domestica*

As concentration ranges in soil, water, and different foods, along with the critical health levels for human exposure, indicating the potential risks associated with As contamination in different environmental media.

The spatial distribution of arsenic contamination hotspots in Bangladesh is illustrated in Figure 10.1 (Zahid et al., 2022). The discovery of As in Bangladesh's groundwater in the 1980s

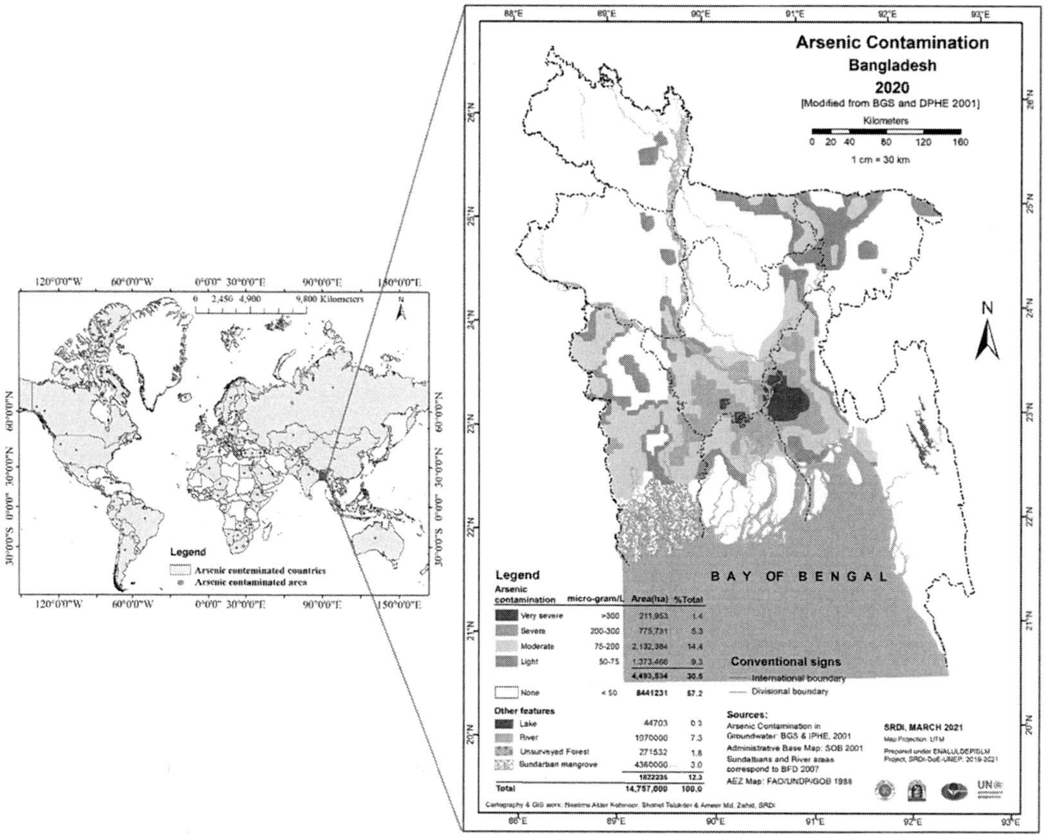

FIGURE 10.1 Arsenic contamination hotspots in Bangladesh.
Reproduced with permission from Zahid et al., 2022.

shocked public health officials and environmental scientists. The initial efforts to combat waterborne diseases in the country, such as cholera and diarrhea, led to the widespread construction of tube wells that drew water from underground sources (Islam et al., 2023). These tube wells were seen as a solution to surface water contamination triggered by biological pathogens. However, it soon became apparent that the groundwater in many areas contained naturally occurring As, a toxic metalloid that can cause a range of severe health problems, including skin lesions, cancers, and cardiovascular diseases (Ahmad et al., 2018). This situation is especially alarming because the health effects of As poisoning, or arsenicosis, often take years or even decades to manifest, meaning that millions of people have been unknowingly consuming contaminated water for extended periods (Smith et al., 2000).

Early studies highlighted the scope of the problem, revealing that the Bengal Basin, covering a significant portion of Bangladesh and parts of India, was among the areas most heavily impacted by As contamination (Ahmed et al., 2004; Ahmad et al., 2018). Studies in the 1990s showed that the As primarily originated from geological sources, where it is trapped in the soil and rock formations and leaches into the groundwater (McArthur et al., 2001). This groundwater, accessed through tube wells, became the main drinking water supply for millions in Bangladesh. The problem was worsened by the fact that As contamination in water is tasteless, colorless, and odorless, making it difficult for people to detect without proper testing (Chowdhury and Jakariya, 1999). By the mid-1990s, the gravity of the crisis was widely acknowledged, prompting the World Health Organization (WHO) to designate the As contamination in Bangladesh as a public health emergency (Smith et al., 2000).

Previous research has significantly contributed to understanding the scope of As contamination and its associated health risks. Numerous studies highlight the high occurrence of As-related illnesses, particularly in rural regions where safe drinking water is scarce. Long-term exposure to As through drinking water has been linked to various severe health conditions, such as cancers of the skin, lungs, bladder, and kidneys, along with cardiovascular diseases and diabetes (Rahman et al., 2009; Islam et al., 2023). Additionally, childhood exposure to As has been connected to developmental delays and growth issues, compounding the long-term effects on impacted communities (Yunus et al., 2011). Beyond drinking water, research has shown that As contamination also affects agricultural systems. Groundwater, frequently contaminated with As and used for irrigation, builds up in soils and is absorbed by crops, particularly rice, a dietary staple in Bangladesh (Meharg and Rahman, 2003). Therefore, As enters the food chain, posing a threat not only to those who drink contaminated water but also to those who consume contaminated food. Table 10.2 summarizes the As concentrations in rice grains under different irrigation and management practices, highlighting the impact of strategies such as alternate wetting and drying (AWD), organic farming, and biofortification on reducing As accumulation.

Despite the wealth of research on the impacts of As contamination, significant gaps remain in scientific understanding of effective mitigation strategies. The health effects and extent of contamination have been the focus of previous studies, but the long-term sustainability of various mitigation efforts has received less attention. For instance, researchers have proposed deep tube wells and household water filtration systems as solutions, but these approaches face numerous challenges such as high costs, maintenance difficulties, and limited access in remote areas (Yunus et al., 2011). Furthermore, the focus on drinking water mitigation often overlooks the broader environmental implications, such as soil contamination and its impact on agricultural productivity (Hossain, 2006). Addressing As contamination in soil and food systems requires an integrated approach that goes beyond water filtration and engages with the agricultural sector, local communities, and sustainable policy frameworks.

The aim of this chapter is to synthesize available research on As contamination mitigation in Bangladesh, with a particular focus on soil and water novel management strategies. This chapter seeks to identify shortcomings in previous efforts and propose sustainable, integrated solutions that address both agricultural and water contamination. The main objectives are to: (1) assess the effectiveness of soil management techniques, such as biochar application and phytoremediation, in reducing As levels in agricultural soils; (2) evaluate the impact of water management strategies, including alternate wetting and drying (AWD) and deep tube wells, on lowering As concentrations in irrigation water and rice crops; and (3) identify knowledge gaps and suggest future research directions for better As mitigation through soil and water management in Bangladesh.

The hypotheses supporting these objectives are: (1) soil management strategies, such as the biochar application and phytoremediation significantly reduces As uptake in crops, resulting in safer agricultural products in As-contaminated regions; (2) water management strategies, such as AWD

TABLE 10.2
Arsenic Content in Rice Grains under Different Irrigation and Management Practices

Management System	Arsenic Content in Rice Grain (mg/kg)	Reference
Conventional Flooding	0.28–1.12	Norton et al., 2019; Harine et al., 2021; Acharjee et al., 2021
Alternate Wetting and Drying	0.24–0.50	Norton et al., 2019; Harine et al., 2021; Acharjee et al., 2021
Organic Farming	0.0012	Hussain et al., 2021
Biofortification with Arsenic-Resistant Varieties	0.195–0.364	Moulick et al., 2022

and deep tube wells, effectively reduce As concentrations in rice fields, minimizing overall As exposure in rice production; and (3) current soil and water management strategies for As mitigation in Bangladesh face challenges in terms of long-term effectiveness, scalability, and regional adaptation, while future research develops innovative, sustainable approaches that address these challenges and enhance As reduction (AR) across diverse agricultural systems.

The novelty of this chapter lies in its comprehensive approach to understanding As mitigation. While much of the existing research has focused on the immediate health impacts of As poisoning, this chapter seeks to broaden the scope by integrating environmental, agricultural, and socio-economic factors. Given the scale of the As crisis in Bangladesh, it is essential to develop novel strategies that not only reduce human exposure but also protect the environment and ensure the long-term viability of the country's food systems. This chapter will also highlight the importance of engaging with local communities in the development and implementation of effective mitigation strategies, as the success of any intervention depends on its acceptance and sustainability at the community level.

METHODOLOGICAL APPROACH

Systematic Literature Search and Selection

A comprehensive literature review of all published studies on As contamination, its impacts, and mitigation strategies in Bangladesh was conducted until 25 November 2024. The literature search included electronic databases such as Web of Science, Scopus, PubMed, and Google Scholar. Keywords used in the search included "arsenic contamination", "Bangladesh", "arsenic mitigation", "biochar", "phytoremediation", "groundwater contamination", "soil amendments", "water management", "irrigation practices", and "arsenic removal technologies". The search was restricted to peer-reviewed studies published in English from 2000 onwards, focusing on agricultural, environmental, and public health impacts. The inclusion criteria for the literature review were: field studies conducted in Bangladesh, or similar geographic regions; quantitative data on As concentrations in soil, water, or crops; evaluation of mitigation strategies; and studies with a clear methodology for As assessment. Exclusion criteria included laboratory-based studies without field validation and non-peer-reviewed literature. This chapter synthesizes findings from studies that meet these criteria to understand current As mitigation strategies and their effectiveness comprehensively.

Data Compilation and Synthesis

All relevant publications identified (84) were managed using Mendeley, an open-source reference management software. The studies were assessed for relevance and classified into thematic categories such as soil management, water filtration, and irrigation practices. Key information from each study, including study area, percentage of AR, intervention type, outcomes, management cost, health effect, etc. was summarized. Geospatial data, including the mapping of arsenic contamination hotspots in Bangladesh (Figure 10.1), were obtained from the Soil Resource Development Institute, Dhaka Divisional Office, Bangladesh.

SYNTHESIS OF FINDINGS

What is Arsenic Poisoning?

Arsenic poisoning, also known as arsenicosis, refers to the harmful health effects caused by the ingestion or inhalation of As, a naturally occurring metalloid found in the earth's crust. It exists in both organic and inorganic forms, with inorganic As compounds being more toxic and commonly associated with environmental contamination, particularly in groundwater. In regions like

Bangladesh, where millions of people rely on groundwater for drinking and irrigation, the contamination of water sources with As poses a severe public health risk. Long-term exposure to As through drinking water has been linked to a variety of acute and chronic health problems, making As poisoning a significant concern for both individuals and communities (Smith et al., 2000). In Figure 10.2, the major routes of As exposure in humans, including links between industrial emissions, agricultural practices, and food consumption, are illustrated.

Arsenic is classified as a Group 1 human carcinogen by the International Agency for Research on Cancer (IARC) due to its high toxicity and potential to cause cancer when consumed over long periods (Rahman et al., 2009). The toxicity of As arises from its ability to interfere with cellular processes, particularly those related to the function of enzymes and cellular respiration. Inorganic As, often found in groundwater, can disrupt the body's biochemical pathways, leading to oxidative stress, cell damage, and eventually cell death. This toxicity manifests in a wide range of symptoms and health effects, depending on the level and duration of exposure (Yunus et al., 2011).

The health effects of As poisoning can be classified into acute and chronic categories. Acute poisoning occurs with high doses of arsenic over a short period and can lead to severe symptoms, including vomiting, abdominal pain, diarrhea, and even death if the exposure is substantial (Rahman et al., 2009). However, the more common and insidious form of As poisoning is chronic exposure, which occurs over long periods at lower doses. Chronic exposure to As-contaminated water can result in a range of health problems that develop gradually, often taking years or decades to manifest. These health problems include skin lesions, hyperpigmentation, and keratosis, which are among the earliest signs of As exposure (Smith et al., 2000).

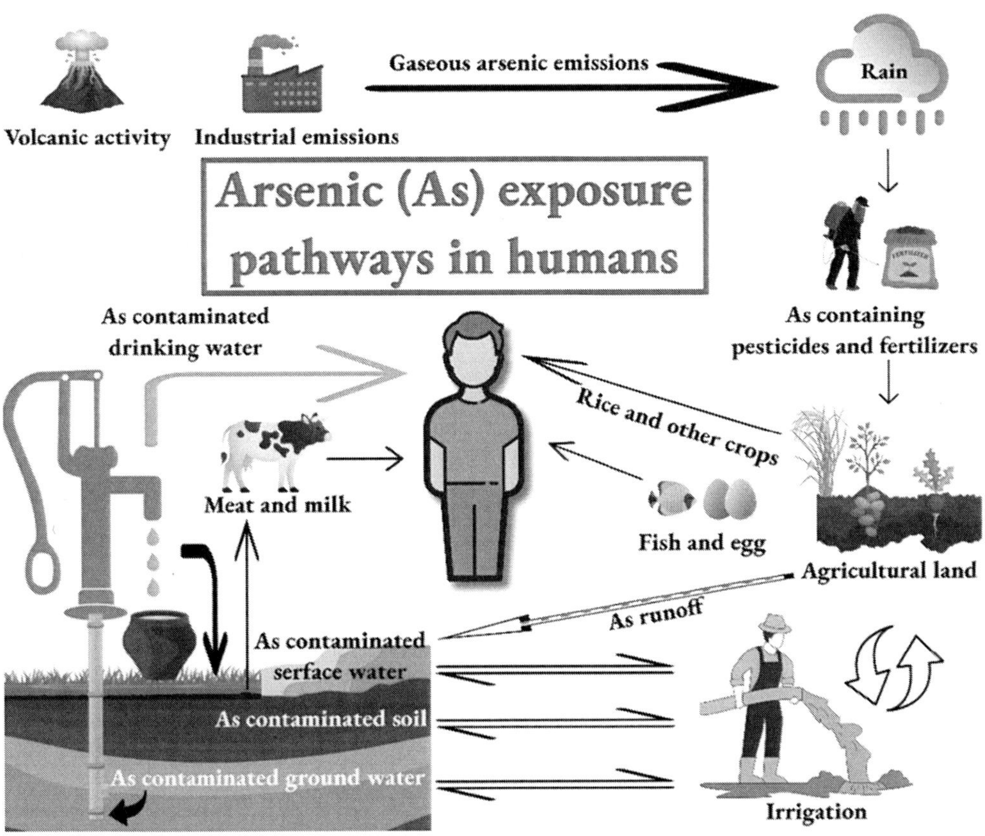

FIGURE 10.2 Human arsenic exposure pathways: linking industrial emissions, agriculture, and food consumption.

One of the most serious long-term effects of chronic As poisoning is its carcinogenic potential. Prolonged exposure to As has been strongly linked to cancers of the skin, lungs, bladder, kidneys, and liver (Rahman et al., 2009). Beyond its carcinogenic effects, As poisoning has also been associated with cardiovascular diseases, diabetes, and neurological problems, including cognitive impairments and developmental delays in children. Arsenic exposure during pregnancy can also lead to adverse birth outcomes, including low birthweight and increased infant mortality (Yunus et al., 2011).

The symptoms and health impacts of As poisoning vary depending on the level of exposure and individual susceptibility. While skin conditions such as lesions and keratosis are often the first visible signs, more severe internal effects, such as organ damage and cancer, may take years to develop. This delayed onset of symptoms makes it difficult for affected individuals to associate their health problems with As exposure, which complicates diagnosis and treatment efforts (Smith et al., 2000). Furthermore, once symptoms appear, the damage may already be irreversible, underscoring the need for early detection and preventive measures.

Sources and Causes of Arsenic Poisoning

Arsenic poisoning, particularly in Bangladesh, is a multifaceted issue stemming from both natural and anthropogenic sources. Understanding the origins of As contamination is critical for mitigating its harmful effects on human health and the environment. Geological and environmental factors, along with human activities, influence the presence of As, a naturally occurring metalloid, in the environment. The risks associated with As contamination are particularly severe in regions like Bangladesh, where groundwater heavily relies on for drinking water and irrigation. This section examines the sources and causes of As poisoning, focusing on both natural and anthropogenic factors, as well as specific case studies that illustrate the problem's scope.

The natural presence of As in the environment is largely due to geological formations and processes. In Bangladesh, the source of As contamination is primarily geological. The Bengal Basin, which spans Bangladesh and parts of India, contains large deposits of As within its sediment layers. These sediments were deposited over millennia by rivers draining from the Himalayas, carrying with them As-containing minerals. Over time, these minerals have become embedded in the aquifers that supply groundwater to millions of people (McArthur et al., 2001). The As present in these sediments is primarily in its inorganic form, which is highly toxic and more easily mobilized under certain environmental conditions.

One of the key geological processes that contribute to As mobilization is the anaerobic conditions found in many aquifers, the reduction of iron oxides can release arsenic into groundwater (Smith et al., 2000). A variety of factors, including the organic matter (OM) present in the sediments and the availability of certain minerals, influence this process, either promoting or inhibiting its mobility. For example, in Bangladesh, the high OM content of the sediments in the Bengal Basin creates reducing conditions that promote the dissolution of As-rich iron oxides. Tube wells used for drinking water and irrigation can then draw up the released As into the groundwater.

While geological factors are the primary source of As in Bangladesh's groundwater, anthropogenic activities can exacerbate the problem. Agricultural practices, particularly the use of As-laden groundwater for irrigation, contribute to the contamination of soil and crops (Meharg and Rahman, 2003). Groundwater containing high levels of As often irrigates rice paddies in Bangladesh, where rice is a staple food. Over time, this leads to the accumulation of As in the soil, which rice plants can absorb and eventually consumed. This creates a dual pathway for As exposure: drinking water and consuming contaminated food.

The use of chemical fertilizers and pesticides in agriculture may also contribute to As contamination. While many countries now largely ban As-based pesticides, their widespread use in agricultural practices persists. Residual As from these pesticides can remain in the soil for decades, continuing to pose a risk to human health (Hossain, 2006). Industrial activities like mining and smelting in some regions can release As into the environment as a byproduct. For example, smelting of As-containing ores can release it into the atmosphere, where it can be deposited on land and water bodies, further contributing to environmental contamination (Smith et al., 2000).

One of the most well-documented cases of As poisoning is in Bangladesh, where the widespread use of tube wells has led to mass exposure to As-contaminated groundwater. In the 1970s, tube wells were promoted as a means of providing safe drinking water to rural communities, as surface water was often contaminated with pathogens such as cholera and dysentery. While the use of tube wells significantly reduced the incidence of waterborne diseases, it inadvertently exposed millions of people to As-contaminated groundwater (Chowdhury and Jakariya, 1999). It wasn't until the 1990s that the full scale of the As crisis became apparent, as more and more cases of As-related health problems were reported.

Figure 10.3 illustrates some of the most common dermatological effects observed in patients affected by As poisoning, which has well-documented acute and chronic effects on human health. Acute As poisoning can lead to a range of symptoms, including vomiting, diarrhea, abdominal pain, and even death in extreme cases (Smith et al., 2000). However, the more insidious threat comes from chronic exposure to lower levels of As, which is the case for millions of people in Bangladesh and other affected regions. Chronic exposure to As over a period of years or decades can cause a range of serious health problems, including skin lesions, cancer, cardiovascular disease, and diabetes (Yunus et al., 2011). One of the most well-known health effects of chronic As exposure is the development of skin lesions, which can take the form of dark spots or hardened patches of skin. These lesions are often the first visible sign of As poisoning and can progress to more serious conditions, including skin cancer (Ahmad et al., 2018). Arsenic exposure has also been linked to cancers of the bladder, lung, and kidneys, making it a significant carcinogen. In fact, As is classified as a Group 1 carcinogen by the International Agency for Research on Cancer (IARC), meaning that it is known to cause cancer in humans (Yunus et al., 2011).

In addition to its carcinogenic effects, As has also been shown to have serious cardiovascular and metabolic effects. Studies have found that chronic exposure to As can increase the risks of hypertension, heart disease, and stroke (Rahman et al., 2009). Moreover, there is growing evidence that As exposure is linked to the development of type 2 diabetes. The exact mechanisms by which As causes these diseases are still being studied, but it is thought that As interferes with cellular respiration and causes oxidative stress, leading to damage in various tissues and organs. The case of Bangladesh serves as a stark reminder of the complexities of As poisoning, where natural geological factors and human activities intersect to create a public health crisis. The use of groundwater for

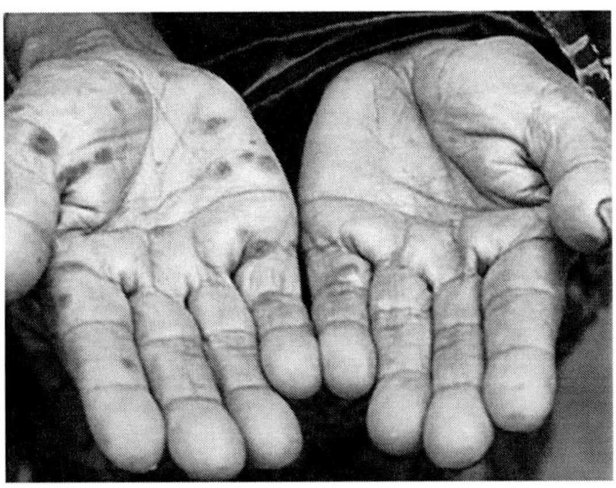

FIGURE 10.3 Dermatological effect of arsenic poisoning: arsenical keratosis on palms.
Photo by Anita Ghosh/reach, via Flickr, licensed under cc by 2.0.

drinking and irrigation, combined with agricultural practices and industrial activities, has resulted in widespread As contamination that affects millions of people. Despite the implementation of mitigation efforts, including the promotion of alternative water sources and the use of As removal technologies, the magnitude of the problem remains daunting (Hossain, 2006).

Mitigation of Arsenic Poisoning Through Soil Management

The effectiveness of various soil management practices in reducing As contamination across South Asian countries is evident, as described in Table 10.3. The practices included biochar, vermicompost, arsenic-resistant bacteria, farmyard manure, and phytoremediation. These methods showed promising results in reducing As levels in soil and crops, with its reduction percentages ranging from 20% to 56%. The studies, conducted primarily in Bangladesh, India, and Pakistan, highlighted the potential of these practices to improve soil structure, increase crop yield, and enhance overall soil fertility while mitigating arsenic contamination.

Biochar

Arsenic contamination in Bangladesh poses severe risks to agriculture and public health, with biochar emerging as a promising soil amendment to mitigate these effects. In Bangladesh, As-contaminated groundwater is frequently utilized for irrigating dry-season rice (boro). While this practice has significantly boosted food security, it is also causing a rise in As accumulation in both soil and rice, potentially leading to reduced crop yields over time due to As toxicity (Dittmar et al., 2010). Biochar, produced via pyrolysis of organic materials, effectively immobilizes As by adsorbing it onto its high surface area, thereby reducing its bioavailability to plants (Alam et al., 2018; Sattar et al., 2022). This is particularly beneficial in rice cultivation, where As contamination from irrigation water is prevalent. Studies have demonstrated significant reductions in As uptake in

TABLE 10.3
Effectiveness of Soil Management Practices in Reducing Arsenic Contamination

SMP	AR (%)	Study Area	Key Findings	Reference
Biochar	25	Bangladesh	Biochar showed potential but needed high-quality organic material.	Alam et al., 2020
Vermicompost and Biochar	56	Bangladesh	Biochar reduced arsenic uptake in rice and improved soil structure.	Roy et al., 2024
Arsenic-resistant Bacteria	50	Bangladesh	Bacteria reduced arsenic mobility and boosted plant resistance.	Bachate et al., 2009
FYM	22	India	Increased crop yield and reduced arsenic accumulation.	Das et al., 2005
Biochar	20	India	Indigenous biochar reduced soil arsenic and improved crop safety.	Srivastava et al., 2021
Phytoremediation	30	India	Phytoremediation lowered arsenic and boosted soil fertility.	Srivastava et al., 2021
Phosphorus (P) and OM Addition	20	India	OM and P reduced arsenic mobility and boosted microbial biomass.	Das and Das, 2020
Biochar	30	Pakistan	Biochar reduced soil arsenic and improved plant growth.	Alsamadany et al., 2022
Compost	20	Pakistan	Lowered arsenic levels and enhanced soil fertility.	Mehmood et al., 2017

Note: SMP = Soil Management Practices; AR = Arsenic Reduction; FYM = Farmyard Manure; OM = organic matter.

rice, leading to improved grain quality and yield (Kabir et al., 2021). Additionally, biochar enhances soil health by improving pH, nutrient retention, and microbial activity, which in turn boosts crop productivity even in contaminated soils (Alam et al., 2018). For instance, the combined use of biochar and silicon has been shown to further reduce As uptake and increase plant resilience, particularly in maize (Sattar et al., 2022). However, the potential health risks associated with biochar, ranging from its production to field application, have raised significant environmental and health concerns, particularly in terms of exposure to volatile organic compounds (VOCs) and heavy metals, which can be mitigated by adopting appropriate safety measures (Mia et al., 2024). Comparative studies from South Asia, including India, Nepal, and Pakistan, corroborate these findings, indicating biochar's broader potential in reducing As toxicity and improving soil fertility across the region (Alam et al., 2018). These insights underscore biochar's role as a sustainable, low-cost solution for managing arsenic contamination in agricultural soils.

Compost

Arsenic contamination in Bangladesh's agricultural soils poses a significant risk to food safety and crop productivity. Compost application has emerged as a promising strategy to mitigate this issue by enhancing soil health and reducing arsenic bioavailability.

The OM in compost adsorbs As, limiting its mobility and uptake by plants while promoting microbial activity that transforms arsenic into less toxic forms. In Bangladesh, the use of well-decomposed compost has proven particularly effective in reducing As accumulation in rice, a staple crop, thereby improving grain quality and minimizing human exposure risks (Das et al., 2016). Compost not only mitigates As toxicity but also enhances soil structure, water retention, and nutrient availability, leading to improved plant growth and higher yields, even in contaminated soils. Studies have demonstrated that combining compost with other amendments, such as zinc, further reduces its bioavailability, boosting crop resilience (Das et al., 2005). Comparative research from India and Nepal supports these findings, showing that compost application, including vermicompost, significantly lowers As concentrations in soils and crops, underscoring its potential as a sustainable, low-cost solution for As mitigation in South Asia (Gunjal, 2021).

Manure

Arsenic contamination in Bangladeshi soils, primarily due to the use of As-laden groundwater for irrigation, presents a significant challenge to agriculture and public health. The application of well-decomposed farmyard manure has been identified as an effective strategy for mitigating As toxicity in soils. Manure's OM adsorbs As, forming stable complexes that reduce its bioavailability and mobility, thus limiting its uptake by plants. In Bangladesh, particularly in rice cultivation, the use of manure has significantly decreased As content in rice grains, enhancing both crop yield and quality (Das et al., 2008). Additionally, manure improves soil structure, nutrient availability, and microbial activity, which are crucial to maintaining soil health and agricultural productivity in As-contaminated areas (Das et al., 2005). Comparative studies from India and Nepal corroborate these findings, demonstrating that manure application effectively reduces As mobility and accumulation in various crops, underscoring its potential as a sustainable, low-cost solution for managing soil contamination in South Asia (Das et al., 2008).

Phytoremediation

Phytoremediation, the use of plants to remove or stabilize environmental contaminants, has shown significant potential for mitigating As contamination in soils, especially in Bangladesh, where As-laden groundwater is frequently used for irrigation. This strategy leverages the natural abilities of certain plants to absorb and detoxify As from soils, with effectiveness varying based on plant species, contamination levels, and soil conditions. Indigenous plants like *Pteris vittata* (Chinese brake fern) have been identified in Bangladesh for their capacity to hyperaccumulate As, with research supporting their use in contaminated soils (Rahman et al., 2010). Additionally, efforts to develop rice varieties with reduced As uptake offer dual benefits of mitigating soil As and lowering dietary exposure (Mitra et al., 2017). Beyond AR, phytoremediation enhances soil health by improving OM content and microbial activity, contributing to better crop yields and sustainable agriculture, particularly in As-affected rice paddies (Rahman et al., 2010). Comparative studies from India, Nepal, and Pakistan further reinforce the effectiveness of phytoremediation as a cost-effective alternative to conventional methods, emphasizing the need for selecting plant species adapted to local conditions to optimize results (Das and Das, 2020).

Microbial Inoculation

Microbial inoculation, the introduction of beneficial microorganisms into soil, is an emerging strategy for mitigating As toxicity in Bangladesh's agricultural soils by reducing its bioavailability, transforming it into less toxic forms, and improving soil health. Various studies demonstrate that microorganisms such as *Pseudomonas* sp. and *Citrobacter* sp. play a crucial role in the biogeochemical cycling of As, using mechanisms like oxidation, reduction, methylation, and demethylation to alter arsenic's chemical forms, which influence its toxicity and mobility. Notably, *Pseudomonas* sp. has been found more effective than *Citrobacter* sp. in decreasing soil As levels, particularly when combined with OM like FYM, which enhances microbial activity and As immobilization by boosting soil OM and microbial biomass carbon (Das and Das, 2020). Organic matter serves as a nutrient source, stimulating microbial growth and activity, while P competes with arsenate for adsorption sites, further reducing its bioavailability. Research indicates that the combined application of OM, P, and microbial inoculants leads to substantial reductions in soil As, as P enhances microbial arsenate reduction, which is then immobilized. However, managing the balance between OM and P is critical, as excessive P can inhibit plant As uptake, leading to elevated soil As if mismanaged. Field studies in Bangladesh have shown the practical benefits of microbial inoculation, with Gustave et al. (2018) reporting a significant reduction in As uptake by rice plants when rice paddies were inoculated with As-resistant bacteria, resulting in lower As concentrations in rice grains – a crucial outcome for Bangladesh, where rice is a dietary staple and high arsenic levels in rice pose significant health risks. The combination of microbial inoculants with organic amendments like compost or biochar has also been found to improve soil structure, water retention, and nutrient availability, all of which contribute to healthier plant growth and reduced As uptake, offering an effective solution for managing As-contaminated and agriculturally degraded soils in Bangladesh.

Mitigation of Arsenic Poisoning Through Water Management

The efficiency of different water management approaches in reducing As exposure in the same region is demonstrated in Table 10.4. The table outlines various methods such as alternate wetting and drying (AWD), sprinkler irrigation, pond sand filters, rainwater harvesting, and advanced filtration techniques. These approaches varied in cost from low to high and showed significant potential in reducing As levels in water sources and crops, particularly rice. The studies emphasized the importance of these practices in improving water quality, enhancing food safety, and ultimately having positive impacts on public health by reducing As-related diseases.

Alternate Wetting and Drying Irrigation

Alternate wetting and drying (AWD) irrigation has emerged as an effective approach for reducing As contamination in rice fields in Bangladesh, where As-laden soils present major health concerns. By periodically allowing the soil to dry out rather than maintaining continuously flooded, AWD lowers As absorption in rice crops. This method changes the soil's redox conditions, which inhibits the transformation of arsenate (AsV) into arsenite (AsIII), the more easily absorbed form of arsenic. Research by Carrijo et al., (2018) has demonstrated that AWD can lower As concentrations in rice grains by 56% to 68%. Similar results have been observed in India and Nepal, though effectiveness

TABLE 10.4
Efficiency of Water Management Approaches in Reducing Arsenic Exposure

WMP	Insights	Location	Cost	Health Impact	Reference
AWD	Conserving water by alternating wetting and drying phases in rice cultivation	Bangladesh	Low to moderate	Reducing arsenic levels in rice, improving food safety	Price et al., 2013
Sprinkler Irrigation	Using sprinklers to reduce water use and prevent arsenic leaching	Bangladesh	Moderate	Lowering arsenic exposure in crops	Harine et al., 2021
AWD	Flooding the field and allowing it to dry before re-flooding	Bangladesh	Moderate	Reducing arsenic exposure in rice	Linquist et al., 2015
PSF	Treating arsenic-contaminated pond water	Bangladesh	Low to moderate	Improving water quality	Hossain et al., 2015
Rainwater Harvesting	Collecting and storing rainwater for drinking and irrigation	Bangladesh	Low to moderate	Providing arsenic-free water, positively impacting health	Hossain et al., 2015
Deep Tube Wells	Using for arsenic-safe groundwater	Bangladesh	Moderate to high	Reducing arsenic exposure significantly	Howard et al., 2007
ICP	Removing arsenic using naturally occurring iron in groundwater	Bangladesh	Low	Offering significant health benefits	Mamtaz and Bache, 2000
CBF	Managing clogging in arsenic-contaminated water	Bangladesh	Moderate	Significantly reducing arsenic-related health issues	Imteaz et al., 2016
NFM	Using low-pressure nanofiltration for arsenic removal from groundwater	Bangladesh	Moderate	Significantly reduces arsenic exposure	Oh et al., 2000
SDMD	Using solar energy to power membrane distillation for arsenic removal	India	High	Removing arsenic from drinking water, improving public health	Pal and Manna, 2010
Intermittent Irrigation	Alternating aerobic and anaerobic irrigation to reduce rice arsenic uptake	India	Moderate	Reducing arsenic in rice grains, enhancing food safety	Shrivastava et al., 2020
Rainwater Harvesting	Collecting rainwater for domestic and agricultural use	India	Low to moderate	Providing arsenic-free water, improving health	Delaire et al., 2017
GPR	Regulating groundwater extraction to prevent arsenic spread.	India	Moderate	Improving drinking water quality	Kavil et al., 2020
Nanoadsorbents	Removing arsenic from water with mixed oxide nanoadsorbents	India	Low to moderate	Significantly reducing waterborne diseases	Dinesha et al., 2017
IL	Installing liners at waste sites to stop arsenic leaching	Pakistan	High	Reducing arsenic-related diseases	Rukh et al., 2024

Note: WMP = Water Management Practices; AWD = Alternate Wetting and Drying; PSF = Pond Sand Filters; ICP = Iron Co-Precipitation; CBF = Clogging and Backwashing Filtration; NFM = Nanofiltration membrane; SDMD = Solar-Driven Membrane Distillation; GPR = Groundwater Pumping Regulation; IL = Impermeable Liners.

varies due to differences in soil types and irrigation practices (Harine et al., 2021; Pal and Manna, 2010). Despite its benefits, AWD's adoption in Bangladesh faces challenges, including potential impacts on rice yield and the need for site-specific management to accommodate varying soil conditions. However, when properly managed, AWD does not compromise yield and can even enhance it by improving root oxygenation.

Alternative Irrigation Sources

Arsenic contamination in groundwater poses a critical threat to agriculture and public health in Bangladesh, particularly through the use of contaminated water for irrigation, leading to the accumulation of As in soils and crops like rice (Ali et al., 2003; Rahman et al., 2009). To mitigate this issue, alternative irrigation sources have been explored. Deep aquifers, which contain lower As levels, have shown promise in reducing soil As content and crop uptake, though concerns about over-extraction persist (Ravenscroft et al., 2014). Similarly, surface water sources such as rivers and lakes offer a viable alternative, aiding in both AR and groundwater conservation (Saha and Ali, 2007). Rainwater harvesting and small reservoirs also present effective solutions, especially in areas with substantial rainfall, helping to alleviate dependence on contaminated groundwater during dry seasons (Ahmed et al., 2004). Comparative experiences from South Asian countries like India, Nepal, and Pakistan underscore the benefits and challenges of these alternative sources, providing valuable insights for Bangladesh's efforts in mitigating As contamination through sustainable water management practices (Bhattacharya et al., 2022).

Alternative Water Sources

Mitigating As poisoning in Bangladesh through alternative water sources is crucial due to the widespread contamination of groundwater, which poses significant risks to agriculture and public health. The use of deep aquifers, generally As-free, has shown promise in reducing its accumulation in soils and crops, though concerns about over-extraction persist (Brammer, 2009). Surface water sources, like rivers and ponds, offer a seasonal alternative that can dilute As concentrations in paddy fields, as evidenced in studies from Bangladesh and Nepal (Shrestha et al., 2003; Hossain et al., 2015). Rainwater harvesting, particularly effective during the monsoon season, provides a sustainable solution that reduces dependency on contaminated groundwater, a practice successfully implemented in Bangladesh (Ahmed et al., 2004). Comparatively, Pakistan has explored similar strategies, emphasizing the integration of alternative water sources in national water management plans (Quadri et al., 2024). These collective insights underscore the importance of adopting diverse irrigation sources to mitigate As risks in Bangladesh.

Arsenic Removal Filters

Arsenic contamination in Bangladesh's groundwater is a critical public health issue, with millions relying on shallow tube wells containing unsafe As levels. Its removal by filters offer a practical and immediate solution, operating through mechanisms such as adsorption, coagulation, and ion exchange to reduce As concentrations in drinking water. Various filters, including bucket filters, community-based systems, and composite-iron matrix filters, have been deployed, showing varying effectiveness depending on the technology and As levels (Maqsood et al., 2021). For instance, laterite filters have proven capable of reducing As to safe levels, making them a viable household option. The effectiveness of these filters is not solely dependent on their technical performance but also on user acceptance and maintenance, with higher adoption rates linked to ease of use (Inauen et al., 2013). Comparatively, in other South Asian countries like India, Nepal, and Pakistan, similar challenges exist. Community-based As removal plants in West Bengal, India, and integrated strategies in Nepal, combining filters with education and alternative water sources, highlight the regional importance of user engagement and sustainable practices for long-term success. The experiences across South Asia emphasize that while As removal filters are effective, their success hinges on proper implementation, community involvement, and consistent maintenance (Howard et al., 2007).

Biosand and Pond Sand Filter

Arsenic contamination in groundwater is a critical public health issue in Bangladesh, necessitating effective mitigation strategies such as biosand (BSFs) and pond sand filters. These filtration systems operate through physical filtration, biological degradation, and chemical adsorption to remove contaminants, including As, from water sources. BSFs, typically employed at the household level, and PSFs, used at the community level, have shown promising results in reducing As concentrations to safe levels (Pokhrel et al., 2009). Despite their effectiveness, challenges such as regular maintenance, user acceptance, cost, and environmental variability pose significant barriers to their sustained use. In South Asia, similar technologies have been implemented with varying success, as seen in Nepal and India, where community engagement and context-specific adaptations have been crucial to their effectiveness (Hossain et al., 2015). These comparative insights highlight the need for tailored solutions and continuous monitoring to ensure long-term success in As mitigation efforts across the region.

Deep Tube Wells

Deep tube wells, accessing water from aquifers typically deeper than 150 m and generally free from As contamination, have become a crucial mitigation strategy. Studies, such as those by Hoque et al., (2004), confirm the effectiveness of deep tube wells in providing As-safe drinking water, with their geochemical stability being a significant factor. Supported by both government and non-governmental organizations, the implementation of deep tube wells in rural Bangladesh has markedly reduced As exposure. However, challenges such as high installation and maintenance costs, and the risk of cross-contamination if seals between aquifers are compromised have been noted (Hossain et al., 2015). Comparative insights from South Asia reveal similar successes in India, where deep tube wells have substantially reduced As-related health issues in West Bengal (Chatterjee et al., 2022). In contrast, the adoption in Nepal and Pakistan has been limited, emphasizing the necessity for comprehensive geological surveys to ensure the effectiveness of deep tube wells in these regions.

Drip Irrigation

Drip irrigation, a water-efficient technique delivering water directly to plant roots, has emerged as a promising strategy to mitigate arsenic uptake, particularly in rice cultivation. Unlike traditional flood irrigation, which promotes anaerobic soil conditions and the conversion of less soluble arsenate (AsV) to more toxic arsenite (AsIII), drip irrigation maintains aerobic soil conditions, limiting As mobilization and reducing plant uptake (Basu et al., 2015). Comparative studies, such as those conducted in West Bengal, India, have demonstrated a reduction in As accumulation in rice grains under controlled soil moisture conditions similar to those achieved with drip irrigation (Basu et al., 2015). While not yet widely adopted for rice in Bangladesh, drip irrigation has shown significant potential in reducing As levels in other crops by minimizing the volume of As-contaminated water applied to soils (Mlangeni, 2023). This method also enhances water-use efficiency, crucial in regions like Bangladesh facing water scarcity due to population growth and climate change. Comparative insights from South Asian countries, including India and Nepal, further underscore the effectiveness of drip irrigation in reducing As and other contaminants, though its application in rice remains limited due to cultural preferences for flood irrigation (Chowdhury et al., 2017). The adoption of drip irrigation in Bangladesh aligns with broader sustainable agriculture goals and offers a viable pathway to mitigate As poisoning while conserving water resources.

Filtration and Backwashing

Filtration, combined with backwashing, is a crucial technology for mitigating As poisoning. Filtration removes As by passing contaminated water through a medium that traps the particles (Mamtaz and Bache, 2000). Regular maintenance through backwashing, as highlighted by (Imteaz et al., 2016), is essential for sustaining filtration efficiency, preventing clogging, and restoring filter performance, making these systems more sustainable for long-term use. Backwashing, which reverses the water flow to clean the filter, is vital for maintaining high As removal rates and preventing bio-clogging from bacterial growth. Despite the effectiveness of these techniques, challenges in Bangladesh include the cost and accessibility of systems, user awareness, the quality of filter media, and environmental factors such as iron and ammonium concentrations, which necessitate more frequent backwashing (Imteaz et al., 2016). Addressing these challenges through education and community-based training is crucial for the successful adoption of these As mitigation strategies.

Nanofiltration Membrane and Nanoadsorbents

Nanotechnology offers promising solutions for As mitigation in Bangladesh, particularly through the use of low-pressure nanofiltration membrane and nanoadsorbents. In Bangladesh, nano-filtration membrane (NFM) has proven to be a highly effective method for reducing As contamination in groundwater, achieving significant As removal with moderate implementation costs (Oh et al., 2000). This method has been shown to significantly reduce As exposure, particularly in rural areas where access to clean water is limited.

In comparison, India has successfully utilized mixed oxide nanoadsorbents for As removal, with a focus on both groundwater and surface water treatment. These nanoadsorbents have demonstrated low to moderate costs and have effectively reduced waterborne diseases (Dinesha et al., 2017). Overall, while nanotechnology has shown great potential in both India and Pakistan, its application in Bangladesh, particularly through NFM, remains crucial for long-term As mitigation.

CHALLENGES AND FUTURE DIRECTIONS

The challenges of mitigating As poisoning in Bangladesh are multifaceted, involving gaps in both scientific research and regulatory frameworks (Islam et al., 2023). While various strategies have been proposed and implemented, including biochar application, composting, phytoremediation, and alternative irrigation methods, each approach faces obstacles that hinder its widespread adoption and effectiveness. A comprehensive understanding of these challenges, coupled with targeted research and innovation, is necessary to develop sustainable solutions for managing As poisoning in the country's soil and water.

Research priorities must focus on addressing knowledge gaps related to the long-term effects of remediation techniques. While biochar, compost, and manure have shown promise in immobilizing As in soil, significant gaps exist in understanding their persistence and impact over time. For instance, the effectiveness of biochar depends on its specific properties – such as surface area, porosity, and the feedstock used for production (Munna et al., 2025). However, optimizing biochar to maximize As adsorption has not yet been fully explored (Alam et al., 2018), and deserves further investigation. Likewise, the long-term effects of compost and manure on As dynamics in the soil are poorly understood. The mobility of As and how it interacts with other soil elements can change over time, potentially leading to recontamination or reduced effectiveness (Das et al., 2005; Das et al., 2008). Long-term application of organic amendments such as compost, manure, and biochar has been shown to substantially modify soil physical quality through changes in aggregation, porosity, and hydraulic properties, which can indirectly influence contaminant mobility and nutrient dynamics in agricultural soils (Munna and Lal, 2025).

To address these knowledge gaps, future research should prioritize field trials across different agroecological regions in Bangladesh. Much of the current research has been conducted in controlled environments such as greenhouses, which may not fully reflect the real-world conditions faced by farmers (Sattar et al., 2022). Field studies will help determine the practical feasibility of these remediation techniques and allow for the development of region-specific guidelines for their application. For instance, biochar's effectiveness may vary depending on local soil composition and

As levels, which necessitates regionally tailored solutions (Alam et al., 2018). Similarly, the success of compost and manure amendments will depend on factors such as the type of OM used, decomposition rates, and the specific As concentration in the affected areas (Das et al., 2005; Das et al., 2016).

Phytoremediation also presents an important area for future research. This technique relies on plants that can absorb As from the soil, offering a potentially low-cost and environmentally friendly solution to As contamination. However, the process is slow, often requiring several growing seasons to reduce As levels in the soil significantly.

Moreover, suitable plant species must be identified, particularly those that can be integrated into existing agricultural systems in Bangladesh (Meharg and Rahman, 2003). Research should focus on discovering or engineering plants with high As uptake capabilities, as well as investigating the long-term effects of phytoremediation on soil health. Moreover, the safe disposal of As-enriched plant biomass remains a critical issue that has yet to be adequately addressed (Rahman et al., 2009).

One area where information is lacking is the integration of remediation techniques with broader agricultural practices. For instance, while As removal filters and deep tube wells have been implemented to provide safe drinking water, these interventions do not address soil contamination and its impact on food safety. Future studies should explore how soil remediation techniques such as biochar and phytoremediation can be combined with water filtration technologies to create a more holistic approach to As mitigation (Rahman et al., 2006). Furthermore, comparative studies across South Asia could provide valuable insights into how different regions are addressing As contamination, helping to develop region-specific strategies that are both effective and sustainable (Meharg and Rahman, 2003).

Innovations in As mitigation could involve emerging technologies such as microbial inoculation, which uses arsenic-resistant bacteria to reduce As levels in the soil. Although promising, microbial inoculation faces challenges related to strain selection and environmental adaptation. Identifying microbial strains that are both effective in As immobilization and well-suited to Bangladesh's unique environmental conditions will be crucial for the success of this approach (Das et al., 2005). (Figure 10.4).

Moreover, the scalability of microbial inoculation remains a significant barrier. While laboratory studies have demonstrated its potential, translating these results to field applications requires rigorous testing and refinement (Das et al., 2005). Another promising direction is the development of enhanced irrigation strategies such as AWD, which has been shown to reduce As uptake in rice cultivation. However, long-term studies are needed to assess its sustainability and effectiveness across various soil types and climatic conditions in Bangladesh (Rahman et al., 2009).

AWD can be integrated with soil amendments like biochar to further reduce As uptake and improve crop yields. Farmer education and training are also essential for the success of AWD, as widespread adoption will depend on farmers' understanding of the technique and its benefits (Das et al., 2005).

As for infrastructure challenges, alternative water sources such as deep tube wells and surface water irrigation systems offer viable solutions to As contamination but face significant barriers related to cost and accessibility. The development of these systems requires substantial investment, particularly for smallholder farmers who may not have the financial resources to install and maintain deep wells (Yunus et al., 2016). Additionally, water quality and availability must be monitored to ensure that these systems do not introduce new contaminants into the soil or water supply (Smith et al., 2000). Robust water management strategies that include regular testing and maintenance are critical to the long-term success of alternative irrigation systems (Hossain, 2006).

Regarding policy support, government incentives and subsidies will be crucial for promoting the adoption of As mitigation technologies. Financial assistance for the installation of deep wells, water filtration systems, and soil amendments can help overcome cost barriers and ensure that these technologies reach the most affected populations (Rahman et al., 2006). Additionally, education and

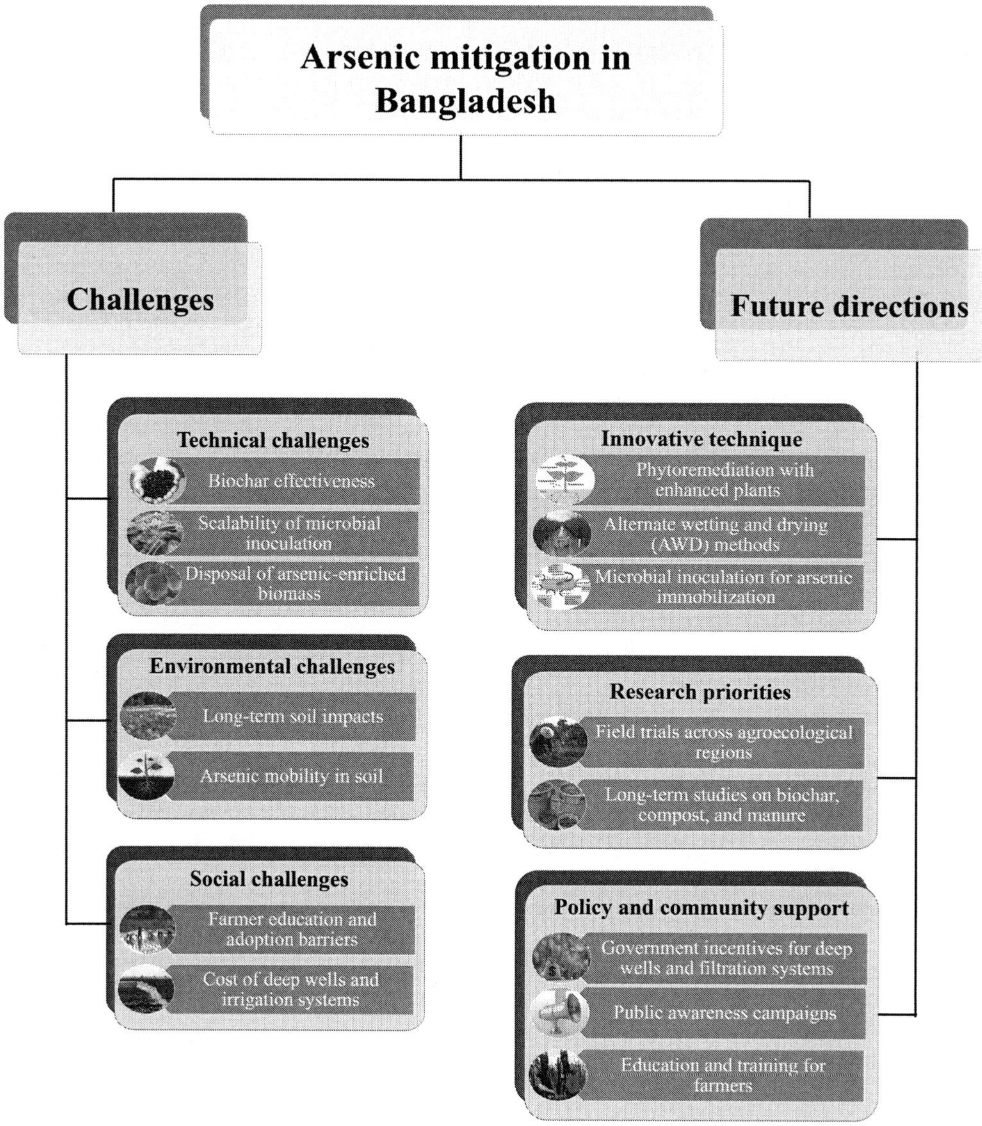

FIGURE 10.4 Key challenges and future directions for arsenic mitigation in Bangladesh.

training programs should be expanded to help farmers understand and implement arsenic mitigation strategies. Public awareness campaigns that emphasize the health risks of As exposure and the importance of safe water and soil management are also necessary for driving behavioral change at the community level (Yunus et al., 2016).

Future research should also prioritize the environmental impacts of As mitigation technologies. For example, introducing biochar, compost, or microbial inoculants into the soil could have unintended consequences on native microbial communities and overall soil health. Long-term environmental monitoring is needed to assess the sustainability of these approaches and ensure that they do not cause additional harm (Das et al., 2016). Similarly, the environmental risks associated with As-laden waste from water filtration systems must be carefully managed to prevent secondary contamination (Rahman et al., 2009).

CONCLUSIONS

The mitigation of As contamination in Bangladesh remains a significant environmental and public health challenge. This chapter has synthesized findings from over 84 studies that highlight the problem's scale and potential solutions. Key strategies, such as biochar application, which has demonstrated As reductions of up to 56% in contaminated soils, offer promising approaches for improving both soil health and crop safety. Similarly, phytoremediation and microbial inoculation, with reductions in As uptake of up to 50%, provide sustainable and cost-effective solutions for long-term management. Research has shown that water management techniques like alternate wetting and drying can reduce As levels in rice by 56–68%, and the implementation of deep tube wells has significantly reduced arsenic exposure for millions of individuals. However, the high cost and maintenance of these technologies, as well as their limited reach in rural areas, present significant barriers to broader implementation. The data emphasize the importance of an integrated approach that addresses both soil and water contamination, engages local communities, and tailors solutions to the specific geographic and socioeconomic conditions of Bangladesh. Future research should focus on the long-term sustainability of these interventions, particularly in real-world agricultural settings, and explore the scalability of emerging technologies such as microbial inoculation. With ongoing efforts and continued innovation, it is possible to significantly reduce As exposure in Bangladesh, improving health outcomes and ensuring the viability of the country's food systems for future generations.

ACKNOWLEDGMENTS

The Foundation for Food and Agriculture Research (Grant-ID: 22-000279; Title: Enhanced Soil Carbon Farming as a Climate Solution) provided financial support with matching contributions from AgMission, Bayer Crop Science US, Corteva Agriscience, Cotton Incorporated, Fondo Regional de Technología Agropecuaria (FONTAGRO), Herbert W. Hoover Foundation, Inter-American Institute for Cooperation on Agriculture (IICA), Kansas Corn, Microsoft, National Sorghum Producers, Ohio Corn and Wheat, Ohio Soybean Council, and The Ohio State University. In-kind support was provided by Agricultural Research Service of the US Department of Agriculture (USDA), Sandia National Laboratory of the US Department of Energy (DOE), and Western Geographic Science Center of the US Geological Survey (USGS).

ACRONYMS

AR	Arsenic Reduction
AWD	Alternate Wetting and Drying
BSF	Biosand Filter
CBF	Clogging and Backwashing Filtration
FYM	Farmyard Manure
GPR	Groundwater Pumping Regulation
ICP	Iron Co-Precipitation
NFM	Nanofiltration Membrane
OM	Organic Matter
PSF	Pond Sand Filter
SDMD	Solar-Driven Membrane Distillation
SMP	Soil Management Practices
WMP	Water Management Practices

REFERENCES

Acharjee, P.U., K. Bhattacharyya, R. Poddar, A. Pari, K. Ray, S.K. Patra, and S. Halder. 2021. Water management and varietal selection approach in mitigation of arsenic in inceptisols of West Bengal, India. *Communications in Soil Science and Plant Analysis* 52:1008–1022.

Adeloju, S.B., S. Khan, and A.F. Patti. 2021. Arsenic contamination of groundwater and its implications for drinking water quality and human health in under-developed countries and remote communities – A review. *Applied Sciences* 11:1926.

Ahmad, A., and P. Bhattacharya. 2019. Arsenic in drinking water: Is 10 μg/L a safe limit? *Current Pollution Reports* 5:1–3.

Ahmad, S.A., M.H. Khan, and M. Haque. 2018. Arsenic contamination in groundwater in Bangladesh: Implications and challenges for healthcare policy. *Risk Management and Healthcare Policy* 11:251–261.

Ahmed, K.M., P. Bhattacharya, M.A. Hasan, S.H. Akhter, S.M.M. Alam, M.A.H. Bhuyian, M.B. Imam, A.A. Khan, and O. Sracek. 2004. Arsenic enrichment in groundwater of the alluvial aquifers in Bangladesh: An overview. *Applied Geochemistry* 19:181–200.

Alam, M.A., W.A. Shaikh, M.O. Alam, T. Bhattacharya, S. Chakraborty, B. Show, and I. Saha. 2018. Adsorption of As (III) and As (V) from aqueous solution by modified Cassia fistula (golden shower) biochar. *Applied Water Science* 8:198.

Alam, M.Z., L. Carpenter-Boggs, M.A. Hoque, and G.J. Ahammed. 2020. Effect of soil amendments on antioxidant activity and photosynthetic pigments in pea crops grown in arsenic-contaminated soil. *Heliyon* 6:e05475.

Ali, M.M., H. Ishiga, and T. Wakatsuki. 2003. Influence of soil type and properties on distribution and changes in arsenic contents of different paddy soils in Bangladesh. *Soil Science & Plant Nutrition* 49:111–123.

Alsamadany, H., H.F. Alharby, H.S. Al-Zahrani, Y.M. Alzahrani, A.A. Almaghamsi, G. Abbas, and M.A. Farooq. 2022. Silicon-nanoparticles doped biochar is more effective than biochar for mitigation of arsenic and salinity stress in Quinoa: Insight to human health risk assessment. *Frontiers in Plant Science* 13:989504.

Bachate, S.P., L. Cavalca, and V. Andreoni. 2009. Arsenic-resistant bacteria isolated from agricultural soils of Bangladesh and characterization of arsenate-reducing strains. *Journal of Applied Microbiology* 107:145–156.

Baker, D.E., and L. Chesnin. 1975. Chemical monitoring of soils for environmental quality and animal and human health. *Advances in Agronomy* 27:305–374.

Basu, B., M. Kundu, M. Hedayatullah, C.K. Kundu, P. Bandyopadhyay, K. Bhattacharya, and S. Sarkar. 2015. Mitigation of arsenic in rice through deficit irrigation in field and use of filtered water in kitchen. *International journal of Environmental Science and Technology* 12:2065–2070.

Berg, M., H.C. Tran, T.C. Nguyen, H.V. Pham, R. Schertenleib, and W. Giger. 2001. Arsenic contamination of groundwater and drinking water in Vietnam: A human health threat. *Environmental Science & Technology* 35:2621–2626.

Bhattacharya, A., S. Sahu, N. Sahu, A. Das, C. Das, V. John, S. Sarkar, S. Duttagupta, J. Bhattacharya, A. Mukherjee, and P.S. Ghosal. 2022. Influence of mass-awareness campaign on community behavior pattern changes for safe drinking water availability in a groundwater arsenic-affected area of South Asia. *Groundwater for Sustainable Development* 18:100766.

Bowen, H.J.M. 1979. *Environmental chemistry of the elements*. Academic Press, New York.

Brammer, H. 2009. Mitigation of arsenic contamination in irrigated paddy soils in South and South-East Asia. *Environment International* 35:856–863.

Bundschuh, J., M.I. Litter, F. Parvez, G. Román-Ross, H.B. Nicolli, J.-S. Jean, C.-W. Liu, D. López, M.A. Armienta, L.R.G. Guilherme, A.G. Cuevas, L. Cornejo, L. Cumbal, and R. Toujaguez. 2012. One century of arsenic exposure in Latin America: A review of history and occurrence from 14 countries. *Science of the Total Environment* 429:2–35.

Carrijo, D.R., N. Akbar, A.F.B. Reis, C. Li, A.C.M. Gaudin, S.J. Parikh, P.G. Green, and B.A. Linquist. 2018. Impacts of variable soil drying in alternate wetting and drying rice systems on yields, grain arsenic concentration and soil moisture dynamics. *Field Crops Research* 222:101–110.

Chatterjee, R., R. Islam, S.K. Acharya, and A. Biswas. 2022. Conservation agriculture: Analysis and prioritization of socio-ecological factors operating at farm levels in Ohio, USA. *Environmental Science & Policy* 138:1–10.

Chen, Q.Y., and M. Costa. 2021. Arsenic: A global environmental challenge. *Annual Review of Pharmacology and Toxicology* 61:47–63.

Chowdhury, A.M., and M. Jakariya. 1999. Testing of water for arsenic in Bangladesh. *Science* 284 (5420):1622.

Chowdhury, M.T.A., C.M. Deacon, G.D. Jones, S.M.I. Huq, P.N. Williams, A.F.M.M. Hoque, L.H.E. Winkel, A.H. Price, G.J. Norton, and A.A. Meharg. 2017. Arsenic in Bangladeshi soils related to physiographic region, paddy management, and micro- and macro-elemental status. *Science of the Total Environment* 590:406–415.

Das, D.K., T.K. Garai, S. Sarkar, and P. Sur. 2005. Interaction of arsenic with zinc and organics in a rice (Oryza sativa L.)-cultivated field in India. *Scientific World Journal* 5:646–651.

Das, D.K., P. Sur, and K. Das. 2008. Mobilisation of arsenic in soils and in rice (Oryza sativa L.) plants affected by organic matter and zinc application in irrigation water contaminated with arsenic. *Plant, Soil and Environment* 54:30–37.

Das, I., S.K. Sanyal, K. Ghosh, and D.K. Das. 2016. Arsenic mitigation in soil-plant system through zinc application in West Bengal soils. *Bioremediation Journal* 20:24–37.

Das, S.K., and S.K. Das. 2020. Influence of phosphorus and organic matter on microbial transformation of arsenic. *Environmental Technology and Innovation* 19:100930.

Delaire, C., A. Das, S. Amrose, A. Gadgil, J. Roy, and I. Ray. 2017. Determinants of the use of alternatives to arsenic-contaminated shallow groundwater: An exploratory study in rural West Bengal, India. *Journal of Water and Health* 15:799–812.

Dinesha, B.L., H. Sharanagouda, N. Udaykumar, C.T. Ramachandr, and A.B. Dandekar. 2017. Removal of pollutants from water/waste water using nanoadsorbents: A potential pollution mitigation. *International Journal of Current Microbiology and Applied Sciences* 6:4868–4872.

Dittmar, J., A. Voegelin, L.C. Roberts, S.J. Hug, G.C. Saha, M.A. Ali, A.B.M. Badruzzaman, and R. Kretzschmar. 2010. Arsenic accumulation in a paddy field in Bangladesh: Seasonal dynamics and trends over a three-year monitoring period. *Environmental Science & Technology* 44:2925–2931.

Doerge, D., E. Dogliotti, L. Edler, P. Farmer, M. Filipič, J. Fink-Gremmels, P. Fürst, T. Guerin, H.K. Knutsen, M. Machala, A. Mutti, J.R. Schlatter, R.V. Leeuwen, M.R. Karagas, T. Schwerdtle, M. Vahter, P. Verger, and B.V. Voet. 2009. Scientific opinion on arsenic in food. *EFSA Journal* 7:1351.

Gunjal, A. 2021. Study of Langmuir kinetics for removal of heavy metals by the fungal biomass. *Proceedings of the Indian National Science Academy* 87:107–109.

Gupta, D.K., and S. Chatterjee (Eds.). 2017. *Arsenic contamination in the environment*. Springer International Publishing.

Gustave, W., Z.-F. Yuan, R. Sekar, H.-C. Chang, J. Zhang, M. Wells, Y.-X. Ren, and Z. Chen. 2018. Arsenic mitigation in paddy soils by using microbial fuel cells. *Environmental Pollution* 238:647–655.

Harine, I.J., M.R. Islam, M. Hossain, H. Afroz, R. Jahan, A.B. Siddique, S. Uddin, M.A. Hossain, S. Alamri, M.H. Siddiqui, and R.J. Henry. 2021. Arsenic accumulation in rice grain as influenced by water management: Human health risk assessment. *Agronomy Basel* 11:1741.

Hoque, B.A., M.M. Hoque, T. Ahmed, S. Islam, A.K. Azad, N. Ali, M. Hossain, and M.S. Hossain. 2004. Demand-based water options for arsenic mitigation: An experience from rural Bangladesh. *Public Health* 118:70–77.

Hossain, M., S.N. Rahman, P. Bhattacharya, G. Jacks, R. Saha, and M. Rahman. 2015. Sustainability of arsenic mitigation interventions – an evaluation of different alternative safe drinking water options provided in Matlab, an arsenic hot spot in Bangladesh. *Frontiers in Environmental Science* 3:30.

Hossain, M.F. 2006. Arsenic contamination in Bangladesh – An overview. *Agriculture, Ecosystems and Environment* 113:1–16.

Howard, G., M.F. Ahmed, S.G. Mahmud, P. Teunis, A. Davison, and D. Deere. 2007. Disease burden estimation to support policy decision-making and research prioritization for arsenic mitigation. *Journal of Water and Health* 5:67–81.

Hussain, M.M., I. Bibi, N.K. Niazi, M.F. Nawaz, and J. Rinklebe. 2021. Impact of organic and inorganic amendments on arsenic accumulation by rice genotypes under paddy soil conditions: A pilot-scale investigation to assess health risk. *Journal of Hazardous Materials* 420:126620.

Imteaz, M.A., M.Y. Ahmed, M.S. Khan, and A. Ahsan. 2016. A simple clogging and backwashing efficiency model for filtration of arsenic-contaminated water. *Desalination and Water Treatment* 57:12237–12243.

Inauen, J., R. Tobias, and H.-J. Mosler. 2013. Predicting water consumption habits for seven arsenic-safe water options in Bangladesh. *BMC Public Health* 13:417.

Islam, M.S., K. Phoungthong, A.R.M.T. Islam, M.M. Ali, A. Sarker, M.H. Kabir, and A.M. Idris. 2023. Present status and mitigation approaches of arsenic in the environment of Bangladesh: A critical review. *International Journal of Environmental Science and Technology* 20:13883–13894.

Kabir, M.H., G. Brodie, D. Gupta, and A. Pang. 2021. Microwave soil treatment along with biochar application alleviates arsenic phytotoxicity and reduces rice grain arsenic concentration. *Energies* 14:8140.

Kabir, M.S., M.A. Salam, D.N.R. Paul, M.I. Hossain, N.M.F. Rahman, A. Aziz, and M.A. Latif. 2016. Spatial variation of arsenic in soil, irrigation water, and plant parts: A microlevel study. *The Scientific World Journal* 2016:1–14.

Kavil, S.P., D. Ghosh, I. Pasic, and J. Routh. 2020. Temporal dynamics of arsenic uptake and distribution: Food and water risks in the Bengal basin. *Toxicological and Environmental Chemistry* 102:62–77.

Linquist, B.A., M.M. Anders, M.A.A. Adviento-Borbe, R.L. Chaney, L.L. Nalley, E.F.F. Da Rosa, and C. Van Kessel. 2015. Reducing greenhouse gas emissions, water use, and grain arsenic levels in rice systems. *Global Change Biology* 21:407–417.

Mamtaz, R., and D.H. Bache. 2000. Low-cost separation of arsenic from water: With special reference to Bangladesh. *Water Environment Journal* 14:260–269.

Maqsood, H., S. Rukh, M. Imran, M. Ayaz, W. Ahmad, A. Matloob, H. Ahmad, A. Khan, and S. Butt. 2021. Evaluation of laterite as filter media to remove arsenic from groundwater. *Journal of the Serbian Chemical Society* 86:195–207.

Matos-Reyes, M.N., M.L. Cervera, R.C. Campos, and M. De La Guardia. 2010. Total content of As, Sb, Se, Te, and Bi in Spanish vegetables, cereals, and pulses and estimation of the contribution of these foods to the Mediterranean daily intake of trace elements. *Food Chemistry* 122:188–194.

McArthur, J.M., P. Ravenscroft, S. Safiulla, and M.F. Thirlwall. 2001. Arsenic in groundwater: Testing pollution mechanisms for sedimentary aquifers in Bangladesh. *Water Resources Research* 37:109–117.

Meharg, A.A., and M.M. Rahman. 2003. Arsenic contamination of Bangladesh paddy field soils: Implications for rice contribution to arsenic consumption. *Environmental Science and Technology* 37:229–234.

Mehmood, T., I. Bibi, M. Shahid, N.K. Niazi, B. Murtaza, H. Wang, Y.S. Ok, B. Sarkar, M.T. Javed, and G. Murtaza. 2017. Effect of compost addition on arsenic uptake, morphological and physiological attributes of maize plants grown in contrasting soils. *Journal of Geochemical Exploration* 178:83–91.

Mia, S., S.Y. Bristy, K. Jindo, M.N.H. Munna, M.M. Rahman, K. Uddin, S.B. Kasim, and M.S. Rahman. 2024. Potential health risks associated with biochar – from production to field application. In P.V. Nidheesh, M. Vithanage, V. Sreedharan, N. Bolan, B. Gao, and A. Bhatnagar (eds.), *Biochar amendments for environmental remediation*, pp. 293–304. CRC Press.

Mitra, A., S. Chatterjee, R. Moogouei, and D. Gupta. 2017. Arsenic accumulation in rice and probable mitigation approaches: A review. *Agronomy* 7:67.

Mlangeni, A.T. 2023. Methylation of arsenic in rice: Mechanisms, factors, and mitigation strategies. *Toxicology Reports* 11:295–306.

Mondal, D., M.M. Rahman, S. Suman, P. Sharma, A.B. Siddique, M.A. Rahman, A.S.M.F. Bari, R. Kumar, N. Bose, S.K. Singh, A. Ghosh, and D.A. Polya. 2021. Arsenic exposure from food exceeds that from drinking water in endemic area of Bihar, India. *Science of the Total Environment* 754:142082.

Moulick, D., D. Ghosh, M. Skalicky, Y. Gharde, M.K. Mazumder, S. Choudhury, J.K. Biswas, S.C. Santra, M. Brestic, P. Vachova, and A. Hossain. 2022. Interrelationship among rice grain arsenic, micronutrients content and grain quality attributes: An investigation from genotype × environment perspective. *Frontiers in Environmental Science* 10:857629.

Munna, M.N.H., and R. Lal. 2025. Impacts of cover cropping and organic amendments on soil physical quality under temperate climate. *Cogent Food and Agriculture* 11:2467452.

Munna, M.N.H., F.Z. Tanu, S. Mia, N.A. Shapna, A. Hakim, and R. Lal. 2025. Reclaiming saline soil by using acid-modified pineapple biochar. *Archives of Agronomy and Soil Science* 71:1–16.

Norton, G.J., A.J. Travis, P. Talukdar, M. Hossain, M.R. Islam, A. Douglas, and A.H. Price. 2019. Genetic loci regulating arsenic content in rice grains when grown flooded or under alternative wetting and drying irrigation. *Rice* 12:54.

Oh, J.I., T. Urase, H. Kitawaki, M.M. Rahman, M.H. Rhahman, and K. Yamamoto. 2000. Modeling of arsenic rejection considering affinity and steric hindrance effect in nanofiltration membranes. *Water Science and Technology* 42:173–180.

Pal, P., and A.K. Manna. 2010. Removal of arsenic from contaminated groundwater by solar-driven membrane distillation using three different commercial membranes. *Water Research* 44:5750–5760.

Pokhrel, D., B. Bhandari, and T. Viraraghavan. 2009. Arsenic contamination of groundwater in the Terai region of Nepal: An overview of health concerns and treatment options. *Environment International* 35:157–161.

Price, A.H., G.J. Norton, D.E. Salt, O. Ebenhoeh, A.A. Meharg, C. Meharg, M.R. Islam, R.N. Sarma, T. Dasgupta, A.M. Ismail, K.L. McNally, H. Zhang, I.C. Dodd, and W.J. Davies. 2013. Alternate wetting and drying irrigation for rice in Bangladesh: Is it sustainable and has plant breeding something to offer? *Food and Energy Security* 2:120–129.

Quadri, S.S.A., R. Ali, A. Mesidor, I.A. Shah, N. Raqeeb, M. Omer Afzal, M. Rizwan, T. Rafique, and S. Nosheen. 2024. Arsenic in the food chain in Pakistan: Assessing risks to human health and ensuring food security through comprehensive contamination mitigation strategies. *Cureus* 16:e54069.

Rahman, M., M. Vahter, M.A. Wahed, N. Sohel, M. Yunus, P.K. Streatfield, S. El Arifeen, A. Bhuiya, K. Zaman, A.M.R. Chowdhury, E.-C. Ekström, and L.A. Persson. 2006. Prevalence of arsenic exposure and skin lesions: A population based survey in Matlab, Bangladesh. *Journal of Epidemiology and Community Health* 60:242–248.

Rahman, M.M., J.C. Ng, and R. Naidu. 2009. Chronic exposure of arsenic via drinking water and its adverse health impacts on humans. *Environmental Geochemistry and Health* 31:189–200.

Rahman, M.S., M.A.M. Miah, H.M. Khaled, A. Islam, and G.M. Panaullah. 2010. Arsenic concentrations in groundwater, soils, and irrigated rice in southwestern Bangladesh. *Communications in Soil Science and Plant Analysis* 41:1889–1895.

Ravenscroft, P., A. Kabir, S.A. Ibn Hakim, A.K.M. Ibrahim, S.K. Ghosh, M.S. Rahman, F. Akhter, and M.A. Sattar. 2014. Effectiveness of public rural waterpoints in Bangladesh with special reference to arsenic mitigation. *Journal of Water Sanitation and Hygiene for Development* 4:545–562.

Roy, R., A. Hossain, M.O. Sharif, M. Das, and T. Sarker. 2024. Optimizing biochar, vermicompost, and duckweed amendments to mitigate arsenic uptake and accumulation in rice (Oryza sativa L.) cultivated on arsenic-contaminated soil. *BMC Plant Biology* 24, 545.

Rukh, S., M.S. Akhtar, F. Alshehri, A. Mehmood, K.M. Malik, S. Almadani, A. Khan, and M. Shahab. 2024. Modeling the transport of inorganic arsenic species through field soils: Irrigation and soil structure effect. *Water* 16:386.

Saha, G.C., and M.A. Ali. 2007. Dynamics of arsenic in agricultural soils irrigated with arsenic contaminated groundwater in Bangladesh. *Science of the Total Environment* 379:180–189.

Sattar, A., A. Sher, M.A.S. Abourehab, M. Ijaz, M. Nawaz, S. Ul-Allah, T. Abbas, A.N. Shah, M.S. Imam, N.R. Abdelsalam, M.E. Hasan, A. Abbas, and M.M. Javaid. 2022. Application of silicon and biochar alleviates the adversities of arsenic stress in maize by triggering the morpho-physiological and antioxidant defense mechanisms. *Frontiers in Environmental Science* 10:979049.

Shandana, A. Khan, M. Waqas, J. Nawab, M. Idress, M. Kamran, and S. Khan. 2024. Total arsenic contamination in soil, vegetables, and fruits and its potential health risks in the Chitral Valley, Pakistan. *International Journal of Sediment Research* 39:257–265.

Shrestha, R.R., M.P. Shrestha, N.P. Upadhyay, R. Pradhan, R. Khadka, A. Maskey, M. Maharjan, S. Tuladhar, B.M. Dahal, and K. Shrestha. 2003. Groundwater arsenic contamination, its health impact, and mitigation program in Nepal. *Journal of Environmental Science and Health, Part A* 38:185–200.

Shrivastava, A., A. Barla, A. Majumdar, S. Singh, and S. Bose. 2020. Arsenic mitigation in rice grain loading via alternative irrigation by proposed water management practices. *Chemosphere* 238:124988.

Smith, A.H., E.O. Lingas, and M. Rahman. 2000. Contamination of drinking water by arsenic in Bangladesh: A public health emergency. *Bulletin of the World Health Organization* 78:1093–1103.

Song, B., M. Lei, T. Chen, Y. Zheng, Y. Xie, X. Li, and D. Gao. 2009. Assessing the health risk of heavy metals in vegetables to the general population in Beijing, China. *Journal of Environmental Sciences* 21:1702–1709.

Srivastava, S., A. Shukla, V.D. Rajput, K. Kumar, T. Minkina, S. Mandzhieva, A. Shmaraeva, and P. Suprasanna. 2021. Arsenic remediation through sustainable phytoremediation approaches. *Minerals* 11:936.

Ulman, C., S. Gezer, Ö. Anal, I.R. Töre, and Ü. Kirca. 1998. Arsenic in human and cow's milk: A reflection of environmental pollution. Water, Air, and Soil Pollution 101:411–416.

UNICEF. 2010. Towards an arsenic safe environment in Bangladesh. United Nations Children's Fund (UNICEF), Dhaka, Bangladesh. Executive Summary, World Water Day, 22 March 2010.

Vinogradov, A.P. 1959. *The geochemistry of rare and dispersed chemical elements in soils*. 2nd ed. Consultants Bureau, New York.

Yunus, F.M., S. Khan, P. Chowdhury, A.H. Milton, S. Hussain, and M. Rahman. 2016. A review of groundwater arsenic contamination in Bangladesh: The millennium development goal era and beyond. *International Journal of Environmental Research and Public Health* 13:215.

Yunus, M., N. Sohel, S.K. Hore, and M. Rahman. 2011. Arsenic exposure and adverse health effects: A review of recent findings from arsenic and health studies in Matlab, Bangladesh. *The Kaohsiung Journal of Medical Sciences* 27:371–376.

Zahid, A.M., N.A. Kohinoor, M.A. Hossain, and J.U.M. Shoaib. 2022. *Land degradation in Bangladesh 2020*. DoE Coordinated and GEF Supported Research Report. SRDI, Dhaka.

Zavala, Y.J., and J.M. Duxbury. 2008. Arsenic in rice: I. Estimating normal levels of total arsenic in rice grain. *Environmental Science and Technology* 42:3856–3860.

11 Managing Soil Health for Enhancing Plant Disease Defensiveness

Chinmayee Priyadarshini, Rattan Lal, and Ye Xia

INTRODUCTION

Soil health is defined as the "capacity of soil to function as a vital living system to sustain biological productivity, maintain environmental quality, and promote plant, animal, and human health" (Doran et al., 1996; Doran and Zeiss, 2000; Larkin, 2015; Lal, 2016). Crop growth occurs through the interaction between plant roots and their environment, which includes the soil or other planting medium. This medium not only provides structural support but also supplies water and nutrients to the plants (Ortiz & Sansinenea, 2022). The concept of soil health inherently implies that healthy soils naturally maintain low levels of plant pathogens and diseases, particularly those that are soil-borne.

Soilborne pathogens, which impact crops directly, also have indirect consequences on human health, as the consumption of contaminated crops can lead to various diseases. For instance, plant pathogens such as *Fusarium*, *Rhizoctonia*, and *Verticillium* can produce mycotoxins, which when ingested, may affect human health by causing toxicity, immune system suppression, or even cancer (Larkin, 2015). Therefore, plant diseases are not just an agricultural concern; they represent a public health risk through the food chain.

Conventional management of soil-borne diseases often relies on disease-resistant cultivars or the use of synthetic pesticides (Weller et al., 2002; Shafique et al., 2016). However, these chemical pesticides not only pose risks to the soil microclimate but also harm the environment and the consumers of the crops (Larkin, 2015). These chemicals can destroy a large group of non-target soil microorganisms, which are essential for maintaining soil health. Thus, management strategies aimed at enhancing soil health logically lead to reduced levels of pathogens and diseases, while ensuring safe food production for humans and animals (Larkin, 2015), and promoting a pollution-free environment.

The key element governing processes in soil is the soil carbon pool. The soil C pool consists of the soil organic carbon (SOC) pool and the soil inorganic carbon (SIC) pool. The SOC pool is a determinant of many soil functions and ecosystem services (Lal, 2016). This pool, which makes up approximately 45–60% of the mass of soil organic matter (SOM), serves as a primary energy source for soil microorganisms. The dynamics of soil health depend heavily on the management of the SOM/SOC pools (Lal, 2016).

In recent years, numerous efforts by agronomists, soil scientists, and ecologists have focused on managing soil health in ways that minimize degradation and are sustainable for the long term. As a result, integrated approaches to soil management that combine physical, chemical, and biological practices have been adopted to sustain soil health. For example, Cardoso et al. (2013) found that physical attributes of soil, such as aggregation, bulk density, porosity, and moisture content, have a slower response compared to biochemical attributes like total carbon (C), nitrogen (N), mineral nutrients, SOM, cation exchange capacity (CEC), biodiversity, soil enzymes, and soil respiration.

DOI: 10.1201/9781003469742-11

Practices such as cover cropping, crop rotation, biological control, the use of growth-promoting microbes, conservation agriculture, mulching, and the application of arbuscular mycorrhizal fungi (AMF) are being increasingly used to enhance soil and plant health. Tahat et al. (2020) observed that AMF, cyanobacteria, and beneficial nematodes improve water-use efficiency (WUE), nutrient availability to plants, phytohormone production, soil nutrient cycling, and plant resistance to environmental stresses.

Since soil is a dynamic system where the physical, chemical, and biological components are interdependent, a blend of these management practices can boost SOC pools and strengthen the inherent disease resistance mechanisms of plants. This, in turn, reduces reliance on pesticides, minimizes pollution, and ensures the production of quality food to meet the needs of a growing global population.

This chapter highlights the importance of soil health in increasing the disease resistance of plants, the linkage between soil health, plant health, and human health, importance of soil C and nitrogen dynamics, and soil management practices for enhancing the inherent disease defensiveness of plants. This chapter aims at testing the hypothesis that maintaining and improving soil health through sustainable management practices enhances plant disease resistance, establishes a strong linkage between soil, plant, and human health, and optimizes soil carbon and nitrogen dynamics, thereby contributing to overall ecosystem resilience and agricultural sustainability.

SOIL HEALTH

Soil health is generally defined as the capacity of soil to function as a living system, supporting plant growth, maintaining biodiversity, and promoting ecosystem services while resisting degradation (Lehman et al., 2015). It is a dynamic and multifaceted concept that includes biological, physical, and chemical properties, all of which contribute to the soil's ability to perform these functions. The definition emphasizes not only productivity but also soil's resilience to stressors such as erosion, pollution, and climate change (Maharjan et al., 2020). Therefore, soil health encompasses soil structure, nutrient cycling, microbial diversity, organic matter content, and its capacity to retain water.

Key Indicators of Soil Health

Identifying reliable indicators of soil health is crucial for monitoring and managing soil conditions. Several studies have proposed a range of indicators that provide insights into the biological, physical, and chemical aspects of soil health. Key biological indicators include soil microbial diversity and activity, such as the abundance of beneficial microorganisms like mycorrhizae, bacteria, and earthworms, which support nutrient cycling and disease resistance (Lehman et al., 2015; Mann et al., 2019). These organisms help degrade organic matter, release essential nutrients, and improve soil structure.

Physical indicators include soil texture, bulk density, porosity, and water-holding capacity. These factors affect soil aeration, drainage, and root penetration, which directly influence plant health and productivity. Karlen et al. (2019) observed that poor physical properties, such as high bulk density, often result in soil compaction, leading to reduced root growth and poor water infiltration, which can degrade soil health over time.

Chemical indicators include soil pH, nutrient content (such as N, phosphorus, and potassium), and the presence of toxic elements like heavy metals. The balance of nutrients in the soil directly affects plant growth, while imbalances or deficiencies can lead to poor yields or crop failure. For instance, elevated N levels can promote soil microbial activity but can also lead to leaching and water pollution (Roper et al., 2017). Therefore, monitoring both macronutrients and micronutrients is essential for ensuring favorable soil health.

Factors Influencing Soil Health

Soil health is influenced by several biotic and abiotic factors. Management practices such as crop rotation, tillage, organic amendments, and the use of cover crops are critical to maintaining or improving soil health. For example, reduced tillage practices and the use of cover crops enhance SOM content, improve soil structure, and promote beneficial microbial activity (Miner et al., 2020). Conversely, intensive tillage, monoculture cropping, and excessive pesticide use can degrade soil health by disrupting soil organisms and reducing SOM content (Idowu et al., 2009).

External factors such as climate, landscape position, and soil parent material also play essential roles. Climate factors, particularly rainfall and temperature, influence soil moisture levels and microbial activity. In drier regions, soil health is often constrained by water availability, while in wetter regions, excessive runoff can lead to soil erosion and high percolation to increase in nutrient leaching. Furthermore, soil health is also strongly influenced by natural disturbances such as floods and droughts, which can temporarily degrade soil conditions but may also contribute to long-term soil resilience (Mann et al., 2019).

Human activities, particularly agricultural practices, have a profound impact on soil health. Urbanization, deforestation, and industrial activities often lead to soil compaction, contamination, and loss of biodiversity. As the human population grows, understanding how to manage soil health sustainably becomes even more critical. Karlen et al. (2019) emphasize that promoting sustainable land management practices, including the adoption of conservation agriculture (CA) and regenerative agriculture (RA) based on agroecological principles (Lal, 2020), is key to restoring and maintaining soil health in agricultural landscapes.

MECHANISMS LINKING SOIL HEALTH TO PLANT DISEASE DEFENSIVENESS

Soil Microbiome and Plant Pathogen Interaction

The soil microbiome is a critical determinant of plant health and disease resistance (Ortiz & Sansinenea, 2022). Beneficial soil microorganisms, including bacteria, fungi, and actinomycetes, play significant roles in the interaction between plants and pathogens. These microorganisms can directly inhibit the growth of harmful pathogens or induce systemic resistance in plants. For example, soil bacteria like *Bacillus* spp. and *Pseudomonas* spp. produce antimicrobial compounds that suppress soil-borne pathogens such as *Fusarium* spp. and *Rhizoctonia* spp. (Ortiz & Sansinenea, 2022). Mycorrhizal fungi such as *Glomus* spp. form symbiotic relationships with plant roots, improving nutrient uptake and enhancing plant resilience against root pathogens (Abawi & Widmer, 2000). A study by Samaddar et al. (2021) found that an increase in microbial diversity in soils is associated with a 30–50% reduction in plant disease incidence, particularly in the case of pathogens like *Verticillium* spp. and *Pythium* spp., which are common causes of root diseases. Samaddar et al. (2021) reported that while increased microbial diversity typically correlates with reduced disease incidence, the degree of suppression varies depending on the environmental conditions (e.g., soil moisture, temperature) and specific microbial community composition. Additionally, while beneficial microbes can suppress pathogens, this is not always guaranteed, especially under conditions of extreme soil degradation.

Soil-Induced Plant Immunity

Soil health can trigger plant immunity through complex interactions with the rhizosphere. Plants can recognize and respond to signals from beneficial microbes in the soil, activating their systemic innate immune systems, a process known as induced systemic resistance (ISR) (Ortiz & Sansinenea, 2022). ISR can trigger broader-spectrum and long-lasting plant disease resistance against diverse pathogens. Certain beneficial soil-borne bacteria and fungi, such as *Trichoderma* spp. and

Rhizobium spp., are known to stimulate ISR by producing elicitors that activate plant systemic defense pathways through the whole plants. Larkin (2015) highlighted that ISR is very effective in preventing root associated diseases caused by pathogens such as *Fusarium* spp. and *Pythium* spp., which often enter plants through the roots. Larkin also provided statistical data showing that the application of beneficial microbes in soils can reduce disease severity by up to 40–60% in crops like tomatoes and potatoes, especially when treated with *Trichoderma* spp. or *Bacillus subtilis*. In trials conducted on soil-infested with *Fusarium oxysporum*, ISR treatments resulted in a 50% reduction in disease symptoms compared to untreated control plants.

Research by Cardoso et al. (2013) indicated that organic amendments like compost can enhance ISR by promoting microbial populations that facilitate these responses. The study reports a 30% increase in the resistance of crops like lettuce and cabbage to *Pythium* and *Rhizoctonia* root rot when organic matter is added to the soil. However, the study also noted that the impact of organic amendments on ISR is contingent on soil texture, with sandy soils showing a 20% weaker response compared to loamy soils.

Nutrient Availability and Disease Resistance

Nutrient availability in soil is a key factor influencing plant disease resistance. A balanced supply of macronutrients and micronutrients is essential for plant growth and immune functions. Nutrient deficiencies, such as N, can weaken defenses and increase susceptibility to pathogens like *Fusarium* spp. and *Rhizoctonia* spp. (Larkin, 2015). In a controlled study, Six et al. (2000) found that N-deficient plants exhibited 40% higher susceptibility to *Rhizoctonia solani* root rot when balanced nutrients were met. Fu et al. (2021) revealed that leguminous plants used in crop rotations can significantly enhance N fixation, enriching the soil with this critical nutrient, which in turn improves disease resistance in subsequent crops. The study found that in a rotation system involving legumes, N levels in the soil increased by 25–30%, resulting in a 15–20% decrease in disease incidence of *Fusarium* spp. and *Pythium* spp. in the following crops.

Kopittke et al. (2024) found that plants in soils with pH levels between 6.0 and 7.5 had 30% higher resistance to soil-borne pathogens compared to those in acidic or alkaline soils, suggesting the significance of soil pH in the nutrient availability and plant disease resistance.

Soil Quality and Plant Health

Six et al. (2000) reported that soils with good aggregation properties have a 30–40% higher microbial activity than those with poor aggregation, which directly correlates with increased disease resistance in crops like corn and soybeans. Larkin (2015) provided statistical data showing that no-tillage (NT) systems, which improve soil aggregation, reduced the incidence of *Rhizoctonia solani* and *Pythium* spp. root diseases by up to 50% in both maize and soybean crops. Additionally, NT systems were associated with a 25% increase in SOM content over five years, which was linked to improved soil health and plant disease resistance. Furthermore, soil management practices such as organic mulching, cover cropping, and composting can significantly improve soil structure. In long-term studies by Liptzin et al. (2022), soils receiving organic amendments had a 40–60% greater microbial biomass and a 25–30% reduction in plant diseases compared to conventionally managed soils. This illustrates the importance of SOM in maintaining soil health and preventing the proliferation of pathogens.

IMPACT OF PLANT DISEASE RESISTANCE ON HUMAN HEALTH

The development of disease-resistant crops is a critical strategy for ensuring food security and reducing the reliance on chemical pesticides. According to Brevik et al. (2020), resistant crops can reduce the need for harmful chemical treatments, which in turn lowers the risks associated with pesticide residues in food. Pesticides have been linked to numerous human health problems, including neurological disorders, endocrine disruption, and cancer (Blair et al., 2015). By enhancing plant resistance to diseases such as blight, rust, and mildew, agricultural systems can minimize pesticide usage, thereby reducing exposure to these harmful chemicals (Rani et al., 2021). Moreover, plant disease resistance can help increase agricultural productivity and crop yields, which is crucial in meeting the nutritional needs of growing global populations (Mena & Angelino, 2020). Ensuring the health of crops through resistance to diseases contributes not only to food security but also to the nutritional quality of the food that reaches consumers. As Brevik et al. (2020) pointed out, soil health plays a significant role in the nutrient content of plants. Healthy soils support diverse microbial communities that enhance nutrient availability and uptake by plant roots. However, diseased plants may be less effective at absorbing these nutrients, which can lead to deficiencies in human diets. This is particularly concerning in regions where populations rely heavily on crops such as rice, wheat, and maize, which may be more susceptible to pathogens and diseases. Plant disease resistance can help mitigate these nutritional losses, ensuring that crops remain rich in vital nutrients like iron, zinc, and vitamins, which are essential for human health (Mena & Angelino, 2020).

Beyond nutrient content, plant disease resistance can also affect the safety of the food supply. Some pathogens, such as fungi and bacteria, produce mycotoxins or other toxic compounds that can contaminate crops and pose serious health risks to humans. Mycotoxin contamination is a significant concern, as these toxins are often present in food even after harvesting, leading to long-term exposure through diet. Disease-resistant plants can be less susceptible to pathogen infections, thus reducing the potential for mycotoxin production and making food safer for consumption (Samaddar et al., 2021). For instance, disease-resistant corn varieties can help mitigate the risk of Aspergillus species infection, which is a primary producer of aflatoxins, highly carcinogenic compounds that pose a significant threat to human health (Pathak et al., 2022). By incorporating disease resistance traits into crops, agricultural systems can lower the risk of mycotoxin contamination and protect human populations from the health risks associated with these harmful substances.

In addition to direct health impacts, plant disease resistance also plays a role in reducing the broader environmental consequences that can affect human health. Soil health, as discussed by Brevik et al. (2020), is essential for maintaining ecosystem services like water filtration, nutrient cycling, and carbon sequestration. Plant diseases can degrade soil health by aggravating risks of soil erosion, reducing soil fertility, and increasing the prevalence of soil-borne pathogens, which can have downstream effects on human populations. For example, diseases that reduce plant cover or root systems can increase losses by soil erosion, which can degrade water quality and lead to sedimentation of water bodies used for drinking or irrigation. This degradation can affect human health by increasing the risks of waterborne diseases (Kopittke et al., 2024). Disease-resistant crops can help preserve soil health by reducing plant mortality and maintaining healthy root systems for higher crop yield and quality, thus protecting both the environment and human health.

Plant disease resistance also intersects with the growing concerns about antimicrobial resistance (AMR), a significant global health threat. Wall et al. (2015) and Samaddar et al. (2021) both noted that the overuse of antibiotics and other antimicrobial agents in agriculture is contributing to the spread of AMR, which can have serious consequences to human health. In agriculture, the use of antibiotics to prevent or treat plant pathogens and diseases may lead to the development of resistant bacteria and fungi in the soil and on crops. These resistant bacteria and fungi can then enter the human food chain, potentially leading to infections that are more difficult to treat. Developing disease-resistant plants can help reduce the need for antimicrobial agents, thereby mitigating the risk of AMR. By focusing on biological resistance mechanisms within plants, farmers can reduce their reliance on chemical treatments and minimize the potential for AMR to spread through agricultural systems (Blair et al., 2015).

Moreover, the hygiene hypothesis, as discussed by Wall et al. (2015), suggests that exposure to diverse microorganisms in the environment, including those found in soil and plants, plays a critical

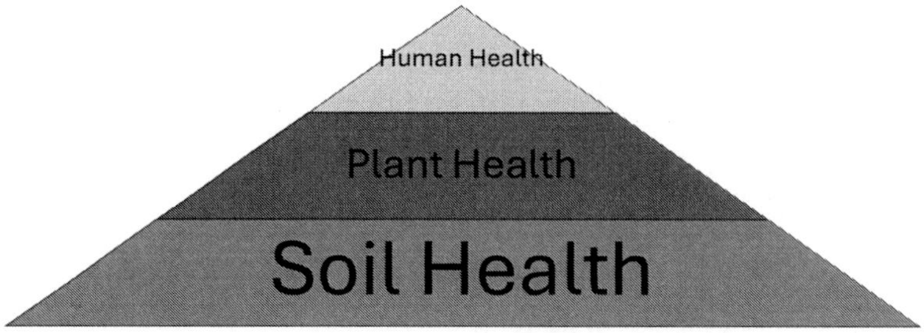

FIGURE 11.1 Pyramid depicting the importance of soil health in a food chain.

FIGURE 11.2 Process model showing the effect of soil-borne diseases on humans.

role in the development of a healthy immune system. While the overuse of pesticides and fungicides in agriculture can limit microbial diversity in soil and on plants, the cultivation of disease-resistant plants can support a balanced microbial ecosystem. By promoting disease resistance in crops through ecological methods, such as crop rotation and agroecological practices, agricultural systems can maintain microbial diversity in the soil and plants or on plant surfaces. This microbial diversity can, in turn, have positive effects on human health by supporting immune function and reducing the incidence of autoimmune diseases and allergies (Wall et al., 2015).

Climate change is expected to exacerbate plant diseases by altering environmental conditions such as temperature and humidity, which can favor the proliferation of certain pathogens (Kopittke et al., 2024). In response, plant breeders are increasingly developing disease-resistant varieties that are better suited to these changing conditions. However, the resilience of these crops will depend on the underlying health of the soil and the broader agricultural ecosystem. As Brevik and Sauer (2015) note, future research should focus on developing integrated strategies that enhance the resilience of both crops and soils to the combined stresses of disease and climate change. By ensuring that plant disease resistance is part of a broader strategy that includes soil health, crop diversity, and sustainable farming practices, human populations can be better prepared for the health challenges posed by climate change (Figures 11.1 and 11.2).

SOIL HEALTH MANAGEMENT FOR ENHANCING PLANT DISEASE RESISTANCE

Soil health management can improve soil structure, lessen soil compaction, increase aeration, and increase SOM content, which will help retain the beneficial soil microflora and microfauna, resulting in better yield and health of crops (Abawi and Widmer, 2000; Larkin, 2015). Management practices such as crop rotations, cover crops, green manures, organic amendments, CA, and use of bio stimulants like AMF have the potential to manage soil-borne diseases through mechanisms like increasing soil microbial biomass, activity, and diversity, resulting in greater biological suppression of pathogens and diseases (Larkin, 2015). The increase in the soil microbial population leads to an increase in SOM content. Approximately 45–60% of the SOM consists of SOC, which is the primary source of energy for the soil biota. The SOM has three prime components: plant/animal residues and microbial biomass, active and labile SOM, and relatively stable SOM (Lal, 2016).

Soil Biodiversity as a Driver of Soil Health

Soil biodiversity is a crucial indicator and driver of soil health. A diverse and balanced soil ecosystem supports the resilience of soil to disturbances and enhances its functionality in nutrient cycling, pest regulation, and organic matter decomposition. The soil microbiome, which includes bacteria, fungi, protozoa, and nematodes, plays an indispensable role in maintaining soil health by promoting nutrient availability and protecting plants from pathogens (Lehman et al., 2015).

Soil biodiversity is also vital for improving soil C sequestration. Higher microbial diversity increases SOM decomposition and C storage in soils, helping to mitigate climate change by reducing atmospheric carbon dioxide levels. According to Norris et al. (2020), healthy soils with a rich microbial community are better at storing C, with soils under sustainable management practices, such as reduced tillage and organic farming, sequestering up to 50% more C than those under conventional management. Thus, soil biodiversity not only enhances soil health but also contributes to global C cycling, making it a key factor in both soil and environmental health.

Furthermore, the diversity of soil organisms enhances soil structure and fertility (Mann et al., 2019). Earthworms, for example, improve soil aeration and water retention through their burrowing activities, while mycorrhizal fungi form symbiotic relationships with plant roots, enhance water and nutrient uptake. Mann et al. (2019) highlight the importance of these organisms in maintaining soil aggregation, which helps prevent erosion and enhances soil's ability to retain moisture during dry periods.

Importance of Soil C-N Dynamics

The key element governing the processes in soil is the soil C-pool. The soil C pool consists of the SOC pool and the SIC pool. SOC pool is the determinant of many soil functions and ecosystem benefits. The fate of soil health heavily relies on SOM and SOC pools (Lal, 2016). SOM is also a major source of N and cations that influence the CEC of the withered soils. The dynamics of SOM and nutrient cycling are closely related through microbial processes that involve both the trapping of nutrients in SOM and their subsequent release through mineralization (Belsky et al., 1989, Mazzarino et al., 1993). The C-N components of the soil are crucial as they govern some of the major soil processes and affect the soil properties. Soils with a higher C:N ratio have an active biota, good soil physical and chemical properties. However, the C-N fraction of the soil is sensitive and dynamic in nature, easily affected by climate change, crop production activities (fertigation, organic amendments, biofertilizers), leaching, soil erosion and environmental pollution (Qaswar et al., 2022).

Impact of C-N Dynamics on Soil Structure

Soil structure plays a crucial role in soil functions, affecting its capacity to support plant and animal life to influence environmental quality, soil water quality, and soil carbon sequestration. Aggregate stability is an indicator of soil structure whereas aggregation is formed due to the flocculation and cementation of the soil particles which is facilitated by binding agents like clays, ionic bridges, SOC, biota, and soil carbonates (Six et al., 2000, Duilker et al., 2003, Bronick & Lal, 2005). Soil biota primarily consists of organic matter which contributes to the SOC pool. The residence time of SOC in soil and the decomposition rate of organic matter influence the formation of soil structure (Bronick and Lal, 2005).

"The C:N ratio is commonly used as an indicator of SOC turnover, although for CR and inert C (IC) fractions, the lignin/N ratio, or other more resistant compounds may be more appropriate" (Vanlauwe et al., 1997). The soil structure influences the decomposition of SOM through factors like gas exchange, porosity, soil moisture, and the physical placement of carbon, such as its depth and occlusion. Turnover dynamics differ across soil regions, individual aggregates, and over time, with SOC, the turnover rate often slows down in macroaggregates towards the end of the growing season (Bronick & Lal, 2005). Since C-N processes affect the soil structure to a greater extent, it is important to maintain an optimum soil C:N ratio. This could be achieved by soil management practices like cover cropping, reduced or zero tillage, organic amendments, biofertilizer and bio stimulant incorporation in soil, green manuring, etc.

C-N Dynamics and Soil Aggregation Affected by Cover Crops and Bio Stimulants Synergy

Management of soil structure is crucial to achieving optimum soil health. A well-aggregated soil has a crumb structure, is porous, contains a higher level organic matter with high water retention properties, and provides improved soil aeration. An integrated soil management practice involving reduced tillage, cover cropping, AMF inoculation, and prescribed crop rotation will help not only to achieve the desired soil structure but also improve the inherent disease defensiveness of plants and result in higher yield. Kumar et al. (2022) studied the effects of biofertilizers, farmyard manure, and green manure on aggregate distribution, C stability in clay organic complexes, aggregate concentration, and labile organic carbon fractions in a long-term rice research experiment on some Inceptisols. They indicated how different types of organic amendments could be coupled to enhance clay organic complexes, soil aggregation, and associated carbon pools in the soil. They evaluated the glomalin-related protein that could be exploited to derive insights into the soil carbon pools. Kohler et al. (2017) examined the effects of three species of AMF on the formation of stable soil aggregates in the semi-arid regions of Mazarron and Rellano in Spain with variable soil physical characteristics such as soil textures and soil calcium carbonates. Their results indicated that the AMF strain *R. irregularis* increased the soil aggregates significantly in the two different types of soils, independent of the presence of calcium carbonates and soil textures. Similarly, in a study conducted by Siddiky et al. (2012) on the interaction between AMF mycelium and *Collembola* and their effect on soil aggregation in the absence of plant roots, they found a higher percentage of soil aggregation with a greater mean weight diameter in the soil hyphal compartments having both AMF and *Collembola* compared to individual AMF and *Collembola* controls. These studies explore not only a vital interaction between the soil structures and soil microbes but also emphasizes their interdependent roles.

Like AMF and biofertilizers, cover crops also help to sustain soil health by adding organic matter to the soil, which enhances the soil aggregation. Moore et al. (2014) studied the effects of winter cover crops on soil quality in a nine-year-old soybean and corn silage crop rotation system in the upper Midwest. Their experiment indicated a significant increase of 5% in SOM, 44% of particulate organic matter (POM), and 38% of potentially mineralizable nitrogen (PMN) in the soil after the incorporation of cover crops following the main crops harvest. This study by Moore et al. (2014) showed how cover crops can improve the soil carbon and soil N fractions. An increase in the soil carbon and N pools improve soil structure by increasing its moisture retention ability. Moreover, well-drained soils act as a reservoir of beneficial organisms to benefit both soil and plants.

Other Approaches to Increase Plant Disease Resistance and Enhance Soil Health

Crop rotation is a key practice for managing soil-borne pathogens. It disrupts pathogen life cycles by alternating host plants, which reduces pathogen populations that are specialized to crops (Peters et al., 2004). They reported that the rotation with non-host crops can significantly reduce the severity of diseases caused by pathogens such as *Verticillium* and *Fusarium*. Additionally, they highlighted that proper rotation can suppress diseases through changes in soil microflora, making the environment less conducive to pathogen survival. Similarly, Ghorbani et al. (2008) emphasized that rotation with legumes or grasses can help break pathogen cycles, particularly in the case of root rot pathogens, by altering the microbial community structure and promoting beneficial soil organisms.

The addition of organic amendments, such as compost, green manure, and animal manures, is another effective strategy to suppress soil-borne diseases (Bailey and Lazarovits, 2003). They discuss how these amendments can improve soil structure, enhance microbial diversity, and stimulate the growth of antagonistic microorganisms that inhibit pathogens. The incorporation of organic matter promotes the establishment of a healthy soil microbiome, which competes with or directly antagonizes harmful pathogens. Bailey and Lazarovits found that soils amended with compost showed a significant reduction in diseases caused by *Rhizoctonia solani* and *Pythium* spp. This approach is supported by Linderman (1989), who noted that organic amendments, particularly those rich in nutrients, fostered populations of beneficial fungi and bacteria that suppress soil-borne pathogens by outcompeting them or producing antimicrobial metabolites (Linderman, 1989).

Residue management is another essential component of soil disease management. Proper handling of crop residues through practices like tilling or mulching can affect the persistence of soil-borne pathogens (Katan, 2000) and they identified that burying or burning crop residues reduced the inoculum levels of certain pathogens, but excessive tillage disrupted soil structure and promoted pathogen spread by disturbing the microbial community. However, residue retention, as advocated by Peters et al. (2004), can enhance soil health by contributing organic matter, fostering microbial diversity, and improving soil structure, which in turn helps reduce pathogen prevalence. Peters and colleagues observed that NT or CA systems, when combined with appropriate residue management, can reduce the severity of diseases like black scurf in potatoes.

RESEARCH GAPS

Studies on soil management practices, such as crop rotation, organic amendments, and residue management, have demonstrated their effectiveness in controlling soil-borne pathogens and diseases (Bailey & Lazarovits, 2003; Peters et al., 2004). However, much of the research has been region-specific or limited to specific crops. For example, while the role of crop rotation in suppressing *Verticillium* and *Fusarium* species is well-documented in temperate climates, its effectiveness in tropical or arid regions remains underexplored, given the distinct soil and climatic conditions in these areas. Similarly, the response of different crops such as vegetables, legumes, and cereals to organic amendments and residue management has not been comprehensively assessed. This highlights the need for broader studies that investigate the efficacy of these soil management practices across diverse ecosystems and crop types, particularly in regions where soil-borne diseases are emerging due to climate change.

Further, while the role of soil microbial communities in disease suppression is well-established (Ghorbani et al., 2008; Bailey & Lazarovits, 2003), the specific microbial mechanisms underlying this process are still not fully understood. While organic amendments are known to increase microbial diversity, the exact microorganisms responsible for pathogen suppression and their interactions with different beneficial microbes and pathogens remain unclear. More detailed studies are required to identify beneficial microbes and explore how their populations are influenced by soil management practices. Understanding the complex interactions between these microbial communities and pathogens is critical for developing sustainable disease control strategies.

Another research gap lies in the long-term effects of soil management practices. Most studies focused on the short-term outcomes of disease suppression and crop yield improvements, but the long-term impacts on soil health, microbial biodiversity, and pathogen resistance remain less understood. Longitudinal studies tracking changes in soil properties, microbial dynamics, and disease resistance over several growing seasons would provide valuable insights into the sustainability of these practices. For example, practices like NT farming and organic amendments may suppress disease in the short term, but there is limited data on whether these benefits are sustained in the long run, particularly in terms of soil fertility and microbial resilience. Such studies could also inform when and how soil management practices need to be adjusted to maintain soil health and disease suppression. The studies describing the effect of soil carbon on the disease suppression is also limited.

Furthermore, the integration of soil management with other disease control strategies, such as chemical treatments, biocontrol agents, and resistant crop varieties, has not been sufficiently explored. While integrated pest management (IPM) has been extensively studied, integrated disease management (IDM) focusing on soil-borne pathogens is underdeveloped. Research into how soil management practices can be combined with other control methods to create holistic disease management systems could lead to more sustainable practices that reduce the reliance on chemical inputs and lower the environmental impact of agriculture.

Additionally, the scalability and economic feasibility of these practices, particularly for smallholder and resource-poor farmers, have not been sufficiently addressed. While many studies show promising results under controlled conditions (Peters et al., 2004; Ghorbani et al., 2008), real-world adoption often faces economic and practical constraints. Cost-benefit analyses of soil management practices, considering different farm sizes and resource availability, are needed to assess the broader applicability and economic sustainability of these practices.

Finally, the impact of climate change on soil-borne diseases and the efficacy of soil management practices under changing environmental conditions represents an urgent area of research. Shifting temperature and precipitation patterns may alter the prevalence of soil-borne pathogens, necessitating adjustments to current management strategies. Further research is needed to predict how soil-borne pathogens will respond to climate change and how management practices can be adapted to mitigate these risks, especially as new pathogens emerge.

CONCLUSION

This chapter provides evidence that improving soil health through sustainable management practices enhances plant disease resistance, establishes a strong linkage between soil, plant, and human health, and optimizes soil carbon and nitrogen dynamics. The findings confirm that these factors collectively contribute to overall ecosystem resilience and agricultural sustainability. Soil health management is crucial for sustainable agriculture, as it promotes strong plant growth, boosts disease resistance, and enhances resilience to pests. Strategies such as cover crops, bio-stimulants, biofertilizers, and NT farming play a key role in improving soil health. Cover crops enhance soil structure, increase SOM content, and support beneficial microbes that help control soil-borne pathogens and reduce disease. Bio-stimulants, like AMF, strengthen plant root system, improve water and nutrient uptake, and enhance plant defense mechanisms. By integrating these practices, farmers can create a healthier soil ecosystem that naturally supports plant health, reduces reliance on agrochemicals, and fosters more sustainable, resilient agricultural systems. However, while plant disease resistance is essential for food security and human health, significant research gaps remain, including better integration of soil health with disease resistance, understanding the impact of disease resistance on nutrient density, and addressing the intersection of plant disease resistance with climate change and antimicrobial resistance. Bridging these gaps will help develop more sustainable farming practices that protect both human and environmental health.

BIBLIOGRAPHY

Abawi, G. S., Widmer, T. L. Impact of Soil Health Management Practices on Soilborne Pathogens, Nematodes and Root Diseases of Vegetable Crops. *Applied Soil Ecology* 2000, *15* (1), 37–47. https://doi.org/10.1016/S0929-1393(00)00070-6

Al-Maliki, S., Ebreesum, H. Changes in Soil Carbon Mineralization, Soil Microbes, Roots Density and Soil Structure Following the Application of the Arbuscular Mycorrhizal Fungi and Green Algae in the Arid Saline Soil. *Rhizosphere* 2020, *14*, 100203. https://doi.org/10.1016/j.rhisph.2020.100203

Bailey, K. L., Lazarovits, G. Suppressing Soil-Borne Diseases with Residue Management and Organic Amendments. *Soil and Tillage Research* 2003, *72* (2), 169–180. https://doi.org/10.1016/S0167-1987(03)00086-2

Belsky, A. J., Amundson, R. G., Duxbury, J. M., Riha, S. J., Ali, A. R., Mwonga, S. M. The Effects of Trees on Their Physical, Chemical and Biological Environments in a Semi-arid Savanna in Kenya. *The Journal of Applied Ecology*, 1989, *26* (3), 1005. https://doi.org/10.2307/2403708

Bimüller, C., Kreyling, O., Kölbl, A., Von Lützow, M., Kögel-Knabner, I. Carbon and Nitrogen Mineralization in Hierarchically Structured Aggregates of Different Size. *Soil and Tillage Research* 2016, *160*, 23–33. https://doi.org/10.1016/j.still.2015.12.011

Blair, A., Ritz, B., Wesseling, C., Beane Freeman, L. Pesticides and Human Health. *Occupational and Environmental Medicine* 2015, *72* (2), 81–82. https://doi.org/10.1136/oemed-2014-102454

Bonilla, N., Gutiérrez-Barranquero, J., Vicente, A., Cazorla, F. Enhancing Soil Quality and Plant Health Through Suppressive Organic Amendments. *Diversity* 2012, *4* (4), 475–491. https://doi.org/10.3390/d4040475

Brevik, E. C., Sauer, T. J. The Past, Present, and Future of Soils and Human Health Studies. *The Soil* 2015, *1* (1), 35–46. https://doi.org/10.5194/soil-1-35-2015

Brevik, E. C., Slaughter, L., Singh, B. R., Steffan, J. J., Collier, D., Barnhart, P., Pereira, P. Soil and Human Health: Current Status and Future Needs. *Air, Soil and Water Research*, 2020, 13, 1178622120934441. https://doi.org/10.1177/1178622120934441

Bronick, C. J., Lal, R. Soil Structure and Management: A Review. *Geoderma*, 2005, *124* (1–2), 3–22. https://doi.org/10.1016/j.geoderma.2004.03.005

Cardoso, E. J. B. N., Vasconcellos, R. L. F., Bini, D., Miyauchi, M. Y. H., Santos, C. A. D., Alves, P. R. L., Paula, A. M. D., Nakatani, A. S., Pereira, J. D. M., Nogueira, M. A. Soil Health: Looking for Suitable Indicators. What Should Be Considered to Assess the Effects of Use and Management on Soil Health? *Scientia Agricola*, 2013, *70* (4), 274–289. https://doi.org/10.1590/S0103-90162013000400009

Carrera, L. M., Buyer, J. S., Vinyard, B., Abdul-Baki, A. A., Sikora, L. J., Teasdale, J. R. Effects of Cover Crops, Compost, and Manure Amendments on Soil Microbial Community Structure in Tomato Production Systems. *Applied Soil Ecology* 2007, *37* (3), 247–255. https://doi.org/10.1016/j.apsoil.2007.08.003

Cavagnaro, T. R., Barrios-Masias, F. H., Jackson, L. E. Arbuscular Mycorrhizas and Their Role in Plant Growth, Nitrogen Interception and Soil Gas Efflux in an Organic Production System. *Plant and Soil* 2012, *353* (1–2), 181–194. https://doi.org/10.1007/s11104-011-1021-6

Chahal, I., Van Eerd, L. L. Cover Crop and Crop Residue Removal Effects on Temporal Dynamics of Soil Carbon and Nitrogen in a Temperate, Humid Climate. *PLoS One* 2020, *15* (7), e0235665. https://doi.org/10.1371/journal.pone.0235665

Chahal, I., Vyn, R. J., Mayers, D., Van Eerd, L. L. Cumulative Impact of Cover Crops on Soil Carbon Sequestration and Profitability in a Temperate Humid Climate. *Scientific Reports* 2020, *10* (1), 13381. https://doi.org/10.1038/s41598-020-70224-6

Chen, W., Modi, D., Picot, A. Soil and Phytomicrobiome for Plant Disease Suppression and Management under Climate Change: A Review. *Plants* 2023, *12* (14), 2736. https://doi.org/10.3390/plants12142736

Doran, J., Sarrantonio, M., Liebig, M., Sparks, D. 1996. Soil Health and Sustainability. *Advances in Agronomy 56*, 1–54.

Doran, J. W., Zeiss, M. R. Soil Health and Sustainability: Managing the Biotic Component of Soil Quality. *Applied Soil Ecology*, 2000, *15* (1), 3–11. https://doi.org/10.1016/S0929-1393(00)00067-6

Duiker, S.W., Rhoton, F.E., Torrent, J., Smeck, N.E., Lal, R., 2003. Iron (hydr)oxide Crystallinity Effects on Soil Aggregation. *Soil Science Society of America Journal*. 67, 606–611.

Fall, A. F., Nakabonge, G., Ssekandi, J., Founoune-Mboup, H., Apori, S. O., Ndiaye, A., Badji, A., Ngom, K. Roles of Arbuscular Mycorrhizal Fungi on Soil Fertility: Contribution in the Improvement of Physical, Chemical, and Biological Properties of the Soil. *Frontiers in Fungal Biology* 2022, *3*, 723892. https://doi.org/10.3389/ffunb.2022.723892

Frey, S. D. Mycorrhizal Fungi as Mediators of Soil Organic Matter Dynamics. *Annual Review of Ecology, Evolution, and Systematics* 2019, *50* (1), 237–259. https://doi.org/10.1146/annurev-ecolsys-110617-062331

Fu, B., Chen, L., Huang, H., Qu, P., Wei, Z. Impacts of Crop Residues on Soil Health: A Review. *Environmental Pollutants and Bioavailability* 2021, *33* (1), 164–173. https://doi.org/10.1080/26395940.2021.1948354

García-González, I., Quemada, M., Gabriel, J. L., Alonso-Ayuso, M., Hontoria, C. Legacy of Eight-year Cover Cropping on Mycorrhizae, Soil, and Plants. *Journal of Plant Nutrition and Soil Science* 2018, *181* (6), 818–826. https://doi.org/10.1002/jpln.201700591

Geisseler, D., Horwath, W. R. Short-Term Dynamics of Soil Carbon, Microbial Biomass, and Soil Enzyme Activities as Compared to Longer-Term Effects of Tillage in Irrigated Row Crops. *Biology and Fertility of Soils* 2009, *46* (1), 65–72. https://doi.org/10.1007/s00374-009-0400-0

He, T., Zhang, X., Du, J., Gilliam, F. S., Yang, S., Tian, M., Zhang, C., Zhou, Y. Arbuscular Mycorrhizal Fungi Shift Soil Bacterial Community Composition and Reduce Soil Ammonia Volatilization and Nitrous Oxide Emissions. *Microbial Ecology* 2023, *85* (3), 951–964. https://doi.org/10.1007/s00248-023-02172-3

Idowu, O. J., Van Es, H. M., Abawi, G. S., Wolfe, D. W., Schindelbeck, R. R., Moebius-Clune, B. N., Gugino, B. K. Use of an Integrative Soil Health Test for Evaluation of Soil Management Impacts. *Renewable Agriculture and Food Systems* 2009, *24* (3), 214–224. https://doi.org/10.1017/S1742170509990068

Janvier, C., Villeneuve, F., Alabouvette, C., Edel-Hermann, V., Mateille, T., Steinberg, C. Soil Health through Soil Disease Suppression: Which Strategy from Descriptors to Indicators? *Soil Biology and Biochemistry* 2007, *39* (1), 1–23. https://doi.org/10.1016/j.soilbio.2006.07.001

Jayaraman, S., Naorem, A. K., Lal, R., Dalal, R. C., Sinha, N. K., Patra, A. K., Chaudhari, S. K. Disease-Suppressive Soils – Beyond Food Production: A Critical Review. *Journal of Soil Science and Plant Nutrition* 2021, *21* (2), 1437–1465. https://doi.org/10.1007/s42729-021-00451-x

Jian, J., Du, X., Reiter, M. S., Stewart, R. D. A Meta-Analysis of Global Cropland Soil Carbon Changes Due to Cover Cropping. *Soil Biology and Biochemistry* 2020, *143*, 107735. https://doi.org/10.1016/j.soilbio.2020.107735

Karlen, D. L., Veum, K. S., Sudduth, K. A., Obrycki, J. F., Nunes, M. R. Soil Health Assessment: Past Accomplishments, Current Activities, and Future Opportunities. *Soil and Tillage Research* 2019, *195*, 104365. https://doi.org/10.1016/j.still.2019.104365

Katan, J. Physical and Cultural Methods for the Management of Soil-Borne Pathogens. *Crop Protection* 2000, *19* (8–10), 725–731. https://doi.org/10.1016/S0261-2194(00)00096-X

Kohler, J., Roldán, A., Campoy, M., Caravaca, F. Unraveling the Role of Hyphal Networks from Arbuscular Mycorrhizal Fungi in Aggregate Stabilization of Semiarid Soils with Different Textures and Carbonate Contents. *Plant and Soil*, 2017, *410* (1–2), 273–281. https://doi.org/10.1007/s11104-016-3001-3

Kopittke, P. M., Menzies, N. W., Wang, P., McKenna, B. A., Lombi, E. Soil and the Intensification of Agriculture for Global Food Security. *Environment International*, 2019, *132*, 105078. https://doi.org/10.1016/j.envint.2019.105078

Kopittke, P. M., Minasny, B., Pendall, E., Rumpel, C., McKenna, B. A. Healthy Soil for Healthy Humans and a Healthy Planet. *Critical Reviews in Environmental Science and Technology* 2024, *54* (3), 210–221. https://doi.org/10.1080/10643389.2023.2228651

Krupek, F. S., Mizero, S. M., Redfearn, D., Basche, A. Assessing How Cover Crops Close the Soil Health Gap in On-farm Experiments. *Agricultural & Environmental Letters* 2022, *7* (2), e20088. https://doi.org/10.1002/ael2.20088

Kumar, D., Purakayastha, T. J., Das, R., Yadav, R. K., Shivay, Y. S., Jha, P. K., Singh, S., Aditi, K., Prasad, P. V. V. Long-Term Effects of Organic Amendments on Carbon Stability in Clay–Organic Complex and Its Role in Soil Aggregation. *Agronomy* 2022, *13* (1), 39. https://doi.org/10.3390/agronomy13010039

Lal, R. Soil Health and Carbon Management. *Food and Energy Security* 2016, *5* (4), 212–222. https://doi.org/10.1002/fes3.96

Lal, R. Regenerative Agriculture for Food and Climate. *Journal of Soil and Water Conservation.*, Aug.2020. https://doi.org/10.2489/jswc.2020.0620A

Larkin, R. P. Soil Health Paradigms and Implications for Disease Management. *Annual Review of Phytopathology*, 2015, *53* (1), 199–221. https://doi.org/10.1146/annurev-phyto-080614-120357

Lazarovits, G. Management of Soil-Borne Plant Pathogens with Organic Soil Amendments: A Disease Control Strategy Salvaged from the Past. *Canadian Journal of Plant Pathology* 2001, *23* (1), 1–7. https://doi.org/10.1080/07060660109506901

Lehman, R., Cambardella, C., Stott, D., Acosta-Martinez, V., Manter, D., Buyer, J., Maul, J., Smith, J., Collins, H., Halvorson, J., Kremer, R., Lundgren, J., Ducey, T., Jin, V., Karlen, D. Understanding and Enhancing Soil Biological Health: The Solution for Reversing Soil Degradation. *Sustainability* 2015, *7* (1), 988–1027. https://doi.org/10.3390/su7010988

Lehman, R. M., Taheri, W. I., Osborne, S. L., Buyer, J. S., Douds, D. D. Fall Cover Cropping Can Increase Arbuscular Mycorrhizae in Soils Supporting Intensive Agricultural Production. *Applied Soil Ecology* 2012, *61*, 300–304. https://doi.org/10.1016/j.apsoil.2011.11.008

Linderman, R. G. Organic Amendments and Soil-Borne Diseases. *Canadian Journal of Plant Pathology* 1989, *11* (2), 180–183. https://doi.org/10.1080/07060668909501136

Lundquist, E. J., Jackson, L. E., Scow, K. M., Hsu, C. Changes in Microbial Biomass and Community Composition, and Soil Carbon and Nitrogen Pools after Incorporation of Rye into Three California Agricultural Soils. *Soil Biology and Biochemistry* 1999, *31* (2), 221–236. https://doi.org/10.1016/S0038-0717(98)00093-5

Maharjan, B., Das, S., Acharya, B. S. Soil Health Gap: A Concept to Establish a Benchmark for Soil Health Management. *Global Ecology and Conservation* 2020, *23*, e01116. https://doi.org/10.1016/j.gecco.2020.e01116

Mann, C., Lynch, D., Fillmore, S., Mills, A. Relationships between Field Management, Soil Health, and Microbial Community Composition. *Applied Soil Ecology* 2019, *144*, 12–21. https://doi.org/10.1016/j.apsoil.2019.06.012

Mazzarino, M. J., Szott, L., Jimenez, M. Dynamics of Soil Total C and N, Microbial Biomass, and Water-Soluble C in Tropical Agroecosystems. *Soil Biology and Biochemistry* 1993, *25* (2), 205–214. https://doi.org/10.1016/0038-0717(93)90028-A

Mena, P., Angelino, D. Plant Food, Nutrition, and Human Health. *Nutrients* 2020, *12* (7), 2157. https://doi.org/10.3390/nu12072157

Miner, G. L., Delgado, J. A., Ippolito, J. A., Stewart, C. E. Soil Health Management Practices and Crop Productivity. *Agricultural & Environmental Letters* 2020, *5* (1), e20023. https://doi.org/10.1002/ael2.20023

Moonilall, N. I., Sklenka, K. A., Nocco, M. A., Lal, R. Rehabilitative Capacity of Amendments to Restore Maize Productivity Following Artificial Topsoil Erosion/ Deposition. *Field Crops Research* 2023, *304*, 109178. https://doi.org/10.1016/j.fcr.2023.109178

Moore, E. B., Wiedenhoeft, M. H., Kaspar, T. C., Cambardella, C. A. Rye Cover Crop Effects on Soil Quality in No-till Corn Silage-soybean Cropping Systems. *Soil Science Society of America Journal*, 2014, *78* (3), 968–976. https://doi.org/10.2136/sssaj2013.09.0401

Nicolopoulou-Stamati, P., Maipas, S., Kotampasi, C., Stamatis, P., Hens, L. Chemical Pesticides and Human Health: The Urgent Need for a New Concept in Agriculture. *Frontiers in Public Health* 2016, *4*. https://doi.org/10.3389/fpubh.2016.00148

Nieder, R., Benbi, D. K., Reichl, F. X. *Soil Components and Human Health*. Springer Netherlands: Dordrecht, 2018. https://doi.org/10.1007/978-94-024-1222-2

Norris, C. E., Bean, G. M., Cappellazzi, S. B., Cope, M., Greub, K. L. H., Liptzin, D., Rieke, E. L., Tracy, P. W., Morgan, C. L. S., Honeycutt, C. W. Introducing the North American Project to Evaluate Soil Health Measurements. *Agronomy Journal* 2020, *112* (4), 3195–3215. https://doi.org/10.1002/agj2.20234

Oliver, M. A. Soil and Human Health: A Review. *European Journal of Soil Science* 1997, *48* (4), 573–592. https://doi.org/10.1111/j.1365-2389.1997.tb00558.x

Ortiz, A., Sansinenea, E. The Role of Beneficial Microorganisms in Soil Quality and Plant Health. *Sustainability*, 2022, *14* (9), 5358. https://doi.org/10.3390/su14095358

Pathak, V. M., Verma, V. K., Rawat, B. S., Kaur, B., Babu, N., Sharma, A., Dewali, S., Yadav, M., Kumari, R., Singh, S., Mohapatra, A., Pandey, V., Rana, N., Cunill, J. M. Current Status of Pesticide Effects on Environment, Human Health and It's Eco-Friendly Management as Bioremediation: A Comprehensive Review. *Frontiers in Microbiology* 2022, *13*, 962619. https://doi.org/10.3389/fmicb.2022.962619

Pauwels, R., Graefe, J., Bitterlich, M. An Arbuscular Mycorrhizal Fungus Alters Soil Water Retention and Hydraulic Conductivity in a Soil Texture Specific Way. *Mycorrhiza*, 2023, *33* (3), 165–179. https://doi.org/10.1007/s00572-023-01106-8

Peters, R. D., Sturz, A. V., Carter, M. R., Sanderson, J. B. Influence of Crop Rotation and Conservation Tillage Practices on the Severity of Soil-Borne Potato Diseases in Temperate Humid Agriculture. *Canadian Journal of Soil Science* 2004, *84* (4), 397–402. https://doi.org/10.4141/S03-060

Qaswar, M., Li, D., Huang, J., Han, T., Ahmed, W., Ali, S., Khan, M. N., Khan, Z. H., Xu, Y., Li, Q., Zhang, H., Wang, B., Tauqeer, A. Dynamics of Organic Carbon and Nitrogen in Deep Soil Profile and Crop Yields Under Long-term Fertilization in Wheat-maize Cropping System. *Journal of Integrative Agriculture*, 2022, *21* (3), 826–839. https://doi.org/10.1016/S2095-3119(20)63501-8

Rani, L., Thapa, K., Kanojia, N., Sharma, N., Singh, S., Grewal, A. S., Srivastav, A. L., Kaushal, J. An Extensive Review on the Consequences of Chemical Pesticides on Human Health and Environment. *Journal of Cleaner Production* 2021, *283*, 124657. https://doi.org/10.1016/j.jclepro.2020.124657

Rillig, M. C., Mummey, D. L. Mycorrhizas and Soil Structure. *New Phytologist* 2006, *171* (1), 41–53. https://doi.org/10.1111/j.1469-8137.2006.01750.x

Roper, W. R., Osmond, D. L., Heitman, J. L., Wagger, M. G., Reberg-Horton, S. C. Soil Health Indicators Do Not Differentiate among Agronomic Management Systems in North Carolina Soils. *Soil Science Society of America Journal* 2017, *81* (4), 828–843. https://doi.org/10.2136/sssaj2016.12.0400

Sagova-Mareckova, M., Omelka, M., Kopecky, J. The Golden Goal of Soil Management: Disease-Suppressive Soils. *Phytopathology* 2023, *113* (4), 741–752. https://doi.org/10.1094/PHYTO-09-22-0324-KD

Sainju, U. M., Whitehead, W. F., Singh, B. P. Cover Crops and Nitrogen Fertilization Effects on Soil Aggregation and Carbon and Nitrogen Pools. *Canadian Journal of Soil Science* 2003, *83* (2), 155–165. https://doi.org/10.4141/S02-056

Samaddar, S., Karp, D. S., Schmidt, R., Devarajan, N., McGarvey, J. A., Pires, A. F. A., Scow, K. Role of Soil in the Regulation of Human and Plant Pathogens: Soils' Contributions to People. *Philosophical Transactions of the Royal Society B* 2021, *376* (1834), 20200179. https://doi.org/10.1098/rstb.2020.0179

Sequeira, C. H., Alley, M. M. Soil Organic Matter Fractions as Indices of Soil Quality Changes. *Soil Science Society of America Journal* 2011, *75* (5), 1766–1773. https://doi.org/10.2136/sssaj2011.0067

Shafique, H. A., Sultana, V., Ehteshamul-Haque, S., Athar, M. Management of Soil-borne Diseases of Organic Vegetables. *Journal of Plant Protection Research*, 2016, *56* (3), 221–230. https://doi.org/10.1515/jppr-2016-0043

Siddiky, Md. R. K., Kohler, J., Cosme, M., Rillig, M. C. Soil Biota Effects on Soil Structure: Interactions between Arbuscular Mycorrhizal Fungal Mycelium and Collembola. *Soil Biology and Biochemistry*, 2012, *50*, 33–39. https://doi.org/10.1016/j.soilbio.2012.03.001

Six, J., Elliott, E.T., Paustian, K. Soil Structure and Soil Organic Matter: II. A Normalized Stability Index and the Effect of Mineralogy. *Soil Science Society of America Journal*, 2000, *64*, 1042–1049.

Tahat, M. M. Alananbeh, K., Othman, Y., Leskovar, D. Soil Health and Sustainable Agriculture. *Sustainability*, 2020, *12* (12), 4859. https://doi.org/10.3390/su12124859

Tosi, M., Ogilvie, C. M., Spagnoletti, F. N., Fournier, S., Martin, R. C., Dunfield, K. E. Cover Crops Modulate the Response of Arbuscular Mycorrhizal Fungi to Water Supply: A Field Study in Corn. *Plants* 2023, *12* (5), 1015. https://doi.org/10.3390/plants12051015

Vanlauwe, B., Diels, J., Sanginga, N. Merckx, R. Residue quality and decomposition: an unsteady relationship? In G. Cadisch, K. Giller, *Driven by Nature: Plant Litter Quality and Decomposition* (pp. 141–146). CAB International: Wallingford, Oxon, UK, 1997.

Vincent, I. R., Rosskopf, E. N., Brecht, J. K., Dufault, N. S., Sandoya-Miranda, G., Zhao, X. Effects of Anaerobic Soil Disinfestation for Soilborne Disease and Weed Management on Baby Leaf Lettuce Performance in a High Tunnel Organic Production System. *Agronomy* 2024, *14* (4), 764. https://doi.org/10.3390/agronomy14040764

Wall, D. H., Nielsen, U. N., Six, J. Soil Biodiversity and Human Health. *Nature* 2015, *528* (7580), 69–76. https://doi.org/10.1038/nature15744

Weller, D. M., Raaijmakers, J. M., Gardener, B. B. M., Thomashow, L. S. Microbial Populations Responsible for Specific Soil Suppressiveness to Plant Pathogens. *Annual Review of Phytopathology*, 2002, *40* (1), 309–348. https://doi.org/10.1146/annurev.phyto.40.030402.110010

Woods, M., McAlister, J. A., Geddes-McAlister, J. A One Health Approach to Overcoming Fungal Disease and Antifungal Resistance. *WIREs Mechanisms of Disease* 2023, *15* (4), e1610. https://doi.org/10.1002/wsbm.1610

Zhang, J., Zhao, R., Li, X., Zhang, J. Potential of Arbuscular Mycorrhizal Fungi for Soil Health: A Review. *Pedosphere* 2024, *34* (2), 279–288. https://doi.org/10.1016/j.pedsph.2024.02.002

12 Interdisciplinary System Strategy to Human Health By Operationalization of the "One Health Concept"

Rattan Lal

THE NEED FOR A PARADIGM SHIFT

Soil, the super living organism capable of self-regulation, is increasingly being subject to land misuse and soil mismanagement. Furthermore, the super organism is prone to widespread soil degradation, pollution, and contamination. Addressing this serious global problem necessitates innovative cooperation among a range of disciplines including pedology, earth sciences, agronomy, plant physiology, pedochemistry, human ecology, health sciences, and the medical profession. Soil degradation, pollution, and contamination create a public health emergency that calls for a coherent and well-conceived intedisciplinary approach.

During the first quarter of the 21st century, primarily driven by a growing world population (8.2 billion people in 2025 and projected to reach 9.5 billion by 2050), and by people's aspirations to achieve affluent lifestyles, evolving global scenarios include: (1) accelerating anthropogenic climate change (ACC), (2) extremely fragile food systems, (3) increasing shortages of water, (4) aggravating political instability, (5) a growing risk of pandemic, (6) degrading soils, (7) polluted environments, (8) excessive resource use, (9) loss of biodiversity, and (10) endangered human health and well-being. These and other issues are being addressed from the human rather than nature's benefit. Soil health is the vital link in this entire chain of interconnectivity in the circular health paradigm (Figure 12.1). Just as soil can be useful or harmful to humans, anthropogenic activities can also restore or destroy soil health, leading to beneficial or adverse impacts on human health and well-being. Thus, a systematic and long-term study on land use and soil management is needed for principal ecoregions (biomes) with regard to some basic questions such as: What heals humans? What infects humans? What nourishes humans? What contaminates humans? (Pepper et al., 2009). The One Health strategy is based on the ideology that there is a strong interconnectivity between the health of soil, plants, animals, people, and the ecosystem is interconnected (Figure 12.2, Van Brussen et al., 2019). Indeed, soil (along with its biodiversity and other pedological processes) has an important influence on human health and is justifiably considered as "Nature's contribution to people" (Samaddar et al., 2021). Thus, there is a strong need to establish integrated, efficient, economic, cultural, and multifunctional approaches to managing and sustaining soil health. This integrated approach emphasizes the need for judicious management of diverse but interconnected domains (soil, plant, livestock, human, ecosystem, and planetary) because each of these domains affects human health and well-being (Vieweger and Döring, 2015). The United Nations recognized the importance of these domains in instigating the International Year of Soil (2015) and the International Year of Plants (2020). Thus, there is a growing interest in the interdisciplinary approach to human health in relation to soil health (Montgomery et al., 2024; El-Ramady et al., 2019; Gullino et al., 2024; Kopittke et al., 2024; Oliver and Gregory, 2015). In this context,

DOI: 10.1201/9781003469742-12

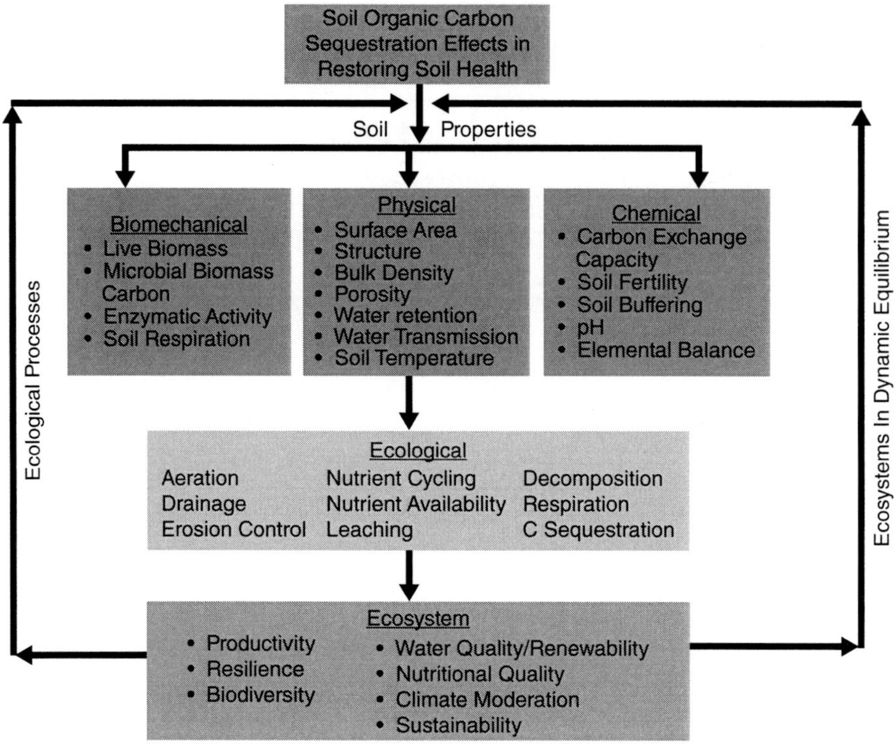

FIGURE 12.1 Restoring soil health and ecological processes by sequestration of organic carbon in degraded lands.

assessment of soil health must be shifted from exclusively physicochemical approaches to a holistic and interdisciplinary strategy (Zeng et al., 2025). The latter may include soil-health indicators based on ecological, biological, microbiological, pedological, and agronomic indicators. The growing interest in planetary health is attributed to the acknowledgement of a strong link between ecosystems, biodiversity, and human health (Zhu et al., 2019). Because food quality is strongly related to soil health (Alshaal et al., 2019), the objective of this chapter is to review some adverse effects of degraded, contaminated, and polluted soils on human health. Approaches to soil management for producing safe and healthy food are also discussed.

SOIL CONTAMINATION BY HEAVY METALS AND OTHER POLLUTANTS

Soil contamination by heavy metal pollution is a global issue which has adverse effects on functions of the super living organism (Hegedus et al., 2023). Although heavy metal contamination is a major cause of soil pollution, the use of plastic in agriculture (mulch) is also a significant cause of soil pollution (Table 12.1). Soil contamination by different sources is a worldwide issue and has emerged as a major challenge to global soil security (Carre et al., 2017). The trend of increasing nanoparticles, aggravated by industrial and waste-disposal activities (Nathanail, 2005) and rapid urbanization (Durdu et al., 2023; Adimalla, 2020) has exacerbated the problem of soil contamination with heavy metals.

APPROACHES TO SOIL MANAGEMENT IN THE CONTEXT OF HUMAN HEALTH

In some large-scale production systems, soil has become a more inert medium to grow crops (Krupenikov et al., 2011) with adverse effects on soil functions and human health. However, there is an increasing interest in soil health and its management. The success of any strategy of soil

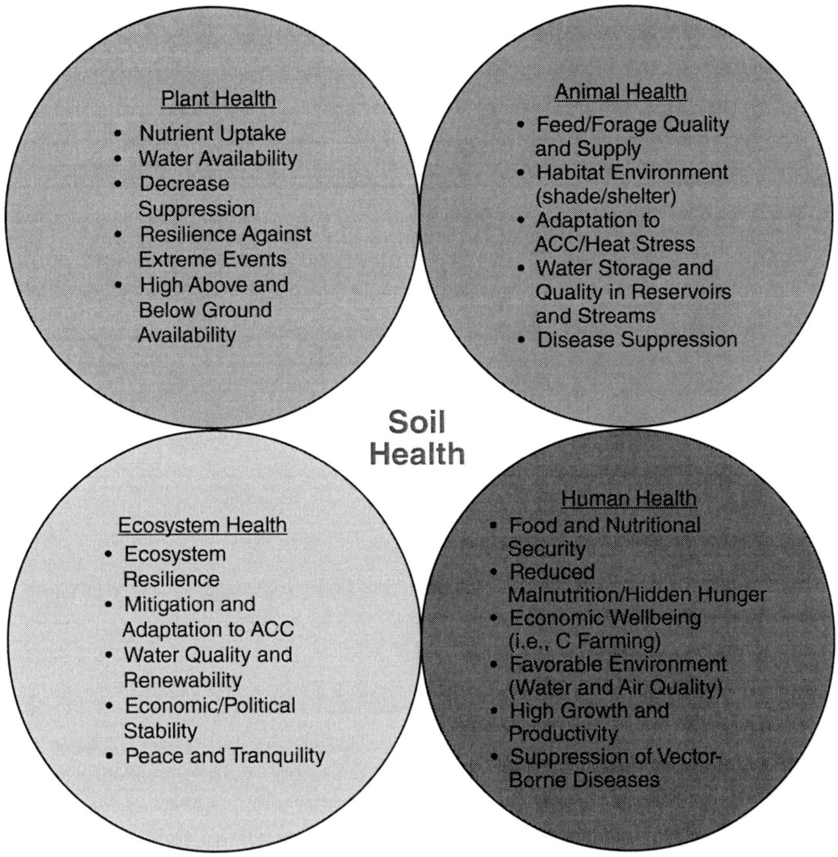

FIGURE 12.2 Principles of the One Health concept based on interconnectivity between soil, plant, animal, human, and ecosystem health.

decontamination or strategy for human health requires an interdisciplinary approach. Managing soil health (i.e., tilth, nutrient availability, available water capacity, activity and species diversity of soil biota, elemental imbalance, and toxicity) is relevant to several scientific disciplines (Wood and Blankinship, 2022) working together as an interdisciplinary team to address this issue of global significance. Human health depends on a safe, proper, and sufficient supply of plant nutrients. A systemic and in-depth understanding of soil health as affected by land use and management is critical in addressing emerging issues of human health in relation to heavy metal toxicity, micro-plastic pollution, water and air quality, micronutrient deficiency, anti-microbial resistance, growing incidence of zoonotic diseases, abrupt climate change, and food security, among others. Thus, a focus on managing soil health offers innovative options and opportunities for preventive interventions (Montgomery et al., 2024), saving costs and reducing the loss of human life, as is stated in the One Health concept (Figure 12.2).

Biofortification with deficient nutrients is a widely observed strategy. The use of zinc (Zn) fertilizer in Zn-deficient soils leads to high-yielding maize (*zea mays*) varieties, with the aim of increasing Zn content in maize grains for improving human health (El-Ramady et al., 2022). Phytoremediation is a widely recognized process for the decontamination of soil from heavy metals and other pollutants. Plant cyanobacteria interaction can be used to benefit human health and well-being through the bioactive compounds that have some plant-growth promoting attributes, and can be used as a biofertilizer (Nowruzi et al., 2021). Nature-based fertilizers are also a beneficial option because of the adverse effects of chemical fertilizers, especially with excessive and indiscriminate use. However, although nature-based fertilizers are an eco-friendly option (Ebanna et al., 2025),

TABLE 12.1
Soil Contamination by Heavy Metals, Microplastics, and Other Pollutants with Adverse Effects on Human Health

Parameter	Specific Health Factor	Reference
Heavy metals in soil and health risks	Hg transfer from soil to plants	
	Toxic metals in soil and human health	
	Soil contamination from mining by Fe, Ca, Zn, Cu, Cd	
	Bioaccumulation and transfer of Zn in soil, plants and animal systems	Chen et al. (2022); Rajput et al. (2020)
	High concentration of Cr in plants, animals and humans by industrial procedures	Kapoor et al. (2022)
	Toxic levels of Se have adverse effects on human disorders and diseases	Natasha et al. (2018)
	As poisoning affects about 500 million people	
	Five million sites worldwide have soil contamination	Duckworth et al. (2022)
	Pb contamination can cause damage to human health	Moldoveanu and Moldoveanu (2010)
	Urbanization and soil contamination with As, Cr, Cu, Zn, Ni and Pb	Adimalla (2020)
Microplastic pollutants and health risks	Micro- and nanoplastics and human health	Perkovic et al. (2022))
Other pollutants in soil with adverse effects on humans	Effects of herbicides (Glyphosate)	
	Use of antibiotic-laden manure on vegetables and human health	Kumar et al. (2005)

some (e.g., biochar, compost, and manure) are needed in large amounts. In addition, the introduction of plant-growth promoting rhizobacteria may also improve soil biological fertility which can increase crop resilience and advance sustainable agriculture.

NATURE-BASED FERTILIZERS

Among several strategies for the decontamination of polluted soils, the homesis concept refers to positive biological effects caused by low doses of a stressor that is known to have adverse effects at high doses, similar to that of a vaccine for some diseases in pandemics and other health contexts (Agathokieou et al., 2023. Homesis may be a pertinent strategy in relation to the soil–human health nexus. In this context, utilizing soil chemistry (Duckworth et al., 2022) and other disciplines is important. Furthermore, nano-biofertilizers are also appropriate for the protection of soil health and environment.

Site-specific cropping systems are another pertinent option. In Florida, USA, Celestin et al. (2025) identified site-specific cropping systems based on soil properties. Clestin and colleagues used the propensity score weighting to model each sample's probability of belonging to a specific cropping system, and highlighted the importance of considering variability in inherent properties.

SOIL STRUCTURE AND PHYSICAL PROPERTIES

There exists an intricate relationship between SOM content, soil structure and the ecosystem C budget. Consequently, soil structure strongly affects nutrient and water supply, buffering capacity (Hatano et al., 2024), and adaptation to and mitigation of climate change effects (Lal, 2004). Soil structure and physical health also affects water retention, nutrient cycling, and emission of greenhouse gases (GHGs). Earthworms, their activity and species diversity, can play an important role in improving soil structure and physical properties. Earthworms are rightfully considered as engineers of soil and human health (Agbenyeku et al., 2015). Management of soil organic carbon (SOC) content is critical to restoring soil health (Lal, 2004) and improving crop yield. In addition to conservation agriculture (including no-till, residue mulch, and cover cropping), crop rotation is also critical to restoring SOC content and enhancing agronomic productivity because of its effects on nutrient dyanamics and input of above-ground and below-ground biomass/

SOIL LAW AND PROTECTION POLICY

There is a strong need to develop and implement soil protection policies to promote the "One Health" concept for improving soil functions and for advancing human health and well-being. Presently, there are limited, if any, legal protocols to protect, restore, and sustainably manage soil health with the aim of improving the environment that supports human health. There must be specific laws to protect specific soil functions. The European Union has established an interdisciplinary team to identify key challenges for providing legal protection of soil. Abbas and colleagues emphasized that while laws must have uniformity and legal certainty, soils are inherently heterogenous with regard to specific functions and capacity to provide ecosystem services. They further emphasized that while laws are aimed at a specific property, or a commodity that it produces, soils are interconnected spatially and temporally and it is challenging to connect the two. Certainly, there is a need for better communication between those involved in science and those in policy. The goal is to develop soil protection policies at local, regional, and global levels. Strengthening soil literacy in schools can help advance the understanding of soil–human health interconnectivity. The increased literacy can connect the general public with soil to improve human health by increased understanding of soil security Soil health science must be used to promote benefits to agronomy, the environment (Wood and Blankinship, 2022), and human health (Pepper et al., 2009; Montgomery et al., 2024).

SOIL HEALTH AND MICRONUTRIENT DEFICIENCY

Human health depends on the safe, proper, and sufficient supply of nutrients, most of which come from healthy and fresh soils. Thus, quality of food depends on the physical, chemical, biological, and ecological properties of soil (Alshaal et al., 2019). Micronutrient availability in soil plays a significant role in crop production and human health, especially in degraded/depleted soils of developing countries (Table 12.2). However, about half of the world's population is prone to micronutrient malnutrition. Zinc deficiency affects 2 billion people globally (Hussain et al., 2022). Iron and iodine deficiency is the most common problem. Table 12.3 provides technological options for remediation and Table 12.4 summarizes the relevance of soil health to human health.

TABLE 12.2
Effects of Nutrient Deficiency in Soil on Human Health

Plant Nutrient	Deficiency Symptoms in Human	Reference
Se	Cancer, heart disease and other chronic conditions	Huang et al. (2025)
Zn, Fe, Cu	Are essential for formation of carbohydrates	García-Latorre et al. (2024)
I	Iodine deficiency disorder (IDD)	Caffagni et al. (2011)
Ni	It is a component of urease and hydronesa	
Fe	Anemia, fatigue and impaired cognitive function	
Mg	Contributes to muscle cramps and irregular heartbeat, hypertension and neurological disorders	Prasad (2013)
Zn	Impaired immune functions, delayed wound healing, cognitive impairment, impaired development in children, organ destruction	Hussain (2022)
Ca	Muscle cramps, brittle bones, neurological issues	Shlisky et al. (2022)

TABLE 12.3
Examples of Some Technological Options to Restore Soil Health for Human Well-being under the "One Health" Concept

Technology	Description	References
A. Restoring Soil Health	Managing soil fertility	
i. Biofortification	Mitigation of pathogenic organisms in soil	El-Ramady et al. (2021)
ii. Biochar	Improving use efficiency of inputs	
iii. Bioremediation	The importance of cyanobacteria	
B. Improving Plant Nutrition by Feeding Soil	Restoring soil quality and soil health	Krupenikov et al. (2011);
i. Micronutrient fertilization	To alleviate deficiency of Se, Zn	El-Ramady et al. (2022)
ii. Restoring soil biological fertility	Increasing soil-plant-micro inter and intra actions	
iii. Organo-mineral fertilizers	To minimize environmental risks of chemical fertilizers	
iv. Nano-biofertilizers	To improve use efficiency	
C. Soil Carbon Sequestration	Managing global warming, conservation agriculture and agroforestry	Pepper et al. (2009); Lal (2004)
i. Soil structure	For soil health intensification	Hatano et al. (2024)
ii. Soil biota	Role of earthworms	Agbenyeku et al. (2015)
iii. Cropping systems	Linking cropping systems to soil health	Haney et al. (2018)
D. Feeding the Urban Population		
i. Reducing contamination	Managing ecological risks to human health	Adimalla (2020)

TABLE 12.4
Effects of Soil Health on Environmental Parameters Affecting Human Health

Factor	Effect on Human Health	Reference
I. Air Quality	• Soil plays a critical role in air quality from GHG emission to ozone depletion including odor, particulates, pathogen transport	Kalimitchenko et al. (2023); Giltap et al. (2021)
	• Pollutants from air can be removed by vegetation (e.g., trees, grasslands, shrublands	Gopala Krishman et al. (2018)
	• Adoption of conservation practices and establishment of vegetation cover have beneficial impacts on	
II. Water Quality	• Soil plays a critical role in water cycle and its quality especially in agroecosytems	
	• Water quality is adversely affected by soil organic P in arable lands	Dodd and Sharpley (2015)
	• Application of manure pollutes water	
	• Natural resource management practices (conservation agriculture, mulch farming, cover cropping) can reduce risks of water pollution	Sahrawat et al. (2010)

BIBLIOGRAPHY

Adimalla, N. (2020). Heavy metals pollution assessment and its associated human health risk evaluation of urban soils from Indian cities: A review. *Environmental Geochemistry and Health*, *42*(1), 173–190. https://doi.org/10.1007/s10653-019-00324-4

Agathokleous, E., Liu, C.-J., & Calabrese, E. J. (2023). Applications of the hormesis concept in soil and environmental health research. *Soil & Environmental Health*, *1*(1), 100003. https://doi.org/10.1016/j.seh.2023.100003

Agbenyeku, E. E.-O., Muzenda, E., & Msibi, I. M. (2015). Earthworms as engineers of soil and human health.

Alblas, E., Edlinger, A., Helfenstein, J., Diaz Bernal, M. J., Stomph, D., Berendsen, B., Briones, M. J. I., Creamer, R., van Egmond, F., Grealish, G., Hacke, M., Jongedijk, E., Lawson, A., Lima, C., Moloney, J., Mommer, L., Nussbaum, M., Roberton, S., Ros, M., & Tempels, B. (2025). Bridging law and soil science to promote soil health. *European Journal of Soil Science*, *76*(4), e70127. https://doi.org/10.1111/ejss.70127

Alshaal, T., El-Ramady, H., Elhawat, N., El-Nahrawy, S., Omara, A. E.-D., Elsakhawy, T., Ghazi, A., Abbas, M. H., Farid, I. M., Abdalla, N., Fári, M., & Domokos-Szabolcsy, É. (2019). Soil health and its biology. In H. El-Ramady, T. Alshaal, N. Bakr, T. Elbana, E. Mohamed, & A.-A. Belal (Eds.), *The Soils of Egypt* (pp. 175–185). Springer International Publishing. https://doi.org/10.1007/978-3-319-95516-2_10

Angon, P. B., Islam, M. S., Kc, S., Das, A., Anjum, N., Poudel, A., & Suchi, S. A. (2024). Sources, effects and present perspectives of heavy metals contamination: Soil, plants and human food chain. *Heliyon*, *10*(7). https://doi.org/10.1016/j.heliyon.2024.e28357

Aziz, M. Z., Yaseen, M., Abbas, T., Naveed, M., Mustafa, A., Hamid, Y., Saeed, Q., & Xu, M. (2019). Foliar application of micronutrients enhances crop stand, yield and the biofortification essential for human health of different wheat cultivars. *Journal of Integrative Agriculture*, *18*(6), 1369–1378. https://doi.org/10.1016/S2095-3119(18)62095-7

Bahareh, Nowruzi, Bouaïcha, N., Metcalf, J. S., Porzani, S. J., & Konur, O. (2021). Plant-cyanobacteria interactions: Beneficial and harmful effects of cyanobacterial bioactive compounds on soil-plant systems and subsequent risk to animal and human health. *Phytochemistry*, *192*, 112959. https://doi.org/10.1016/j.phytochem.2021.112959

Bana, R. C., Gupta, A. K., Bana, R. S., Shivay, Y. S., Bamboriya, S. D., Thakur, N. P., Puniya, R., Gupta, M., Jakhar, S. R., Kailash Choudhary, R. S., Bochalya, R. S., Bajaya, T., Kumar, V., Kumar, P., & Choudhary, A. K. (2022). Zinc-coated urea for enhanced zinc biofortification, nitrogen use efficiency and yield of basmati rice under typic fluvents. *Sustainability*, *14*(1), 104. https://doi.org/10.3390/su14010104

Banuelos, G. S., Lin, Z.-Q., & Caton, J. (2023). Editorial: Selenium in soil-plant-animal systems and its essential role for human health. *Frontiers in Plant Science*, *14*. https://doi.org/10.3389/fpls.2023.1237646

Beelman, R. B., Richie, J. P., Phillips, A. T., Kalaras, M. D., Sun, D., & Duiker, S. W. (2021). Soil disturbance impact on crop ergothioneine content connects soil and human health. *Agronomy*, *11*(11), 2278. https://doi.org/10.3390/agronomy11112278

Bolan, S., Sharma, S., Mukherjee, S., Isaza, D. F. G., Rodgers, E. M., Zhou, P., Hou, D., Scordo, F., Chandra, S., Siddique, K. H. M., & Bolan, N. (2025). Wildfires under changing climate, and their environmental and health impacts. *Journal of Soils and Sediments*. https://doi.org/10.1007/s11368-025-04020-y

Brevik, E. C., Steffan, J. J., Rodrigo-Comino, J., Neubert, D., Burgess, L. C., Cerdà, A. (2018). Connecting the public with soil to improve human health. *European Journal of Soil Science 70*, 898–910. https://doi.org/10.1111/ejss.12764

Caffagni, A., Arru, L., Meriggi, P., Milc, J., Perata, P., & Pecchioni, N. (2011). Iodine fortification plant screening process and accumulation in tomato fruits and potato tubers. *Communications in Soil Science and Plant Analysis*, *42*(6), 706–718. https://doi.org/10.1080/00103624.2011.550372

Carré, F., Caudeville, J., Bonnard, R., Bert, V., Boucard, P., & Ramel, M. (2017). Soil contamination and human health: A major challenge for global soil security. In D. J. Field, C. L. S. Morgan, & A. B. McBratney (Eds.), *Global Soil Security* (pp. 275–295). Springer International Publishing. https://doi.org/10.1007/978-3-319-43394-3_25

Celestin, F., Maltais-Landry, G., Dubeux, J. C. B., Mylavarapu, R. S., & Lin, Y. (2025). Cropping system is a key determinant of soil health after accounting for environmental and edaphic variability. *Geoderma*, *458*, 117330. https://doi.org/10.1016/j.geoderma.2025.117330

Chang, X., Fang, Y., Wang, Y., Wang, F., Shang, L., & Zhong, R. (2022). Microplastic pollution in soils, plants, and animals: A review of distributions, effects and potential mechanisms. *Science of The Total Environment*, *850*, 157857. https://doi.org/10.1016/j.scitotenv.2022.157857

Chaudhari, S. K., Biswas, P. P., & Kapil, H. (2020). Soil health and fertility. In B. B. Mishra (Ed.), *The Soils of India* (pp. 215–231). Springer International Publishing. https://doi.org/10.1007/978-3-030-31082-0_11

Chen, F., Muhammad, F. G., Khan, Z. I., Ahmad, K., Malik, I. S., Ashfaq, A., Naeem, M., Nadeem, M., Ma, J., Awan, M. U. F., Mahpara, S., & Mehmood, S. (2022). Bioaccumulation and transfer of zinc in soil plant and animal system: A health risk assessment for the grazing animals. *Environmental Science and Pollution Research*, *29*(2), 2718–2727. https://doi.org/10.1007/s11356-021-15808-z

Costantini, E. A. C., & Mocali, S. (2022). Soil health, soil genetic horizons and biodiversity. *Journal of Plant Nutrition and Soil Science*, *185*(1), 24–34. https://doi.org/10.1002/jpln.202100437

Dhillon, K. S., Dhillon, S. K., & Bijay-Singh. (2019). Chapter one—Genesis of seleniferous soils and associated animal and human health problems. In D. L. Sparks (Ed.), *Advances in Agronomy* (Vol. 154, pp. 1–80). Academic Press. https://doi.org/10.1016/bs.agron.2018.11.001

Dodd, R. J., & Sharpley, A. N. (2015). Recognizing the role of soil organic phosphorus in soil fertility and water quality. *Resources, Conservation and Recycling*, *105*, 282–293. https://doi.org/10.1016/j.resconrec.2015.10.001

Duckworth, O. W., Polizzotto, M. L., & Thompson, A. (2022). Bringing soil chemistry to environmental health science to tackle soil contaminants. *Frontiers in Environmental Science*, *10*. https://doi.org/10.3389/fenvs.2022.981607

Durdu, B., Gurbuz, F., Koçyiğit, H., & Gurbuz, M. (2023). Urbanization-driven soil degradation; ecological risks and human health implications. *Environmental Monitoring and Assessment*, *195*(8), 1002. https://doi.org/10.1007/s10661-023-11595-x

Elbanna, B. A., Darweesh, M., Elbltagy, H. M., Al Anany, F. S., & Elbasiouny, H. (2025). Nature-based fertilizers for sustainable soil fertility and environmental health. *Egyptian Journal of Soil Science*. https://doi.org/10.21608/ejss.2025.358830.1989

Elemike, E. E., Uzoh, I. M., Onwudiwe, D. C., & Babalola, O. O. (2019). The role of nanotechnology in the fortification of plant nutrients and improvement of crop production. *Applied Sciences*, *9*(3), 499. https://doi.org/10.3390/app9030499

El-Ramady, H., Alshaal, T., Elsakhawy, T., Omara, A. E.-D., Abdalla, N., & Brevik, E. C. (2019). Soils and humans. In H. El-Ramady, T. Alshaal, N. Bakr, T. Elbana, E. Mohamed, & A.-A. Belal (Eds.), *The Soils of Egypt* (pp. 201–213). Springer International Publishing. https://doi.org/10.1007/978-3-319-95516-2_12

El-Ramady, H., Brevik, E. C., Elbasiouny, H., Elbehiry, F., El-Henawy, A., Faizy, S. E.-D., Elsakhawy, T., Omara, A. E.-D., Amer, M., & Eid, Y. (2021). Soils, biofortification, and human health under COVID-19: Challenges and opportunities. *Frontiers in Soil Science*, *1*. https://doi.org/10.3389/fsoil.2021.732971

El-Ramady, H., Hajdú, P., Tör⊠s, G., Badgar, K., Llanaj, X., Kiss, A., Abdalla, N., Omara, A. E.-D., Elsakhawy, T., Elbasiouny, H., Elbehiry, F., Amer, M., El-Mahrouk, M. E., & Prokisch, J. (2022). Plant nutrition for human health: A pictorial review on plant bioactive compounds for sustainable agriculture. *Sustainability*, *14*(14), 8329. https://doi.org/10.3390/su14148329

Frontiers | Indirect Effects of the Herbicide Glyphosate on Plant, Animal and Human Health Through its Effects on Microbial Communities | Indirect Effects of the Herbicide Glyphosate on Plant, Animal and Human Health Through its Effects on Microbial Communities. (n.d.). Retrieved August 8, 2025, from https://doi.org/10.3389/fenvs.2021.763917

Full article: The Soil Health-Human Health Nexus. (n.d.). Retrieved August 8, 2025, from https://doi.org/10.1080/10643389.2012.694330

Galbraith, J. M. (2018). Human-altered and human-transported (HAHT) soils in the U.S. soil classification system. *Soil Science and Plant Nutrition*, *64*(2), 190–199. https://doi.org/10.1080/00380768.2018.1442682

García-Latorre, C., Reynolds-Marzal, M. D., De la Peña-Lastra, S., Pinheiro, N., & Poblaciones, M. J. (2024). Soil and foliar zinc biofortification of triticale (x Triticosecale) under mediterranean conditions: Effects on forage yield and quality. *Plants*, *13*(14), 1917. https://doi.org/10.3390/plants13141917

Giltrap, D., Cavanagh, J., Stevenson, B., & Ausseil, A.-G. (2021). The role of soils in the regulation of air quality. *Philosophical Transactions of the Royal Society B: Biological Sciences*, *376*(1834), 20200172. https://doi.org/10.1098/rstb.2020.0172

Gopalakrishnan, V., Hirabayashi, S., Ziv, G., & Bakshi, B. R. (2018). Air quality and human health impacts of grasslands and shrublands in the United States. *Atmospheric Environment*, *182*, 193–199. https://doi.org/10.1016/j.atmosenv.2018.03.039

Gullino, M. L., Pasquali, M., Pugliese, M., & Capua, I. (2024). Positioning plant health within the evolving human-animal-environmental health paradigms. *One Health*, *19*, 100931. https://doi.org/10.1016/j.onehlt.2024.100931

Haney, R. L., Haney, E. B., Smith, D. R., Harmel, R. D., & White, M. J. (2018). The soil health tool—Theory and initial broad-scale application. *Applied Soil Ecology*, *125*, 162–168. https://doi.org/10.1016/j.apsoil.2017.07.035

Hatano, R., Mukumbuta, I., & Shimizu, M. (2024). Soil Health Intensification through Strengthening Soil Structure Improves Soil Carbon Sequestration. *Agriculture*, *14*(8), 1290. https://doi.org/10.3390/agriculture14081290

Hegedus, C., Paşcalău, S.-N., Andronie, L., Rotaru, A.-S., Cucu, A.-A., & Dezmirean, D. S. (2023). The journey of 1000 leagues towards the decontamination of the soil from heavy metals and the impact on the soil–plant–animal–human chain begins with the first step: Phytostabilization/phytoextraction. *Agriculture, 13*(3), 735. https://doi.org/10.3390/agriculture13030735

Huang, S., Han, Y., Song, R., Wang, X., Zhou, Y., Luo, H., Ren, X., & Yu, K. (2025). Effects of bio-nano-selenium on wheat grain morphology, selenium transport enrichment and antioxidant enzyme activities. *Frontiers in Plant Science, 16*. https://doi.org/10.3389/fpls.2025.1516005

Hussain, A., Jiang, W., Wang, X., Shahid, S., Saba, N., Ahmad, M., . . . & Mustafa, A. (2022). Mechanistic impact of zinc deficiency in human development. *Frontiers in Nutrition, 9*, 717064.

Imran, & Amanullah. (2021). Phosphorus and boron application optimizing biofortification of p and productivity of french bean (Phaseolus vulgaris L.). *Communications in Soil Science and Plant Analysis, 52*(22), 2876–2883. https://doi.org/10.1080/00103624.2021.1971689

Imtiaz, M., Rashid, A., Khan, P., Memon, M. Y., & Aslam, M. (n.d.). The role of micronutrients in crop production and human health.

Kalinitchenko, V. P., Swidsinski, A. V., Glinushkin, A. P., Meshalkin, V. P., Gudkov, S. V., Minkina, T. M., Chernenko, V. V., Rajput, V. D., Mandzhieva, S. S., Sushkova, S. N., Okolelova, A. A., & Shestakova, A. A. (2023). New approach to soil management focusing on soil health and air quality: One earth one life (critical review). *Environmental Geochemistry and Health, 45*(12), 8967–8987. https://doi.org/10.1007/s10653-023-01550-7

Kapoor, R. T., Bani Mfarrej, M. F., Alam, P., Rinklebe, J., & Ahmad, P. (2022). Accumulation of chromium in plants and its repercussion in animals and humans. *Environmental Pollution, 301*, 119044. https://doi.org/10.1016/j.envpol.2022.119044

Kopittke, P. M., Minasny, B., Pendall, E., Rumpel, C., & McKenna, B. A. (2024). Healthy soil for healthy humans and a healthy planet. *Critical Reviews in Environmental Science and Technology, 54*(3), 210–221. https://doi.org/10.1080/10643389.2023.2228651

Kouhounde, S., Adéoti, K., Mounir, M., Giusti, A., Refinetti, P., Otu, A., Effa, E., Ebenso, B., Adetimirin, V. O., Barceló, J. M., Thiare, O., Rabetafika, H. N., & Razafindralambo, H. L. (2022). Applications of probiotic-based multi-components to human, animal and ecosystem health: Concepts, methodologies, and action mechanisms. *Microorganisms, 10*(9), 1700. https://doi.org/10.3390/microorganisms10091700

Krupenikov, I. A., Boincean, B. P., & Dent, D. (2011). Soil health and soil quality. In D. Dent, B. P. Boincean, & I. A. Krupenikov (Eds.), *The Black Earth: Ecological Principles for Sustainable Agriculture on Chernozem Soils* (pp. 105–107). Springer, Netherlands. https://doi.org/10.1007/978-94-007-0159-5_16

Kumar, K., Gupta, S. C., Baidoo, S. K., Chander, Y., & Rosen, C. J. (2005). Antibiotic uptake by plants from soil fertilized with animal manure. *Journal of Environmental Quality, 34*(6), 2082–2085. https://doi.org/10.2134/jeq2005.0026

Kumar, M., Kumar, R., Sharma, S., Mukherjee, S., Bolan, S., Shrivastava, A., Hettiarachchi, M., Siddique, K. H. M., & Bolan, N. (2025). Microplastics pollution modulating soil biological health – A review. *Soil Use and Management, 41*(1), e70009. https://doi.org/10.1111/sum.70009

Kumleh, A. S. (n.d.). Genetic approaches and nutrient management in rice soil: A review. Pertanika Journal of Tropical Agricultural Science, *42*(3), 973–974.

Lal, R. (2004). Soil carbon sequestration impacts on global climate change and food security. *Science*, 304, 1623–1627.

Larramendy, M., & Soloneski, S. (2016). *Soil Contamination: Current Consequences and Further Solutions*. BoD – Books on Demand.

Moldoveanu, A. M., & Moldoveanu, A. C. (2010). Assessment of the impact of soil pollution on human health. *Journal of Environmental Protection and Ecology, 11*(4), 1239–1245.

Montgomery, D. R., Rabinowitz, P., Sipos, Y., & Wheat, E. E. (2024). Soil health: A common focus for one health and planetary health interventions. *One Health, 18*, 100673. https://doi.org/10.1016/j.onehlt.2023.100673

Muramoto, J., Parr, D. M., Perez, J., & Wong, D. G. (2022). Integrated soil health management for plant health and one health: Lessons from histories of soil-borne disease management in california strawberries and arthropod pest management. *Frontiers in Sustainable Food Systems, 6*. https://doi.org/10.3389/fsufs.2022.839648

Nano-biofertilizers: A promising technology for sustainable soil fertility, soil health, and environmental protection. (n.d.). Retrieved August 8, 2025, from https://www.accscience.com/journal/AJWEP/articles/online_first/5118

Natasha Shahid, M., Niazi, N. K., Khalid, S., Murtaza, B., Bibi, I., & Rashid, M. I. (2018). A critical review of selenium biogeochemical behavior in soil-plant system with an inference to human health. *Environmental Pollution*, *234*, 915–934. https://doi.org/10.1016/j.envpol.2017.12.019

Nathanail, C. P. (2005). Generic and site-specific criteria in assessment of human health risk from contaminated soil. *Soil Use and Management*, *21*(s2), 500–507. https://doi.org/10.1079/SUM2005341

Nouri, M., & Haddioui, A. (2015). Human and animal health risk assessment of metal contamination in soil and plants from Ait Ammar abandoned iron mine, Morocco. *Environmental Monitoring and Assessment*, *188* (1), 6. https://doi.org/10.1007/s10661-015-5012-6

Ochoa-Hueso, R. (2017). Global change and the soil microbiome: A human-health perspective. *Frontiers in Ecology and Evolution*, *5*. https://doi.org/10.3389/fevo.2017.00071

OECD, Turkey, T. S., & T. R. C. of. (2003). *Biological Resource Management in Agriculture Innovative Soil-Plant Systems for Sustainable Agricultural Practices*. OECD Publishing.

Oliver, M. A., & Gregory, P. J. (2015). Soil, food security and human health: A review. *European Journal of Soil Science*, *66*(2), 257–276. https://doi.org/10.1111/ejss.12216

One health: Considerations in the International Year of Plant Health. (n.d.). Retrieved August 8, 2025, from http://www.scielo.edu.uy/scielo.php?pid=S2301-15482020000402307&script=sci_arttext

Penuelas, J., Janssens, I. A., Ciais, P., Obersteiner, M., & Sardans, J. (2020). Anthropogenic global shifts in biospheric N and P concentrations and ratios and their impacts on biodiversity, ecosystem productivity, food security, and human health. *Global Change Biology*, *26*(4), 1962–1985. https://doi.org/10.1111/gcb.14981

Pepper, I. L., Gerba, C. P., Newby, D. T., & Rice, C. W. (2009). Soil: A public health threat or savior? *Critical Reviews in Environmental Science and Technology*, *39*(5), 416–432. https://doi.org/10.1080/10643380701664748

Perković, S., Paul, C., Vasić, F., & Helming, K. (2022). Human health and soil health risks from heavy metals, micro(nano)plastics, and antibiotic-resistant bacteria in agricultural soils. *Agronomy*, *12*(12), 2945. https://doi.org/10.3390/agronomy12122945

Potential management practices of saltwater intrusion impacts on soil health and water quality: A review | Journal of Water and Climate Change | IWA Publishing. (n.d.). Retrieved August 8, 2025, from https://iwaponline.com/jwcc/article/12/5/1327/76653/Potential-management-practices-of-saltwater

Prasad, A. S. (2013). Discovery of human zinc deficiency: Its impact on human health and disease. *Advances in Nutrition*, *4*(2), 176–190. https://doi.org/10.3945/an.112.003210

Rajput, V., Minkina, T., Sushkova, S., Behal, A., Maksimov, A., Blicharska, E., Ghazaryan, K., Movsesyan, H., & Barsova, N. (2020). ZnO and CuO nanoparticles: A threat to soil organisms, plants, and human health. *Environmental Geochemistry and Health*, *42*(1), 147–158. https://doi.org/10.1007/s10653-019-00317-3

Sahrawat, K. L., Wani, S. P., Pathak, P., & Rego, T. J. (2010). Managing natural resources of watersheds in the semi-arid tropics for improved soil and water quality: A review. *Agricultural Water Management*, *97*(3), 375–381. https://doi.org/10.1016/j.agwat.2009.10.012

Sandstead, H. H. (2000). Causes of iron and zinc deficiencies and their effects on brain1. *The Journal of Nutrition*, *130*(2), 347S–349S. https://doi.org/10.1093/jn/130.2.347S

Shi, W., Wu, N., Zhang, Z., Liu, Y., Chen, J., & Li, J. (2024). A global review on the abundance and threats of microplastics in soils to terrestrial ecosystem and human health. *Science of The Total Environment*, *912*, 169469. https://doi.org/10.1016/j.scitotenv.2023.169469

Shlisky, J., Mandlik, R., Askari, S., Abrams, S., Belizan, J. M., Bourassa, M. W., Cormick, G., Driller-Colangelo, A., Gomes, F., Khadilkar, A., Owino, V., Pettifor, J. M., Rana, Z. H., Roth, D. E., & Weaver, C. (2022). Calcium deficiency worldwide: Prevalence of inadequate intakes and associated health outcomes. *Annals of the New York Academy of Sciences*, *1512*(1), 10–28. https://doi.org/10.1111/nyas.14758

Siqueira Freitas, D., Wurr Rodak, B., Rodrigues dos Reis, A., de Barros Reis, F., Soares de Carvalho, T., Schulze, J., Carbone Carneiro, M. A., & Guimarães Guilherme, L. R. (2018). Hidden Nickel deficiency? Nickel fertilization via soil improves nitrogen metabolism and grain yield in soybean genotypes. *Frontiers in Plant Science*, *9*. https://doi.org/10.3389/fpls.2018.00614

Soorya, M., Suganya, S., Elayarajan, M., Davamani, V., & Chitra, P. (2024). Nutrient nexus: Linking soil, plants, animals and humans. *Plant Science Today*, *11*, 5759. https://doi.org/10.14719/pst.5759

Ullah, R., Tsui, M. T.-K., Chow, A., Chen, H., Williams, C., & Ligaba-Osena, A. (2022). Micro(nano)plastic pollution in terrestrial ecosystem: Emphasis on impacts of polystyrene on soil biota, plants, animals, and humans. *Environmental Monitoring and Assessment*, *195*(1), 252. https://doi.org/10.1007/s10661-022-10769-3

van Bruggen, A. H. C., Goss, E. M., Havelaar, A., van Diepeningen, A. D., Finckh, M. R., & Morris, J. G. (2019). One health—cycling of diverse microbial communities as a connecting force for soil, plant, animal, human and ecosystem health. *Science of The Total Environment*, *664*, 927–937. https://doi.org/10.1016/j.scitotenv.2019.02.091

Vieweger, A., & Döring, T. F. (2015). Assessing health in agriculture – towards a common research framework for soils, plants, animals, humans and ecosystems. *Journal of the Science of Food and Agriculture*, *95*(3), 438–446. https://doi.org/10.1002/jsfa.6708

Wood, S. A., & Blankinship, J. C. (2022). Making soil health science practical: Guiding research for agronomic and environmental benefits. *Soil Biology and Biochemistry*, *172*, 108776. https://doi.org/10.1016/j.soilbio.2022.108776

Xue, W., You, J., Su, Y., & Wang, Q. (2019). The effect of magnesium deficiency on neurological disorders: A narrative review article. *Iranian Journal of Public Health*, *48*(3), 379–387.

Yang, H., Yang, X., Ning, Z., Kwon, S. Y., Li, M.-L., Tack, F. M. G., Kwon, E. E., Rinklebe, J., & Yin, R. (2022). The beneficial and hazardous effects of selenium on the health of the soil-plant-human system: An overview. *Journal of Hazardous Materials*, *422*, 126876. https://doi.org/10.1016/j.jhazmat.2021.126876

Zabaloy, María Celina. (2021). Una sola salud: la salud del suelo y su vínculo con la salud humana, *Revista Argentina de Microbiología*, *53*(4), 275–276, ISSN 0325-7541, https://doi.org/10.1016/j.ram.2021.11.001. (https://www.sciencedirect.com/science/article/pii/S0325754121001000)

Zeng, Y., Verhoef, A., Vereecken, H., Ben-Dor, E., Veldkamp, T., Shaw, L., Van Der Ploeg, M., Wang, Y., & Su, Z. (2025). Monitoring and modeling the soil-plant system toward understanding soil health. *Reviews of Geophysics*, *63*(1), e2024RG000836. https://doi.org/10.1029/2024RG000836

Zhu, Y.-G., Zhao, Y., Zhu, D., Gillings, M., Penuelas, J., Ok, Y. S., Capon, A., & Banwart, S. (2019). Soil biota, antimicrobial resistance and planetary health. *Environment International*, *131*, 105059. https://doi.org/10.1016/j.envint.2019.105059

13 Urban Agriculture for Human and Planetary Health

Rattan Lal

INTRODUCTION

Globally, the number of megacities with populations of 10 million or more has increased from 2 in 1950 to 37 in 2025, and this is projected to increase to 48 by 2035 and 67 by 2050. Even the number and population of smaller cities is growing because of the increase in the world population and the trend of migration from rural to urban areas. Out of the world population of 8.2 billion in 2025, ~55% (4.5 billion) live in cities, and two out of three people (~70%% of the world population) is projected to live in cities by 2050 (Ritchie, 2024; Samuel et al., 2022). Some estimates show that the urban population has increased 40-fold since 1870 (Li et al., 2021). The rapid growth in cities, especially in the developing countries of Africa and Asia, is aggravating the risks of global food and nutritional insecurity because of the vulnerability of cities to food shortages (Mietz et al., 2024). Therefore, urban food and nutritional insecurity is the primary driver of the growing interest in urban agriculture (UA) and peri-urban agriculture (PUA) (Lee-Smith, 2014).

The COVID-19 pandemic caused a surge in undernutrition of the affected global population from 690 million to 733 million people worldwide, and the Sustainable Development Goals (SDGs) of the UN's Sustainable Development Agenda 2030 are not on track, and thus, interest in UA and PUA is increasing (Mietz et al., 2023). The urgent crisis of food insecurity in African cities (Crash et al., 2012; Satterwaite et al., 2010) is a complex problem driven by climate change, rapid and ad hoc growth, political unrest, wars, and poverty. Undernutrition and malnutrition in urban centers are affected by factors including food production, processing, distribution, preparation, consumption (Dinku et al., 2023), and retention that depends on quality of water and the overall environment. Rapidly growing cities are also among the principal causes of global environmental pollution. Indeed, integration of urban water management, especially in developing counties, has a number of implications to improving the environment (e.g., air, green space) while feeding the urban population. The problem of undernutrition is affected by technological, socio-economic, and ecological parameters, which need to be understood and appropriately addressed.

Thus, the objective of this chapter is to explain the basic concepts of UA and PUA, identify constraints to upscaling these practices, and identify policy options to advance SDGs while protecting soil, water, and other natural resources in a changing and uncertain climate.

URBAN AND PERI-URBAN AGRICULTURE

Providing adequate, nutritious and safe food to a growing urban population is a major challenge for soil scientists, agronomists, nutritionists, and city planners. The need to grow food within and in the vicinity of large and growing cities is also affecting the environment, soil, and water resources, allocation of large resources to import food, and the overall health of Planet Earth. Yet, conventional agriculture is also a concern because the per capita arable land area (m^2) has decreased from 5,600 in 1950 to 2,300 in the early 2000s, and is projected to be as low as 1,500 by 2050 (Miccoli et al., 2016). Thus, there is a pressing need to identify and implement innovative solutions to feeding a large and increasing urban population.

DOI: 10.1201/9781003469742-13

The term urban agriculture (UA) implies practices of growing crops and raising livestock within urban centers, ranging in size from small gardens to large farms and including systems such as greenhouses, rooftop gardens, and use of unconstructed or vacant land. Vertical farming and rooftop agriculture are also innovative strategies to produce local and reliable food in land-scarce countries (e.g., Singapore). It is widely believed that the top five most productive crops for rooftop agriculture are tomato, chard, lettuce, pepper, and eggplant, and that water consumption is 3 to 4 L m^{-2} d^{-1}. Similar to water use, the productivity of UA (kg of food m^{-2}) can be more than that of conventional agriculture. A meta-analysis by Payen et al. (2022) showed that differences in yields between urban and conventional agriculture were high (17 kg m^{-2} $cycle^{-1}$ vs. 3.8 kg m^{-2} $cycle^{-1}$). Payen and colleagues also documented that some urban spaces and growing systems also had a significant effect on specific crop yields. For example, tomato yields in hydroponic systems were significantly greater than tomato yields in soil-based systems. In comparison with UA, peri-urban agriculture (PUA) refers to growing crops and raising animals in the vicinity of cities, in the transition zone between urban and rural areas. Nonetheless, both UA and PUA are aimed at advancing food security and creating resilient food systems for rapidly growing cities. In addition to consumption within the household, produce from UA and PUA is also used for sale, and the practice supports the creation of green space and community development (Duma et al., 2002); Opitz et al., 2016).

There is a wide range of types of UA (Figures 13.1, 13.2, and Table 13.1) depending on site-specific needs, challenges, and operational management. For example, the challenge of lack of sufficient water for UA can be addressed by usage of roof-harvested rainwater (Amos et al., 2018). The use of indoor smart gardens is another component and a new form of UA. Mihailovic et al. (2023) described the potential of indoor smart gardens in food production, environment sustainability, and human health through the revolutionization of farming practices and addressing the growing demand for food production in megacities. Community gardens are another commonly used practice of UA. Rather than being family-focused, UA can also be commercial part of sustainable cities. Among the options of UA, hydroponic gardening is an important technique (Figure 13.2). Wooten (2018) observed that hydroponic food production can yield the same amount of food as traditional agriculture, but uses up to 95% less water, 80% less land, and has much lower labor costs, due to automation, especially in advanced hydroponic systems. Thus, UA may be a common solution to under- and overnutrition in cities.

Vertical farming (VF) is among the more innovative options of UA, and has a short food supply chain. It involves multistory 3-dimensional glass structures with controlled environments designed to maximize photosynthesis. Thus, VF has lower fertilizer, and pesticide needs while also conserving water and mitigating soil degradation. In this context, innovative UA can be considered another

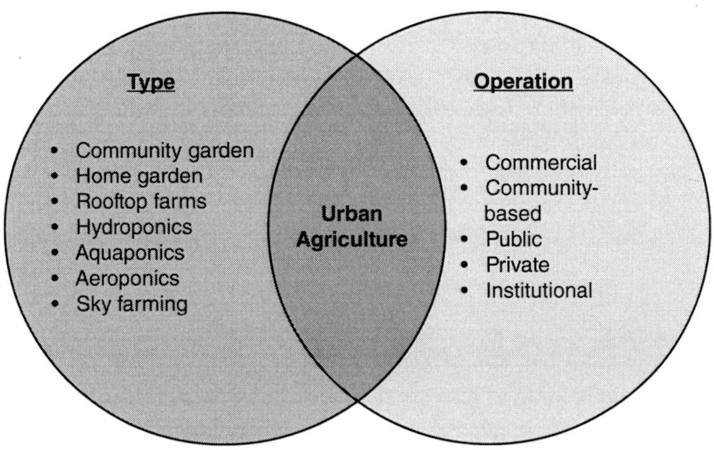

FIGURE 13.1 Type and operational management of home gardens and urban agriculture.

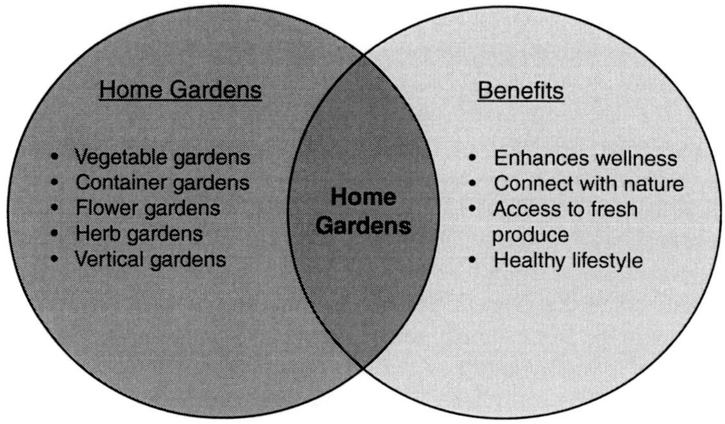

FIGURE 13.2 The usefulness of hydroponic as soil-less culture.

TABLE 13.1
Methods of Urban-Based Food Systems

Mode		Reference
Soil-less	Hydroponics on rooftop	
High-rise building	Vertical spaces for natural-light home gardening	Song et al. (2022)
Crop-livestock integration (CLI)	Integrated urban/crop livestock farming systems	Quaye et al. (2025); Ushimaru et al. (2024)
Hydroponics	Productivity:12 kg m^{-2} yr^{-1}; Land 0.8 m^2 kg^{-1}	Acevedo et al. (2025)

Green Revolution. In land-scarce cities, high-rise apartment buildings can be designed to create vertical spaces for natural-light home gardening along rooftops, corridors, balconies, and facades. However, Song et al. (2022) observed that corridor gardening was limited by light and met only 0.5% demand for vegetables, rooftop gardening with 15 cm growing medium met 3% of demand, while façade gardening met 43% of demand, because of the larger space available.

There is also a growing interest in integrated crop-livestock (ICL) systems in both UA and PUA (Ushimaru et al., 2024). A study by Quaye et al. (2025) observed that household income is negatively correlated with urban food security. Thus, ICL in UA settings can diversify diet and enhance income. Quaye and colleagues also observed that ICL increases household access to food by 84% compared to households without a farm, 48% compared to households only with crops, and 37% compared to households only with livestock. Furthermore, ICL-based UA enhanced household dietary diversity by 12%, 7.6%, and 16% compared to those without a farm, with only crops, or with only livestock, respectively (Quaye et al., 2025). Because of its potential to feed the growing urban population, UA is rightly considered as an innovative form of Green Revolution 2 (Table 13.2).

In addition to food production, UA also has social values, such as education, human health, civic engagement, social and gender equity, and contribution to waste management. Therefore, UA is widely considered a nature-based "One Health" approach. With its focus on interconnectivity and multiple functionality, and helping to resolve the public health crisis by improving the environment (Figure 13.3), UA can meet the challenge of feeding and improving health and quality of life for the teeming populations in cities. Indeed, UA and PUA provide opportunities for sustainable development of rapidly growing cities especially in the Global South.

TABLE 13.2
Types of Urban Agriculture as an Innovative Form of Green Revolution 2.0

Type	Strategy	Reference
Agroecological	Strengthening of community and individual leadership, empowerment, conducive environment	Ribeiro et al. (2015)
Nutrition-sensitive	Micronutrients and health balanced diet, dietary diversity	Gerster-Bentaya (2013); Warren et al. (2015)
Entrepreneurial	Creating employment and income	Dimitri et al. (2016)
Recreational	Mental health and well-being	Pollard et al. (2018); Xu et al. (2024)
Land/water sparing	140 kg vegetable m^{-2}	Armande et al. (2019)
Environmental	Healthy cities	Rao et al. (2022)
Climate smart	Adaptation/mitigation	Chari et al. (2022)

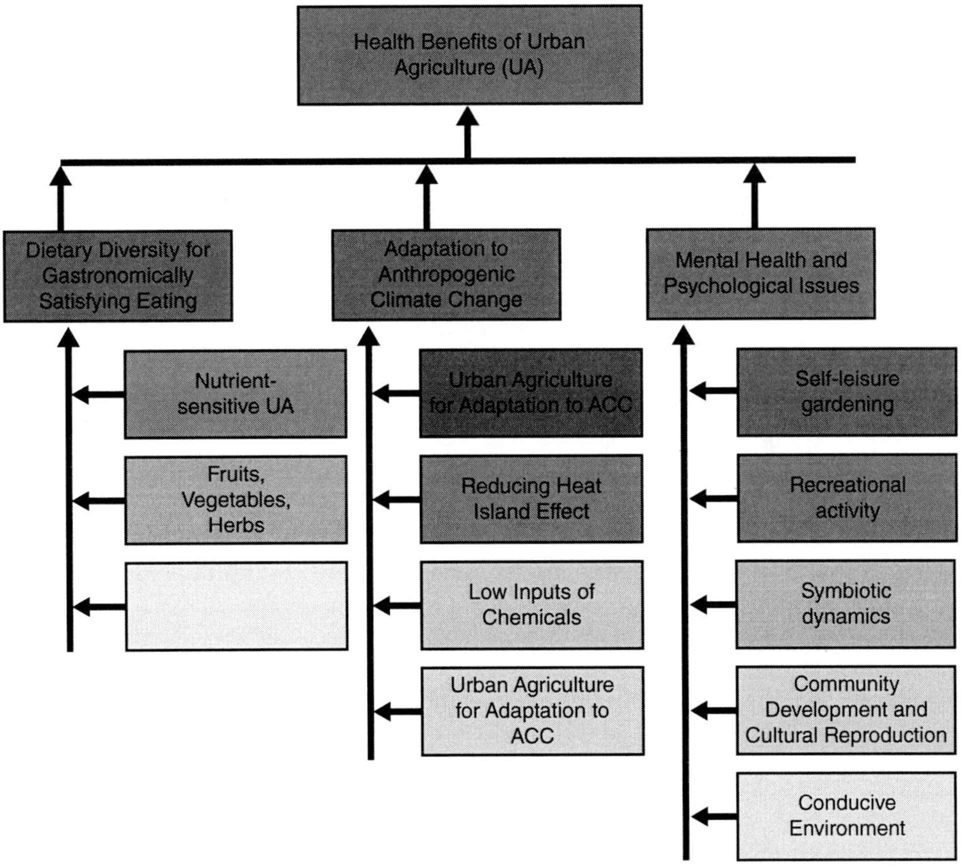

FIGURE 13.3 Addressing public health crises by harnessing health benefits of urban agriculture: beyond hunger and hidden hunger.

ADVANCING SUSTAINABLE DEVELOPMENT GOALS OF THE UN AGENDA 2030

With proper planning, promoting UA and PUA are appropriate means to advancing several SDGs of the Agenda 2030 of the United Nations. especially zero hunger (SDG #2) (Ikundayisi et al., 2025), sustainable development (SDG #11), climate action (SDG #13), and life on land (SDG #16)

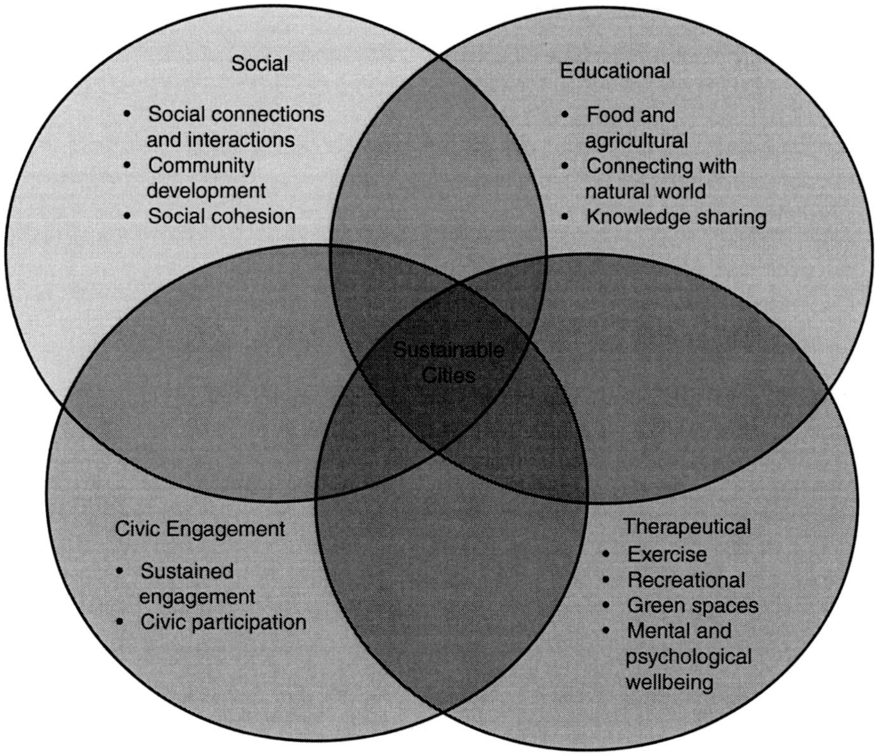

FIGURE 13.4 Urban agriculture and Sustainable Development Goals of the Agenda 2023 of the United Nations.

(Girard, 2021). SDG#11 specifically states the need to "make cities and human settlements inclusive, safe, resilient, and sustainable" (U.N., 2015; Kii, 2021) (Figure 13.4). Because of these and other emerging issues driven by global population, UA may be a solution to advancing food and nutritional security especially in Sub-Saharan Africa (SSA) (Lee-Smith, 2010), and reducing urban pollution because rapidly growing cities are also among principal causes of global environmental pollution. Thus, with Africa's cities projected to double in population by 2050, UA is an important opportunity for sustainable development and minimization of poverty through creation of another income stream (Lee Smith, 2014).

Rapid population growth in cities is also driving the increase in the intercontinental food import (Joshipura, 2024). Thus, there is a strong need for policies to address this crisis (Samuel et al., 2022), and to identify technological options to achieve food self-sufficiency. One of these options is upscaling UA and PUA (Miccoli et al., 2016) in a systematic manner by translating the science of agronomy into action. The annual cost of food imports in SSA was estimated at $43 billion in 2019, and as much as $50 billion in 2025, and increasing. For Africa, some estimates show the cost of food imports may soon reach $90B to $110 billion per year. Food imported into rapidly growing urban centers is aggravating the vicious cycle of losing nutrients and carbon (C) from agro-ecosystems, thereby aggravating soil degradation, on the one hand. and hunger-related urban migration, on the other (Niemczynowicz, 1996). Urban agriculture is a relevant approach to reducing food imports (which may raise the price of food staples), so is directly related to SDG #1 (End Poverty). Dixon et al. (2007) observed that in many cities, thousands of positions of paid employment could be created through the development of sustainable and self-sufficient food systems.

HOW MUCH FOOD CAN BE PRODUCED IN CITIES BY URBAN AGRICULTURE?

Despite its numerous advantages, UA can make only a limited contribution to urban food security in low-income countries (Warren et al., 2015). There are biophysical and socio-economic limits to how much food can be grown in cities. Some crops like grains and fruit trees cannot be grown in cities. However, UA is suited to producing fresh vegetables, while reducing the cost of shipment and developing a resilience system against unexpected disruptions such as the COVID-19 pandemic (Lal, 2020; Lal et al., 2020). Some estimates show that a city like Kampala, Uganda may be growing 25% of food consumed by UA and PUA. In Ilorin, Nigeria, Ola (2020) reported that UA contributed 16.9% meat/fish/eggs, 4.5% of yam/cassava/potato, 0.6% of vegetables and fruits, and 0.5% of grains. In the Global South, hydroponics systems provide the highest productivity of and the lowest land demand, but generate the highest emissions. Furthermore, PUA farms may emit fewer greenhouse gases and are better integrated with existing land use patterns. However, PUA farms may require more water and land), and need decentralized food circulation that can match rooftop farming yields, so that even smaller scale systems could provide 15 to 30% of local food flows. UA could produce 140 kg vegetables m^{-2} yr^{-1} (Armanda et al., 2019).

BENEFITS OF URBAN AND PERI-URBAN AGRICULTURE

Among the numerous benefits of sustainable UA are food and nutritional security, environmental, social and economic dimensions. There is an ever-growing interest in eating healthily, consuming local food, addressing urban hunger, and advancing SDGs (Boukharta et al. 2025). Thus, UA and PUA have a special niche because these innovative food systems have positive impacts on advancing several SDGs, and reducing environmental effects. Therefore, the strategy of edible urbanism 5.0 is pertinent to advancing SDGs, as well as the food revolution 5.0 target of feeding up to 10 billion people. Russo and Cirella (2019) argued that edible urbanism integrates three main principes of sustainability: food security, resilience, and social inclusion. Some benefits of UA are discussed below.

1. **Addressing Malnutrition**
   A high rate of malnutrition (deficiency of micronutrients and protein) in developing countries may be at least partly addressed by UA and PUA (Pantoja-Calderon et al., 2025). Affordability of fresh produce is another advantage (Abdelmenan et al., 2025) for low-income households. UA is an economical source of nutrition-sensitive food (Gerster-Bentaya, 2013) even under harsh environmental conditions. Thus, UA has the health equity dimensions (Dixon et al., 2007), that are related to human health and well-being, along the lines of the One Health approach.
2. **Improving Human Assets**
   One of the benefits of UA is to improve the well-being of the residents in cities. In a study involving community gardens, Rosmiza and Zainal (2021) observed that participants developed strength in terms of human assets such as good health, and determination – not giving up easily.
3. **Saving Energy**
   UA reduces the use of energy-based inputs by supplemental use of compost and recycling of black and grey water, thereby reducing use of chemical fertilizers and pesticides (i.e., herbicides, insecticides, nematicides and fungicides). Decreasing use of water for supplemental irrigation, and reducing transport are also among the major benefits of UA. UA has a short value chain and reduces transport-related energy cost.
4. **Mitigating Global Warming**
   Global warming is one of the major issues of the 21st century. UA is relevant to addressing the issue of anthropogenic climate change (ACC) in cities facing the challenges of global

warming, extreme rainfall, and intensifying urban heat islands (Liu et al., 2025). Being a climatologically diverse continent, cities in SSA may face different ACC challenges, and UA must be specifically designed to address these issues (Chari et al., 2022). Indeed, UA is a smart option to address ACC (Dubbeling et al., 2011) and reduces risks of urban heat islands (Tabrez, 2025).

5. **Developing Communities and Education**
   Community development is another benefit of UA and PUA. In general home gardens, which involve more sustained engagement with other people and the natural world promote opportunities for greater learning, connection, and civic participation. Thus while UA decreases food waste, and reduces air pollution it also fosters social connections, and improves physical or mental health.
6. **Creating Economic Benefits**
   In addition to contributing to the quantity and quality of food, UA can also create another income stream for urban households in rapidly urbanizing countries in SSA and South Asia. UA can contribute to income generation, savings, capital expenditures, and tax revenues.
7. **Reducing Risks of Disruption in Food-Supply Chains**
   Risks of disruption in food-supply chains by natural (COVID19) and human-made crises (war) can be minimized by adoption of science-based UA (Lal, 2020; Lal et al., 2020).

Despite numerous benefits, UA also has some disservices that must be considered. Important among these are soil pollution with heavy metals via industrial waste. The problem is aggravated because soil-testing facilities are not available. The problems of growing food in contaminated soils and in using polluted water for home gardens have serious health issues which must be considered prior to establishing home gardens.

Human Health Benefits of UA

The benefits of UA go beyond the production of food and increasing its access. Indeed, UA provides multiple social, educational, civic, and nutritional services. The social benefits of UA go beyond that of food and UA and PUA have several health benefits which should be known. For example, home gardening is a popular recreational activity promoting mental and psychological well-being, providing outdoor physical exercise and strengthening the connection between humans and nature. Xu et al. (2024) reported that home gardening offers therapeutic benefits by fostering solid bonds through knowledge sharing and produce sharing. Pollard et al. (2018) observed that the present era is an age of concern for mental health and well-being, and that UA is important for health and well-being, social connection, and happiness. Thus, UA may provide a preventative or supportive role for gardeners' health and well-being (Pollard et al., 2018). UA also leads to diet diversity and nutritional improvement (Dixon et al., 2007). Further, agroecological UA is a tool for health promotion through community leadership, empowerment, creation of conducive environment for health, and from income generation (Ribeiro et al., 2015). UA is also a source of nutrition-sensitive food (Gerster-Bentaya, 2013).

In addition to advancing food security. Urban home food gardens also contribute to community development, cultural reproduction, and resilience at multiple scales, conserve agrobiodiversity, and support urban biodiversity (Taylor et al., 2014).

An important merit of UA is connecting producers to consumers. Dimitri et al. (2016) reported that UA practiced in lower-income communities facilitated building communities. These communities donate a higher share of food from their farm and are less likely to own farmland.

In addition to growing food, UA and PUA are also related to public health through urban planning (Brown and Jameton, 2000), so that green space can be demarcated from the beginning, and provisions are made to recycle black and grey water for supplemental irrigation, and create composting areas for converting biowaste into soil improver.

Planetary Health Benefits of Urban Agriculture

The EAT-Lancet Commission has emphasized the need to produce a healthy diet within planetary boundaries by 2050. The Commission established a scientific consensus on a "Planetary Health Diet". It comprises a flexible diet with emphasis on fruits, vegetables, nuts, and legumes while reducing consumption of red meat and unhealthier foods. A major aim of the Commission has been to prevent millions of premature deaths and create a sustainable food system which also protects, restores, and sustains planetary processes. The Planetary Health Diet (PH Diet), under the auspices of the EAT-Lancet Commission, is aimed at harnessing health and environmental benefits. With advances in scientific understanding of UA and PUA in the context of growing cities, a wide range of UA technologies have emerged. UA is highly versatile and also has both productive and aesthetic aims (Gunapala et al., 2025). It addresses the issue of food deserts by enhancing access to fresh produce, empowering communities while reducing transportation costs, strengthening climate resilience, creating environmentally friendly agro-ecosystems, and leading to sustainable cities.

SOME PRECAUTIONS AND BARRIERS FOR URBAN AGRICULTURE

Soil has a great and sacred position worldwide because of its numerous ecosystem services (ESs) such as food, feed, fiber and food for human and animal. Yet, contamination of soil and degradation/deformation of landscape is a worldwide issue (Carre et al., 2017). Soil contamination is a serious problem in industrial areas (Moghtaderi et al., 2018). Soils of urban centers may also be polluted and contaminated with Copper, Lead, Zinc, Manganese, Iron, Nickel (Ayaz et al., 2023). In Pakistan, for example, Ali et al. (2022) observed that Cr contamination of paddy soil is a serious problem.

Similar to contamination, urban soil degradation has severe ecological risks and human health implications. Durdu et al. (2023) reported that urban soils may be contaminated with organochlorine pesticides and heavy metals/metalloids (Ag, As, Si, B, Ca, Cd, Cr, Cu, Fe, Hg, Mn, Ni, Mo, Pb, Se, Zn, and V) which have severe impacts on human health. Furthermore, humans are exposed to heavy metals and microplastics through inhalation, dermal contact, or the consumption of food grown on contaminated soils (Perkovic et al., 2022). In China, Tong et al. (2020) reported serious heavy metal contamination (As, Cd, Cr, Hg, Pb, Cu, Zn, and Ni) in 71 cities during the period from 2003 to 2019, with serious implications to human health and well-being. Levels of heavy metals in agricultural soils (Cr, As, Ni, Cd, Pb, Hg, Zn, and Cu) can harm crop health and yield. Angon et al. (2024) observed that exposure to heavy metals can harm brain, intestines, lungs, liver, and other organs of the human body. The sources of heavy metals in soil can be natural and anthropogenic with adverse effects on soil and plant health, and on human health by entering the food chain (Angon et al., 2024).

Soil contamination can also happen from fertilizers and use of plastic film to conserve water and control weeds. In East Asia it is observed that soil contamination caused by N fertilizer manufacturing is a major concern. Furthermore, the impact and severity of soil microplastics can persist for a long time (Chia et al. (2022).

There are some other barriers to the upscaling and adoption of UA, especially in SSA, In these countries, UA may not be strongly related to urban food security Davies et al. (2021) because of some constraints to adopting UA. Several barriers to adoption of UA (33% households in Zambia and Kenya) include settlement formality, property rights, and proximity to food retailers. The issue of land tenure may also lead to tension between urban developers and resource-rich and resource-poor households. While poverty may be the motivation in some households, entrepreneurial objectives may be the motivation in others (Tevera, 2022). Thus, a lot of research is needed to fill the knowledge gaps in relation to ecological and some human dimensional issues which urban planners in secondary cities should understand to be able to minimize these tensions. In this context, UA land–food systems may be developed in an agent-based approach (Hemerijckx et al., 2025).

RESEARCH PRIORITIES

UA is currently experiencing a significant boom, and it is practiced in terraces, balconies, abandoned plots, and on spaces between the buildings (Richter, 2013). Despite the growing interest in UA and PUA, there is a scarcity of data on soil health and its management. The following are some priorities for research.

1. **Understanding soil health** in UA/PUA is critical to developing efficient management practices. Salomon et al. (2022) observed gradually decreasing soil total N and C concentration, but accumulation of P in soil under UA. Thus, Salomon and colleagues suggested the need for identification of pertinent soil management practices in UA systems. Important among these practices are mulching and nutrient-based compost.
2. Additional research is also needed on **how UA can address poverty and household income**. Zivhave et al. (2025) observed that while UA has a short value chain, it can negatively affect land-use planning decisions, which has implications for developing a long-range income stream for resource-poor landholders.
3. More research is needed on **how biotechnology can be used** to improve the nutritional value of crops grown in UA (Aggarwal et al., 2024).
4. Additional research is needed in the context of physiognomic response of crops (GumisiRiza et al., 2023). It refers to observable physical changes (size, shape,) due to environmental factors reflecting underlying physiological adjustments to cope with stress and revealing adaptations or sensitivities crucial for breeding and management.
5. **Rainwater harvesting from roof tops** can advance food self-sufficiency in UA. More research is needed into how UA can alleviate water shortage in arid regions to achieve sustainable cities (Hume et al., 2022).
6. Additional research is also needed on **the effects of anthropogenic climate change, including** the frequency of extreme events, the heat island effect of UA and vice versa (Chari et al., 2022).
7. Contributions of UA to **urban sustainability and human well-being** need to be studied under diverse conditions with specific focus on environmental sustainability, material well-being, labor and livelihoods, land tenure, urban planning, food and nutritional security, social sustainability and gender issues (Rao et al., 2022)
8. **Human health impacts of UA** may differ across agriculture types (Kirby et al., 2021), and thus must be quantified under site-specific situations and for diverse systems (e.g., with and without livestock).
9. **General perception of the benefits of UA in** terms of ecology, economic impacts, human physical and mental well-being (Dorofiereva et al., 2021) also need to be quantified in different UA/PUA settings. The need for sustainable cities in relation to UA must be critically assessed as well as the role of artificial intelligence in enhancing ecoefficiency and productivity.
10. **Legal measures should be researched and developed** to prevent and manage soil contamination and to enhance food safety for consumer health (Ramon et al., 2019).
11. More research is required to assess the impact of **microplastic contamination on soil health in UA** (Chia et al., 2022), and on human health through alteration of physicochemical properties of soil, microbial population, nutrient cycling, and plant growth and development (Tariq et al., 2024), with specific focus on contamination of urban soils (Panico et al., 2023; Meena et al., 2020).

12. For specific city contexts (culture, population, governance, food habit) there is a need for objective and **critical assessment of the role of UA and PUA in alleviation of hunger and malnutrition** with or without.

CONCLUSIONS

With rapid urbanization; especially in SSA, South Asia and other tropical regions; UA and PUA have an important role in the elimination of hunger and malnutrition. It also has economic benefits in reducing the cost of transportation. Indeed, it is a mechanism of advancing SDGs including #1 (end poverty) and #2 (zero hunger) and others. Among a wide range of UA and PUA strategies, rooftop gardens and multistory glass buildings with controlled environments and soil-less growing media are among some innovative options. Integration of crops with livestock is also a nutrition-positive method of agriculture. Aquaculture and other soil-less production systems in multistory glass buildings along with artificial intelligence are particularly innovative options. Soils of large cities have the potential to produce 10–20% of safe, economic, and locally produced food. UA and PUA have a bright future in feeding the growing urban population while improving the environment.

BIBLIOGRAPHY

Abdelmenan, S., Worku, A., Berhane, H. Y. *et al.* (2025) Affordability of family foods is associated with nutritional status of women with pre-school children in Addis Ababa, Ethiopia. *Scientific Reports 15*, 665. https://doi.org/10.1038/s41598-024-83064-5

Acevedo-De-los-Ríos, A., Dyson, A., Claeys, D., & Cardenas-Mamani, U. (2025). Environmental performance of urban agriculture in the global south: A comprehensive literature review and life cycle analysis approach. *Environmental Impact Assessment Review, 115*, 108040. https://doi.org/10.1016/j.eiar.2025.108040

Aggarwal, B., Rajora, N., Raturi, G., Dhar, H., Kadam, S. B., Mundada, P. S., Shivaraj, S. M., Varshney, V., Deshmukh, R., Barvkar, V. T., Salvi, P., & Sonah, H. (2024). Biotechnology and urban agriculture: A partnership for the future sustainability. *Plant Science, 338*, 111903. https://doi.org/10.1016/j.plantsci.2023.111903

Ali, W., Zhang, H., Mao, K., Shafeeque, M., Aslam, M. W., Yang, X., . . . & Podgorski, J. (2022). Chromium contamination in paddy soil-rice systems and associated human health risks in Pakistan. *Science of the Total Environment, 826*, 153910 https://doi.org/10.1016/j.scitotenv.2022.153910.

Amos, C. C., Rahman, A., Karim, F., & Gathenya, J. M. (2018). A scoping review of roof harvested rainwater usage in urban agriculture: Australia and Kenya in focus. *Journal of Cleaner Production, 202*, 174–190. https://doi.org/10.1016/j.jclepro.2018.08.108

Angon, P. B., Islam, M. S., Das, A., Anjum, N., Poudel, A., & Suchi, S. A. (2024). Sources, effects and present perspectives of heavy metals contamination: Soil, plants and human food chain. *Heliyon, 10*(7). https://doi.org/10.1016/j.heliyon.2024.e28357

Armanda, D. T., Guinée, J. B., & Tukker, A. (2019). The second green revolution: Innovative urban agriculture's contribution to food security and sustainability – A review. *Global Food Security, 22*, 13–24. https://doi.org/10.1016/j.gfs.2019.08.002

Ayaz, H., Nawaz, R., Nasim, I., Irshad, M. A., Irfan, A., Khurshid, I., Okla, M. K., Wondmie, G. F., Ahmed, Z., & Bourhia, M. (2023). Comprehensive human health risk assessment of heavy metal contamination in urban soils: Insights from selected metropolitan zones. *Frontiers in Environmental Science, 2023*. https://doi.org/10.3389/fenvs.2023.1260317

Balfour, E. B. (1943). The Living Soil. In *The Living Soil: Evidence of the Importance to Human Health of Soil Vitality, with Special Reference to Post-war Planning / E. B. Balfour* (248). Faber and Faber, London

Boukharta, O. F., Chico Santamarta, L., Correa Guimaraes, A., & Navas Gracia, L. M. (2025). Assessing citizens' perceptions of urban agriculture and its contribution to food security—Worldwide analysis and specific case studies in Spain. *Urban Science*, 9(5), 150. https://doi.org/10.3390/urbansci9050150

Brown, K., & Jameton, A. (2000). Public health implications of urban agriculture. *Journal of Public Health Policy 21*, 20–39. https://doi.org/10.2307/3343472

Carré, F., Caudeville, J., Bonnard, R., Bert, V., Boucard, P., & Ramel, M. (2017). Soil contamination and human health: A major challenge for global soil security. In Field, D.J., Morgan, C.L.S., McBratney, A. B. (eds) *Global Soil Security. Progress in Soil Science*. Cham: Springer. https://doi.org/10.1007/978-3-319-43394-3_25

Čepić, S., Tomićević-Dubljević, J., & Pálsdóttir, A. M. (2024). Unveiling human-nature interaction: Qualitative exploration of perceived benefits of urban gardening. *Health & Place*, *88*, 103276. https://doi.org/10.1016/j.healthplace.2024.103276

Chari, F., & Ngcamu, B. S. (2022). Climate change and its impact on urban agriculture in Sub-Saharan Africa: A literature review. *Environmental & Socio-Economic Studies*, *10*(3), 22–32. https://doi.org/10.2478/environ-2022-0014

Chia, R. W., Lee, J. Y., Jang, J. *et al.* (2022). Soil health and microplastics: a review of the impacts of microplastic contamination on soil properties. *Journal of Soils and Sediments 22*, 2690–2705. https://doi.org/10.1007/s11368-022-03254-4

Crush, J., Frayne, B., & Pendleton, W. (2012). The crisis of food insecurity in African Cities. *Journal of Hunger & Environmental Nutrition*, *7*(2–3), 271–292. https://doi.org/10.1080/19320248.2012.702448

Davies, J., Hannah, C., Guido, Z., Zimmer, A., McCann, L., Battersby, J., & Evans, T. (2021). Barriers to urban agriculture in Sub-Saharan Africa. *Urban Food Policies for a Sustainable and Just Future*, *103*, 101999. https://doi.org/10.1016/j.foodpol.2020.101999

de Oliveira Alves, D., de Oliveira, L., & Mühl, D. D. (2024). Commercial urban agriculture for sustainable cities. *Cities*, *150*, 105017. https://doi.org/10.1016/j.cities.2024.105017

Dima, S. J., Ogunmokun, A. A., & Nantanga, T. (2002). The status of urban and peri-urban agriculture. *A survey report prepared for integrated support to sustainable development and food security programme in Namibia*, Food and Agriculture Organization of the United Nations, Rome, Italy.

Dimitri, C., Oberholtzer, L., & Pressman, A. (2016). Urban agriculture: Connecting producers with consumers. *British Food Journal*, *118*(3), 603–617. https://doi.org/10.1108/BFJ-06-2015-0200

Dinku, A. M., Mekonnen, T. C., & Adilu, G. S. (2023). Urban food systems: Factors associated with food insecurity in the urban settings evidence from Dessie and Combolcha cities, north-central Ethiopia. *Heliyon*, *9*(3). https://doi.org/10.1016/j.heliyon.2023.e14482

Dixon, J., Omwega, A. M., Friel, S. *et al.* (2007). The health equity dimensions of urban food systems. *Journal of Urban Health 84* (Suppl 1), 118–129. https://doi.org/10.1007/s11524-007-9176-4

Dorofieieva, K., & Vugule, K. (2021). Phenomenon of urban agriculture and its role in shaping sustainable cities. *Landscape Architecture and Art*, *18*, 49–58. https://doi.org/10.22616/j.landarchart.2021.18.05

Dubbeling, M., & De Zeeuw, H. (2011, April). Urban agriculture and climate change adaptation: Ensuring food security through adaptation. In *Resilient Cities: Cities and Adaptation to Climate Change-proceedings of the Global Forum 2010* (pp. 441–449). Dordrecht: Springer Netherlands.

Durdu, B., Gurbuz, F., Koçyiğit, H., & Gurbuz, M. (2023). Urbanization-driven soil degradation; ecological risks and human health implications. *Environmental Monitoring and Assessment*, *195*(8), 1002. https://doi.org/10.22616/j.landarchart.2021.18.05

Ebenso, B., Otu, A., Giusti, A., Cousin, P., Adetimirin, V., Razafindralambo, H., Effa, E., Gkisakis, V., Thiare, O., Levavasseur, V., Kouhounde, S., Adeoti, K., Rahim, A., & Mounir, M. (2022). Nature-based one health approaches to urban agriculture can deliver food and nutrition security. *Frontiers in Nutrition*, *9-2022*, 59–70.

El-Ramady, H., Alshaal, T., Elsakhawy, T., Omara, A. E. D., Abdalla, N., & Brevik, E. C. (2018). Soils and humans. In *The Soils of Egypt* (pp. 201–213). Cham: Springer International Publishing.

Gerster-Bentaya, M. (2013). Nutrition-sensitive urban agriculture. *Food Security*. *5*, 723–737. https://doi.org/10.1007/s12571-013-0295-3

Girard, Tyler, (2021). Explaining the Origins and Evolution of the Global Financial Inclusion Agenda. Electronic Thesis and Dissertation Repository, *7921*. https://ir.lib.uwo.ca/etd/7921

Gunapala, R., Gangahagedara, R., Wanasinghe, W. C. S., Samaraweera, A. U., Gamage, A., Rathnayaka, C., Hameed, Z., Baki, Z. A., Madhujith, T., & Merah, O. (2025). Urban agriculture: A strategic pathway to building resilience and ensuring sustainable food security in cities. *Farming System*, *3*(3), 100150. https://doi.org/10.1016/j.farsys.2025.100150

Hemerijckx, L. M., De Vos, K., Kaunda, J. O., & Van Rompaey, A. (2025). Future scenarios for urban agriculture and food security in sub-Saharan Africa: Modelling the urban land-food system in an agent-based approach. *Computers, Environment and Urban Systems*, *118*, 102258.

Howard, S. A. (1940). *An Agricultural Testament* (p. 266). Oxford: Oxford University Press. ISBN-10, 1849027730. ISBN-13, 978-1849027731.

Hume, I. V., Summers, D. M., & Cavagnaro, T. R. (2022). Lawn with a side salad: Rainwater harvesting for self-sufficiency through urban agriculture. *Sustainable Cities and Society*, *87*, 104249. https://doi.org/10.1016/j.scs.2022.104249

Ikudayisi, A. A., & Adejumo, O. (2025). Urban food security perspective towards sustainable development Goal 2: A bibliometric analysis. *Sustainable Futures*, *9*, 100654. https://doi.org/10.1016/j.sftr.2025.100654

Joshipura, T. (2024). Feeding African cities: Hinterland suitability and urban growth in twentieth-century sub-Saharan Africa. *Economic History of Developing Regions*, *39*(3), 251–278. https://doi.org/10.1080/20780389.2024.2376549

Kanosvamhira, T. P. (2024). Sustainable urban agriculture: Unlocking the potential of home gardens in low-income communities. *The Professional Geographer*, *76*(5), 587–596. https://doi.org/10.1080/00330124.2024.2355179

Kii, M. (2021). Projecting future populations of urban agglomerations around the world and through the 21st century. *NPJ Urban Sustain 1*, 10. https://doi.org/10.1038/s42949-020-00007-5

Kirby, C. K., Specht, K., Fox-Kämper, R., Hawes, J. K., Cohen, N., Caputo, S., Ilieva, R. T., Lelièvre, A., Poniży, L., Schoen, V., & Blythe, C. (2021). Differences in motivations and social impacts across urban agriculture types: Case studies in Europe and the US. *Landscape and Urban Planning*, *212*, 104110. https://doi.org/10.1016/j.landurbplan.2021.104110

Kronenberg, J., Andersson, E., Elmqvist, T., Łaszkiewicz, E., Xue, J., & Khmara, Y. (2024). Cities, planetary boundaries, and degrowth. *The Lancet Planetary Health*, *8*(4), e234–e241. https://doi.org/10.1016/S2542-5196(24)00025-1

Lal, R. (2017). Managing urban soils for food security and adaptation to climate change. In *International Congress on Soils of Urban, Industrial, Traffic, Mining and Military Areas* (pp. 302–319). Cham: Springer International Publishing.

Lal, R. (2020). Home gardening and urban agriculture for advancing food and nutritional security in response to the COVID 19 Pandemic. *Food Security*, 12, 871–876.

Lal, R., Brevik, E. C., Dawson, L., Field, D., Glaser, B., Hartemink, A. E., Hatano, R., Lascelles, B., Monger, C., Scholten, T., Singh, B. R., Spiegel, H., Terribile, F., Basile, A., Zhang, Y., Horn, R., Kosaki, T., & Sánchez, L. B. R. (2020). Managing Soils for Recovering from the COVID-19 Pandemic. *Soil Systems*, *4* (3), 46. https://doi.org/10.3390/soilsystems4030046

Lee-Smith, D. (2010). Cities feeding people: an update on urban agriculture in equatorial Africa. *Environment and Urbanization*, *22*(2), 483–499. https://doi.org/10.1177/0956247810377383

Lee-Smith, D. (2014). The dynamics of urban and peri-urban agriculture. In *Digging Deeper: Inside Africa's Agricultural, Food and Nutrition Dynamics* (pp. 197–216). Brill. Publishers,ISBN:978-90-04-28269-8, https://doi.org/10.1163/9789004282698_010

Li, X., Zhou, Y., Hejazi, M. *et al.* (2021). Global urban growth between 1870 and 2100 from integrated high resolution mapped data and urban dynamic modeling. *Communications Earth & Environment*, *2*, 201. https://doi.org/10.1038/s43247-021-00273-w

Liu, Y., Chanse, V., & Chicca, F. (2025). Enhancing post-disaster food security through urban agriculture in the context of climate change. *Land*, *14*(4), Article 4. https://doi.org/10.3390/land14040799

Meena, V., Dotaniya, M. L., Saha, J. K., Das, H., & Patra, A. K. (2020). Impact of lead contamination on agroecosystem and human health. In: Gupta, D., Chatterjee, S., Walther, C. (eds) *Lead in Plants and the Environment. Radionuclides and Heavy Metals in the Environment*. Cham: Springer. https://doi.org/10.1007/978-3-030-21638-2_4

Miccoli, S., Finucci, F., & Murro, R. (2016). Feeding the cities through urban agriculture the community esteem value. *Florence "Sustainability of Well-Being International Forum". 2015: Food for Sustainability and Not Just Food, FlorenceSWIF2015*, *8*, 128–134. https://doi.org/10.1016/j.aaspro.2016.02.017

Mietz, L. K., Civit, B. M., & Arena, A. P. (2023). Cultivating communities in Mendoza, Argentina: Exploring social aspects of urban agriculture. *Environmental & Socio-Economic Studies*, *11*(4), 12–24. https://doi.org/10.2478/environ-2023-0020

Mietz, L. K., Civit, B. M., & Arena, A. P. (2024). Life cycle assessment to evaluate the sustainability of urban agriculture: opportunities and challenges. *Agroecology and Sustainable Food Systems*, *48*(7), 983–1007. https://doi.org/10.1080/21683565.2024.2344025

Mihailović, B., Radosavljević, K., & Popović, V. (2023). The role of indoor smart gardens in the development of smart agriculture in urban areas. *Економика пољопривреде*, *70*(2), 453–468.

Moghtaderi, T., Mahmoudi, S., Shakeri, A., & Masihabadi, M. H. (2018). Heavy metals contamination and human health risk assessment in soils of an industrial area, Bandar Abbas–South Central Iran. *Human and Ecological Risk Assessment: An International Journal*, *24*(4), 1058–1073.

Mukaila, R., & Enete, A. A. (2025). Achieving sustainable food security: Does urban household garden agriculture matter? *Environment, Development and Sustainability* https://doi.org/10.1007/s10668-024-05956-w

Ndakidemi, P. A., Nampijja, Z., & Mbega, E. R. (2023). Soilless urban gardening as a post covid-19 food security salvage technology: A study on the physiognomic response of lettuce to hydroponics in Uganda. *Scientific African*, *20*, e01643.

Niemczynowicz, J. (1996). Megacities from a Water Perspective. *Water International*, *21*(4), 198–205. https://doi.org/10.1080/02508069608686515

Oh, S., & Lu, C. (2022). Vertical farming - smart urban agriculture for enhancing resilience and sustainability in food security. *The Journal of Horticultural Science and Biotechnology*, *98*(2), 133–140. https://doi.org/10.1080/14620316.2022.2141666

Ola, A. (2020). Building a food-resilient city through urban agriculture: The case of Ilorin, Nigeria. *Town and Regional Planning*, *77*, 89–102.

Opitz, I., Berges, R., Piorr, A., & Krikser, T. (2016). Contributing to food security in urban areas: differences between urban agriculture and peri-urban agriculture in the Global North. *Agriculture and Human Values*, *33*(2), 341–358.

Panico, S. C., Santorufo, L., Memoli, V., Esposito, F., Santini, G., Di Natale, G., … & Maisto, G. (2023). Evaluation of soil heavy metal contamination and potential human health risk inside forests, wildfire forests and urban areas. *Environments*, *10*(8), 146.

Pantoja-Calderon, R., Garcia-Cejudo, D., & Roggema, R. (2025). Addressing the paradox of food and health in Mexico: A landscape urbanism approach. *Land*, *14*(3), 506.

Payen, F. T., Evans, D. L., Falagán, N., Hardman, C. A., Kourmpetli, S., Liu, L., … & Davies, J. A. (2022). How much food can we grow in urban areas? Food production and crop yields of urban agriculture: a meta-analysis. *Earth's Future*, *10*(8), e2022EF002748.

Pérez Camerano, C. E., Borja Barrera, M., & Gómez Pedroza, F. (2023). Model to promote urban agriculture based on SDG 2 and 11 in the district of Barranquilla. *Revista Pensamiento Americano*, *16* (32). 1–11.

Perković, S., Paul, C., Vasić, F., & Helming, K. (2022). Human health and soil health risks from heavy metals, micro (nano) plastics, and antibiotic resistant bacteria in agricultural soils. *Agronomy*, *12*(12), 2945.

Pollard, G., Ward, J., & Roetman, P. (2018). Typically diverse: the nature of urban agriculture in South Australia. *Sustainability*, *10*(4), 945.

Quaye, J., Adams, F., Mensah, A., Ullah, A., Euah, S., & Donkor, E. (2025). Crop-livestock integration in urban agriculture: implication for urban food security in Ghana. *Food and Energy Security*, *14*(3), e70100.

Rajagopalan, S., Ramaswami, A., Bhatnagar, A., Brook, R. D., Fenton, M., Gardner, C., … & Whitsel, L. P. (2024). Toward heart-healthy and sustainable cities: a policy statement from the American Heart Association. *Circulation*, *149*(15), e1067–e1089.

Ramon, F., & Lull, C. (2019). Legal measures to prevent and manage soil contamination and to increase food safety for consumer health: The case of Spain. *Environmental Pollution*, *250*, 883–891.

Rao, N., Patil, S., Singh, C., Roy, P., Pryor, C., Poonacha, P., & Genes, M. (2022). Cultivating sustainable and healthy cities: A systematic literature review of the outcomes of urban and peri-urban agriculture. *Sustainable Cities and Society*, *85*, 104063. https://doi.org/10.1016/j.scs.2022.104063

Ribeiro, S. M., Bógus, C. M., & Watanabe, H. A. W. (2015). Agroecological urban agriculture from the perspective of health promotion. *Saúde e Sociedade*, *24*, 730–743. https://doi.org/10.1590/S0104-12902015000200026

Richter, F. (2013). La agricultura urbana y el cultivo de sí. *Los huertos de ocio a la luz de las dinámicas neorrurales. (Urban agriculture and the growing of the self. Leisure gardening in light of new-rural dynamics).* http://hdl.handle.net/10366/123231

Ritchie, Hannah, Samborska, Veronika, & Roser, Max (2024) Urbanization. Published online at https://OurWorldinData.org. Retrieved from: https://ourworldindata.org/urbanization [Online Resource].

Roscher, E., & Arnholdt, W. (2006). Older soil contaminations and human biomonitoring–two typical examples. *Gesundheitswesen (Bundesverband der Arzte des Offentlichen Gesundheitsdienstes (Germany))*, *68*(12), 787–795.

Rosmiza, M., & Zainal, M. (2021). Exploration of human assets among community garden project participants. *Geografia. Malaysian Journal of Society and Space*, *17*(4), 30–44. CABI Databases. https://doi.org/10.17576/geo-2021-1704-03

Russo, A., & Cirella, G. T. (2019). Edible urbanism 5.0. *Palgrave Commun 5*, 163. https://doi.org/10.1057/s41599-019-0377-8

Salomon, M. J., Watts-Williams, S. J., McLaughlin, M. J., & Cavagnaro, T. R. (2022). Spatiotemporal dynamics of soil health in urban agriculture. *Science of the Total Environment*, *805*, 150224. https://doi.org/10.1016/j.scitotenv.2021.150224

Satterthwaite, D., McGranahan, G., & Tacoli, C. (2010). Urbanization and its implications for food and farming. *Philosophical Transactions of the Royal Society, B: Biological Sciences*, *365*(1554), 2809–2820. https://doi.org/10.1098/rstb.2010.0136

Säumel, I., Reddy, S., Wachtel, T., Schlecht, M., & Ramos-Jiliberto, R. (2022). How to feed the cities? Co-creating inclusive, healthy and sustainable city region food systems. *Frontiers in Sustainable Food Systems*, *2022*. https://doi.org/10.3389/fsufs.2022.909899

Song, S., Cheong, J. C., Lee, J. S. H., Tan, J. K. N., Chiam, Z., Arora, S., Png, K. J. Q., Seow, J. W. C., Leong, F. W. S., Palliwal, A., Biljecki, F., Tablada, A., & Tan, H. T. W. (2022). Home gardening in Singapore: A feasibility study on the utilization of the vertical space of retrofitted high-rise public housing apartment buildings to increase urban vegetable self-sufficiency. *Urban Forestry & Urban Greening*, *78*, 127755. https://doi.org/10.1016/j.ufug.2022.127755

Tabrez, Z. (2025). Sustainable cities: Enhancing food systems with urban agriculture. *Discover Food*, *5*(1). https://doi.org/10.1007/s44187-025-00439-x

Tariq, M., Iqbal, B., Khan, I. *et al.* (2024). Microplastic contamination in the agricultural soil—mitigation strategies, heavy metals contamination, and impact on human health: a review. *Plant Cell Reports*, *43*, 65. https://doi.org/10.1007/s00299-024-03162-6

Taylor, J. R., & Lovell, S. T. (2014).Urban home food gardens in the Global North: research traditions and future directions. *Agriculture and Human Values 31*, 285–305 https://doi.org/10.1007/s10460-013-9475-1

Tevera, D. (2022). Secondary cities and urban agriculture in Sub-Saharan Africa. In *Transforming Urban Food Systems in Secondary Cities in Africa* (pp. 133–147). Cham: Springer International Publishing.

Thwaites, H. J., Hume, I. V., & Cavagnaro, T. R. (2025). The case for urban agriculture: Opportunities for sustainable development. *Urban Forestry & Urban Greening*, *110*, 128861. https://doi.org/10.1016/j.ufug.2025.128861

Tong, S., Li, H., Wang, L., Tudi, M., & Yang, L. (2020). Concentration, spatial distribution, contamination degree and human health risk assessment of heavy metals in urban soils across China between 2003 and 2019—a systematic review. *International Journal of Environmental Research and Public Health*, *17*(9), 3099.

U.N. 2015. *SDG Action Platforms. Department of Economic and Social Affairs, Sustainable Development.* New York, USA: U.N.. https://sdgs.un.org/partnership

Ushimaru, S., Iwata, R., Amrullah, E. R., Utami, A. W., & Ishida, A. (2024). Which households raise livestock in urban and peri-urban areas of eight developing Asian countries? *Agriculture*, *14*(3), Article 3. https://doi.org/10.3390/agriculture14030443

Warren, E., Hawkesworth, S., & Knai, C. (2015). Investigating the association between urban agriculture and food security, dietary diversity, and nutritional status: A systematic literature review. *Food Policy*, *53*, 54–66. https://doi.org/10.1016/j.foodpol.2015.03.004

Wooten, H. (2018). 20193451619, English, Journal article Conference paper, USA, 0886-7283, 131, Goldenrod, Proceedings of the *Florida State Horticultural Society*, (210–211), Florida State Horticultural Society, Growing communities, growing food: hydroponic gardening for urban audiences.

Xu, J., Lin, T., Wang, Y., Jiang, W., Li, Q., Lu, T., Xiang, Y., Jiang, J., & Yu, H. (2024). Home food gardening in modern cities: Advances, issues, and future perspectives. *Frontiers in Sustainable Food Systems*, *2024*. https://doi.org/10.3389/fsufs.2024.1391732

Zivhave, M., & Kornienko, K. (2025). Urban agriculture's 'Invisible' short food value chain: How small-scale farming contributes to Johannesburg food security. *Urban Forum 36*, 91–113. https://doi.org/10.1007/s12132-024-09521-6

14 Conclusion

Marium Husain and Rattan Lal

As we reach the end of this book, it is only part of the beginning of the education and work that is needed to understand and implement the connection between our food and food systems and our health as humans. The process of growing and cultivating our food is complicated and global. There is no one right solution to poor food access and harmful agricultural practices. It will require more research and collaboration to find solutions that are appropriate for each community and people. One important example of the need for cultural competency and region-specific focus is the Green Revolution in lower-income countries in the 1960s. The development of high-yield varieties of crops like wheat and rice increased crop yields by 44% between 1965 and 2010 [1]. This had a significant impact on decreasing hunger and food production keeping pace with the rate of global population growth. However, there are many documented unintended consequences of the practices promoted by the Green Revolution [2] listed below:

1. The large-scale simplification of landscapes negatively impacting national and regional-scale biodiversity through the loss of semi-natural habitats resulting in biotic homogenization and the selection of generalist (including crop-pest) species at the expense of specialists.
2. "Intensive use of agrochemicals and homogenization of cropping systems leading to declines in the biodiversity of the flora and fauna adapted to ruderal habitats through direct effects of pesticides or loss of resources provided by non-crop plants.
3. More frequent, mechanized tillage, intensive use of inorganic fertilizers and reduced organic inputs degrading agricultural soils through erosion, contamination, acidification, salinization, and loss of biological diversity with particular concern around the impact of declines in soil organic carbon on soil function.
4. Reliance on pesticides for crop protection and simplification of crop rotations selecting for a small species pool of pernicious pests, weeds, and diseases, and the overuse of a limited number of pesticide active ingredients leading to the evolution of pesticide resistance.
5. Losses of pesticides and chemical fertilizers through volatilization, leaching or adsorption on transported soil particles polluting air and water courses with implications for the health of ecosystems and humans.
6. The contribution of agriculture to greenhouse gas emissions through land-use change, the reliance on fossil fuels in mechanical operations and the production of inorganic fertilizers plus the emission of nitrous oxides and methane from fertilizers and livestock.
7. Decrease in the nutritional value per unit of food by the so-called dilution effect by increase in atmospheric concentration of CO_2 [2, 3].
8. Reduction in diversity of food caused by cereal-centric (rice-wheat) focus and ignoring traditional food crops which have high nutrient density (e.g., sorghum, millet, chickpeas, pulses, root crops, and leafy vegetables) [4, 5].

These negative consequences have been discussed in the book and the solutions outlined belie the importance of not only recognizing the OneHealth model, but also considering and analyzing the long-term impact of any solution. The words of Sir Albert Howard from 1947 are prescient to this entire discussion:

DOI: 10.1201/9781003469742-14

> Our industries, our trade, and our way of life generally have been based first on the exploitation of the earth's surface and then on the oppression of one another–on banditry pure and simple. The inevitable result is now upon us. The unsuccessful bandits are trying to despoil their more successful competitors. The world is divided into two hostile camps: at the root of this vast conflict lies the evil of spoliation which has destroyed the moral integrity of our generation. While this contest marches to its inevitable conclusion, it will not be amiss to draw attention to a forgotten factor which may perhaps help to restore peace and harmony to a tortured world. We must in our future planning pay great attention to food–the product of sun, soil, plant, and livestock–in other words, to farming and gardening. [6]

We can transform our food and food systems to be a healing force not only for our health as humans but for all living beings on our planet. We leave the reader with action items:

1. Continue to read on the links between food and health.
2. Investigate where your food comes from and how it was cultivated.
3. Purchase more food from local farms and farmers.
4. Decrease meat intake.
5. Advocate with policy groups for sustainable agricultural practices and biodiverse crops for all communities.
6. Increase food diversity.
7. Reinclude traditional crops in the food.

For those who are health professionals, we recommend:

1. Talk to patients about the One Health model.
2. Incorporate the One Health model into medical curricula.
3. Create awareness about the nutritional value of traditional crops.
4. Work with the human nutrition specialists regarding the dietary needs of the growing, and increasingly affluent, human population.
5. Participate in research activities evaluating nutrient density, the promotion of local foods, and its economics in the local community.
6. Advocate with policy groups pushing for nutrient-dense products for all communities.

REFERENCES

1. Balfour, L.E. 1977 Towards a Sustainable Agriculture–The Living Soil. In *International Federation of Organic Agriculture Movements (IFOAM)*. Switzerland.
2. Bhardwaj, R., A. Parashar, H. Parewa, and L. Vyas. 2024. An alarming decline in the nutritional quality of foods: The biggest challenge for future generations' health. *Food* 13: 877. doi: 10.3390/foods13060877
3. Kaspari, M., and E.A.R. Welti. 2024. Nutrient dilution and the future of herbivore populations. *Trends in Ecology & Evolution* 39(9): 809–820. doi: 10.1016/j.tree.2024.05.001
4. Knez, M., M. Ranić, and M. Gurinovic. 2023. Underutilized plants increase biodiversity, improve food and nutrition security, reduce malnutrition, and enhance human health and well-being. Let's put them back on the plate! *Nutrition Reviews*, 82. doi: 10.1093/nutrit/nuad103
5. Soria López, A., P. Garcia-Perez, M. Rodríguez, P. García Oliveira, P. Otero, et al. 2022. Challenges for future food systems: From the Green Revolution to food supply chains with a special focus on sustainability. *Food Frontiers* 4: 1–12. doi: 10.1002/fft2.173
6. Howard, A. (1947) *The Soil and Health: A Study of Organic Agriculture*. University Press of Kentucky. JSTOR. https://doi.org/10.2307/j.ctt2jcp63 (Accessed on January 21, 2026).

Index

Pages in *italics* refer to figures and pages in **bold** refer to tables.

For Product Safety Concerns and Information please contact our
EU representative GPSR@taylorandfrancis.com Taylor & Francis
Verlag GmbH, Kaufingerstraße 24, 80331 München, Germany